Recent Advances in Photocatalytic Treatment of Pollutants in Water

Recent Advances in Photocatalytic Treatment of Pollutants in Water

Guest Editors

Xiufang Chen
Zhexin Zhu
Yinsong Si

MDPI

Basel • Beijing • Wuhan • Barcelona • Belgrade • Novi Sad • Cluj • Manchester

Guest Editors
Xiufang Chen
College of Material Science and Engineering
Zhejiang Sci-Tech University
Hangzhou
China

Zhexin Zhu
College of Material Science and Engineering
Zhejiang Sci-Tech University
Hangzhou
China

Yinsong Si
College of Material Science and Engineering
Zhejiang Sci-Tech University
Hangzhou
China

Editorial Office
MDPI AG
Grosspeteranlage 5
4052 Basel, Switzerland

This is a reprint of the Special Issue, published open access by the journal *Catalysts* (ISSN 2073-4344), freely accessible at: https://www.mdpi.com/journal/catalysts/special_issues/72EH37RN94.

For citation purposes, cite each article independently as indicated on the article page online and as indicated below:

Lastname, A.A.; Lastname, B.B. Article Title. *Journal Name* **Year**, *Volume Number*, Page Range.

ISBN 978-3-7258-6181-1 (Hbk)
ISBN 978-3-7258-6182-8 (PDF)
https://doi.org/10.3390/books978-3-7258-6182-8

Contents

catalysts MDPI

Editorial

Editorial for Special Issue "Recent Advances in Photocatalytic Treatment of Pollutants in Water"

Xiufang Chen, Yinsong Si and Zhexin Zhu *

National & Local Joint Engineering Research Center for Textile Fiber Materials and Processing Technology, College of Material Science and Engineering, Zhejiang Sci-Tech University, Hangzhou 310018, China; chenxf@zstu.edu.cn (X.C.); siys@zstu.edu.cn (Y.S.)
* Correspondence: zzx@zstu.edu.cn

Water is the fundamental source of life on earth. Nowadays, the growth of urban populations and industrial development have resulted in a continuous increase in sewage discharge, which frequently contains pathogenic microorganisms, toxic chemicals, and pharmaceutical pollutants [1,2]. Traditional treatment methods struggle to effectively remove emerging organic micro-pollutants such as drug residues and pesticides [3,4]. When these harmful substances enter surface water and groundwater, they can cause eutrophication, algal blooms, and fish death, ultimately threatening human health. In the face of the urgent demand for safe water, advanced oxidation technologies are proving promising due to their ability to mineralize organic pollutants into non-toxic products, CO_2, and H_2O [5,6]. Among these technologies, photocatalysis—using solar energy as its sole power source—is regarded as the "greenest" solution for treating water containing organic micro-pollutants [7,8].

This Special Issue explores recent advancements in photocatalytic technology for water treatment, featuring diverse perspectives, from material design to practical applications. Contribution 1 presents a "urea-dicyandiamide" copolymerization strategy to synthesize porous V-doped g-C_3N_4 (V/CN), achieving a specific surface area of 64.6 m^2/g and improved catalyst yield. The combination of vanadium doping and nanosheet and hollow tubular structures enhances the separation of photogenerated carriers, boosting the photocatalytic activity of g-C_3N_4 in the peroxymonosulfate (PMS) system. Under simulated sunlight using PMS as an oxidant, carbamazepine is completely degraded within 20 min, effectively balancing the advantages of "high activity" and "low cost". The results reveal that this is an economical and efficient method for degrading pharmaceutical pollutants in aquatic environments.

TiO_2 has emerged as one of the most promising materials for degrading organic pollutants, offering multiple advantages, including non-toxicity, low cost, high photoreactivity, and stability. When TiO_2 is irradiated by UV light, highly reactive oxygen species (ROS) are produced and oxidize and mineralize organic pollutants into harmless products [9]. However, the low photocatalytic efficiency and wide bandgap of TiO_2 limit its practical application [10]. To improve catalytic efficiency, contribution 2 introduces a modified TiO_2 photocatalyst doped with bismuth (Bi) and fluorine (F), along with SnO_2 and SiO_2. This co-modification significantly enhances the photocatalytic degradation efficiency of TiO_2; specifically, RhB degradation efficiency reaches 100% within 20 min under simulated sunlight. The reaction rate constant is 41 times greater than that of Bi/TiO_2 alone, demonstrating that this synergistic strategy effectively improves light absorption and carrier separation efficiency.

To address the issue of the wide bandgap in TiO_2 (about 3.2 eV), which restricts its degradation efficiency under visible light, in Contribution 3, Li et al. successfully extend the light response of TiO_2 to the visible spectrum through interstitial nitrogen doping. Their study compares two nitrogen sources, thiourea and ammonium bicarbonate, finding that $5\%N_T/TiO_2$ prepared with thiourea exhibits superior degradation efficiency for methylene blue (MB) compared to both pristine TiO_2 and substitutive N-doped TiO_2 ($5\%N_{AB}/TiO_2$). The benefit of interstitial doping over substitutional doping is that interstitial nitrogen atoms distort the TiO_2 lattice, reduce interplanar spacing, and enhance electron transport. For contribution 4, a triple-action effect was utilized to enhance the photocatalytic activity of TiO_2. TiO_2 nanoparticles were prepared and transformed into nanotubes during the doping progress, supported by gold nanoparticles to increase charge carriers and active sites while preventing recombination reactions. Consequently, the photocatalytic degradation rate of acid green 1 reached 100% after 17 min of light radiation. This study demonstrated that this triple-action effect addresses titanium oxide's limitations by creating new photo-active sites and pathways for charge carriers, in addition to inhibiting recombination reactions. Contribution 5 introduces another material design concept by combining TiO_2 with phosphoric acid-treated peanut shell biochar (p-BC). The TiO_2/p-BC composite reduces the band gap to 2.73 eV, which significantly enhances tetracycline removal efficiency through synergistic adsorption and photocatalysis. Under optimal conditions, the tetracycline removal efficiency reaches 95.3%, remaining above 86% after five cycles, demonstrating excellent stability and reusability. It represents an interesting approach to developing an efficient TiO_2-based photocatalyst, using low-cost agricultural waste as a catalyst precursor while promoting resource utilization.

The support material is a critical component of the catalyst, significantly influencing its activity, reusability, stability, and recovery. For Contribution 6, a TiO_2/CdS nanocatalyst was immobilized on flexible nickel foam to achieve efficient photocatalytic degradation of antibiotic pollutants in water. Under optimal conditions determined based on response surface methodology—specifically the Box–Behnken design (RSM–BBD)—which included 28 g of catalyst at a pH of 9.04 in 150 min, the removal rate for tetracycline hydrochloride (TCH) reached 53.89% at a scale of 10 L, thereby providing direct parameters for process scaling-up. Furthermore, five consecutive cycling experiments demonstrated remarkable stability with only a minimal catalyst loss of 4.44%.

The practical applications of photocatalytic technology must consider environmental influences; parameters optimized in the laboratory may require adjustment for actual water bodies, which is an important consideration in the transition of photocatalytic technology from the laboratory to engineering applications. Contribution 7 shifts the focus from the material to the application environment, systematically investigating how inorganic anions influence the UV/persulfate advanced oxidation process. This study shows that in pure water, the UV/persulfate system can completely degrade herbicides (terbutylazine and isoproturon) within 30 min. However, the presence of Cl^- and HCO_3^- significantly reduces degradation efficiency. Notably, HCO_3^- effectively scavenges HO· and $SO_4{\cdot}^-$, producing $CO_3{\cdot}^-$ with weaker oxidation capacity, thus inhibiting herbicide degradation.

With the widespread detection of organic pollutants such as neonicotinoid pesticides in water bodies, there is an urgent need for the development of efficient photocatalyst. Contribution 8 summarizes the research progress on various solid catalysts for degrading the pesticide imidacloprid. Metal oxides such as nano-ZnO and black TiO_2 demonstrate excellent degradation performance. Carbon-based materials (e.g., g-C_3N_4 and graphene) and metal–organic frameworks enhance catalytic efficiency due to their high specific surface area and superior photoelectric properties. Additionally, Z-scheme heterojunctions have gained attention for their ability to broaden light response, improve redox capabilities, and reduce carrier recombination, which surpasses that of traditional type-I and type-II heterojunctions. Zeolites possess a high specific surface area, tunable pore structures, and excellent ion exchange capacity. Modified zeolite materials have shown great potential as carriers or composite components of photocatalysts in the photocatalytic treatment of pollutants [11]. Contribution 9 reports that their photocatalytic activity can be significantly enhanced through doping with metals (such as Ag, Fe, or Cu) or combining them with semiconductor materials (such as TiO_2, ZnO, or g-C_3N_4). These composite materials not only extend light absorption into the visible light range but also enhance the degradation efficiency of pollutants through the synergy of adsorption and photocatalysis. High-silica zeolites, due to their hydrophobicity, have enhanced adsorption capacity for organic pollutants. Although modified zeolites have achieved remarkable results at the laboratory stage, their industrial application still poses challenges such as high synthesis costs, unstable regeneration efficiency, and difficulties in large-scale production. Future research should focus on precision material design and the optimization of regeneration technology to promote the application of zeolite-based photocatalytic materials in actual water treatment.

Photocatalytic technology also shows great potential for water disinfection as an alternative to traditional chlorination and ultraviolet treatment. Contribution 10 systematically reviews advancements in red phosphorus and black phosphorus for photocatalytic water disinfection, emphasizing their benefits in visible light response, ROS generation, and pathogen inactivation mechanisms. The authors report that both materials exhibit excellent light absorption and biocompatibility while significantly enhancing the separation efficiency of photogenerated carriers and antibacterial performance through heterostructures with graphitic carbon nitride or TiO_2. Current research primarily focuses on pure water systems; future studies should enhance performance evaluation and long-term stability in actual aquatic environments.

Overall, photocatalytic water treatment technology, a key branch of advanced oxidation processes, shows significant potential in the treatment of emerging pollutants and in disinfecting water. There have been notable advances in two key areas: improving photocatalytic efficiency through novel materials and optimizing processes for complex aquatic environments. However, challenges remain regarding large-scale application, such as the low utilization of visible light and unresolved issues relating to material stability over prolonged use. Additionally, the influence of diverse components in real water bodies on the photocatalytic process requires systematic investigation and effective solutions. In the future, with further integration of materials science, catalytic chemistry, and environmental engineering, it is expected that efficient, cost-effective, and stable photocatalytic water treatment technologies will play a crucial role in the purification of water environments.

Conflicts of Interest: The authors declare no conflicts of interest.

List of Contributions

1. Huang, Y.F.; Pang, R.; Sun, S.S.; Chen, X.F.; Chen, F.T.; Lu, W.Y. One-Pot in Situ Synthesis of Porous Vanadium-Doped g-C_3N_4 with Improved Photocatalytic Removal of Pharmaceutical Pollutants. *Catalysts* **2025**, *15*, 206. https://doi.org/10.3390/catal15030206.
2. Qiu, L.; Li, H.L.; Xu, W.Y.; Zhu, R.S.; Ouyang, F. TiO_2 Catalysts Co-Modified with Bi, F, SnO_2, and SiO_2 for Photocatalytic Degradation of Rhodamine B Under Simulated Sunlight. *Catalysts* **2024**, *14*, 735. https://doi.org/10.3390/catal14100735.
3. Li, D.Z.; Calebe, V.C.; Li, Y.Q.; Liu, H.M.; Lei, Y.M. Interstitial N-Doped TiO_2 for Photocatalytic Methylene Blue Degradation under Visible Light Irradiation. *Catalysts* **2024**, *14*, 681. https://doi.org/10.3390/catal14100681.
4. Saber, O.; Awada, C.; Osama, A.; Shaalan, N.M.; Alshoaibi, A.; Mansour, S.A.; Osama M.; Farha A.H. Boosting the Efficiency of Titanium Oxides and Accelerating Industrial Pollutant

Removal Through Triple-Action Effects to Purify Water by Light. *Catalysts* **2024**, *14*, 772. https://doi.org/10.3390/catal14110772.

5. Liu, J.H.; Zheng, L.W.; Gao, Y.C.; Ji, L.; Yang, Z.F.; Wang, H.L.; Shang, M.; Du, J.H.; Yang, X.D. TiO_2/p-BC Composite Photocatalyst for Efficient Removal of Tetracycline from Aqueous Solutions under Simulated Sunlight. *Catalysts* **2024**, *14*, 357. https://doi.org/10.3390/catal14060357.
6. Zhu, K.F.; Ma, L.Z.; Duan, J.L.; Fang, Z.Y.; Yang Z. Photocatalytic Degradation of Tetracycline Hydrochloride Using TiO_2/CdS on Nickel Foam Under Visible Light and RSM–BBD Optimization. *Catalysts* **2025**, *15*, 113. https://doi.org/10.3390/catal15020113.
7. Pérez-Lucas, G.; Campillo, A.; Navarro, S. Impact of Inorganic Anions on the Photodegradation of Herbicide Residues in Water by UV/Persulfate-Based Advanced Oxidation. *Catalysts* **2024**, *14*, 376. https://doi.org/10.3390/catal14060376.
8. He, D.Y.; Dong, J.Y.; Zhang, Y.N.; Zhang, S.; Zhang Y.N.; Qu, J. Red- and Black-Phosphorus-Based Materials for Photocatalytic Water Disinfection. *Catalysts* **2025**, *15*, 218. https://doi.org/10.3390/catal15030218.
9. Fatin Samara, F.; Hamid, A.A.; Gopal, V.; Dronjak, L.; Feghaly F.; Kanan, S. Modified Zeolites for the Removal of Emerging Bio-Resistive Pollutants in Water Resources. *Catalysts* **2025**, *15*, 138. https://doi.org/10.3390/catal15020138.
10. Gao, S.; Li, S.S.; Sun, S.F.; Chen, M.L. Recent Advances in Photocatalytic Degradation of Imidacloprid in Aqueous Solutions Using Solid Catalysts. *Catalysts* **2024**, *14*, 878. https://doi.org/10.3390/catal14120878.

References

1. Peng, X.Y.; Chen, X.F.; Pang, R.; Cheng, L.L.; Chen, F.T.; Lu, W.Y. The Impact of Polymerization Atmosphere on the Microstructure and Photocatalytic Properties of Fe-Doped g-C_3N_4 Nanosheets. *Catalysts* **2024**, *14*, 520. [CrossRef]
2. Cheng, L.; Sun, S.; Chen, X.; Chen, F.; Chen, X.; Lu, W. Convenient Fabrication of Ultrafine VO_x Decorated on Porous g-C_3N_4 for Boosting Photocatalytic Degradation of Pharmaceuticals with Peroxymonosulfate. *Surf. Interfaces* **2023**, *42*, 103300. [CrossRef]
3. Zhang, B.; Wu, W.Z.; Yin, G.C.; Gong, X. A Multifunctional Synergistic Solar-Driven Interfacial Evaporator for Desalination and Photocatalytic Degradation. *ACS Appl. Mater. Interfaces* **2025**, *17*, 6948–6956. [CrossRef] [PubMed]
4. Badoni, A.; Thakur, S.; Vijayan, N.; Swart, H.C.; Bechelany, M.; Chen, Z.S.; Sun, S.H.; Cai, Q.R.; Chen, Y.; Prakash, J. Recent Progress in Understanding the Role of Graphene Oxide, TiO_2 and Graphene Oxide-TiO_2 Nanocomposites as Multidisciplinary Photocatalysts in Energy and Environmental Applications. *Catal. Sci. Technol.* **2025**, *15*, 1702–1770. [CrossRef]
5. Yang, J.M.; Tian, S.F.; Song, Z.; Hao, Y.G.; Lu, M.H. Recent Advances in Sorption-based Photocatalytic Materials for the Degradation of Antibiotics. *Coordination Chem. Rev.* **2025**, *523*, 216257. [CrossRef]
6. Iqbal, M.A.; Akram, S.; Khalid, S.; Lal, B.; Ul Hassan, S.; Ashraf, R.; Kezembayeva, G.; Mushtaq, M.; Chinibayeva, N.; Hosseini-Bandegharaei, A. Advanced Photocatalysis as a Viable and Sustainable Wastewater Treatment Process: A Comprehensive Review. *Environ. Res.* **2024**, *253*, 118947. [CrossRef] [PubMed]
7. Chen, X.; Cheng, L.; Yang, Y.; Chen, X.; Chen, F.; Lu, W. Construction of High-Density Fe Clusters Embedded in a Porous Carbon Nitride Catalyst with Effectively Selective Transformation of Benzene. *ACS Sustain. Chem. Eng.* **2023**, *11*, 1518–1526. [CrossRef]
8. Hwang, G.I.; Lee, S.; Lee, J.H.; Moon, J.; Shin, Y.R.; Baik, D.H.; Jee, M.H.; Jeong, Y.G. TiO_2-incorporated Poly(m-phenylene isophthalamide) Hollow Fiber Membranes for Superior Ultrafiltration, Mechanical Strength, Photocatalytic Activity in Water Treatment. *Chem. Eng. J.* **2025**, *520*, 165930. [CrossRef]
9. Roslan, N.N.; Lau, H.L.H.; Suhaimi, N.A.A.; Shahri, N.N.M.; Verinda, S.B.; Nur, M.; Lim, J.W.; Usman, A. Recent Advances in Advanced Oxidation Processes for Degrading Pharmaceuticals in Wastewater-A Review. *Catalysts* **2024**, *14*, 189. [CrossRef]
10. Liu, S.Y.; Lou, H.H.; Luo, J.Q.; Albashir, D.; Shi, Y.; Chen, Q.H. A Novel in Situ Biosynthesized Bacterial Cellulose/MoS_2/TiO_2 Composite Film for Efficient Removal of Dyes and Pathogenic Bacteria from Industrial Wastewater under Sunlight Illumination. *ACS Appl. Mater. Interfaces* **2025**, *17*, 19543–19561. [CrossRef] [PubMed]
11. Alakhras, F.; Alhajri, E.; Haounati, R.; Ouachtak, H.; Addi, A.A.; Saleh, T.A. A Comparative Study of Photocatalytic Degradation of Rhodamine B using Natural-based Zeolite Cmposites. *Surf. Interfaces* **2020**, *20*, 100611. [CrossRef]

catalysts

MDPI

Article

TiO_2/p-BC Composite Photocatalyst for Efficient Removal of Tetracycline from Aqueous Solutions under Simulated Sunlight

Jianhui Liu [1], Liwen Zheng [1,*], Yongchao Gao [1,*], Lei Ji [1], Zhongfeng Yang [1], Hailun Wang [1], Ming Shang [1], Jianhua Du [2] and Xiaodong Yang [3]

1 Ecology Institute, Qilu University of Technology (Shandong Academy of Sciences), Jinan 250103, China; 15665891128@163.com (J.L.); jilei1010@qlu.edu.cn (L.J.); yangzf@qlu.edu.cn (Z.Y.); wanghl0215@163.com (H.W.); sm17667939671@163.com (M.S.)

2 WSP Australia Pty Limited, Level 3, Mia Yellagonga Tower 2, 5 Spring Street, Perth 6000, Australia; jason.du@wsp.com

3 Department of Geography & Spatial Information Technology, Ningbo University, Ningbo 315211, China; yangxiaodong@nbu.edu.cn

* Correspondence: zhenglw@qlu.edu.cn (L.Z.); gaoyc@sdas.org (Y.G.)

Abstract: Pollution caused by antibiotics has brought significant challenges to the ecological environment. To improve the efficiency of the removal of tetracycline (TC) from aqueous solutions, a composite material consisting of TiO_2 and phosphoric acid-treated peanut shell biochar (p-BC) has been successfully synthesized in the present study by the sol-gel method. In addition, the composite material was characterized using various techniques, including scanning electron microscopy (SEM), Fourier-transform infrared spectroscopy (FTIR), X-ray diffraction (XRD) spectroscopy, X-ray photoelectron spectroscopy (XPS), photoluminescence (PL) spectroscopy, and ultraviolet–visible diffuse reflectance spectroscopy (UV-vis DRS). The XPS and FTIR analyses revealed the formation of a new Ti–O–C bond, while the XRD analysis confirmed the presence of TiO_2 (with an anatase phase) in the composite material. Also, the PL analyses showed a notable decrease in the recombination efficiency of electrons and holes, which was due to the formation of a composite. This was further supported by the UV-vis DRS analyses, which revealed a decrease in bandgap (to 2.73 eV) of the composite material and led to enhanced light utilization and improved photocatalytic activity. Furthermore, the effects of pH, composite dosage, and initial concentration on the removal of TC were thoroughly examined, which resulted in a maximum removal efficiency of 95.3% under optimal conditions. Additionally, five consecutive cycle tests demonstrated an exceptional reusability and stability of the composite material. As a result of the experiments, the active species verified that $\cdot O_2^-$ played a key role in the photodegradation of TC. Four possible degradation pathways of TC were then proposed. As a general conclusion, the TiO_2/p–BC composite can be used as an efficient photocatalyst in the removal of TC from aqueous solutions.

Keywords: TiO_2; biochar; tetracycline; photocatalytic; removal

1. Introduction

Antibiotics have been widely used to treat various bacterial diseases in humans and livestock all over the world [1]. However, only a small portion of the antibiotics can be absorbed and utilized in the body, and a high portion of antibiotics is discharged into the in vitro environment in the form of original drugs or primary metabolites [2]. Furthermore, residual antibiotics can cause antibiotic resistance bacteria (ARB) and antibiotic resistance genes (ARGs) [3], which threaten human health and ecosystem security. As one of the major broad-spectrum antibiotics, tetracycline (TC) is widely used for the treatment of human and animal infection diseases [4], especially in the livestock industry. In recent years, the abuse and wanton discharge of TC has led to its detection in surface water [5], groundwater [6], and sediment [7]. The physicochemical properties and structure of TC have difficulties regarding their degradation in the environment [8]. The environmental accumulation of TC has aroused widespread concern [9]. It is necessary to remove TC from contaminated water to minimize the environmental and ecological risks.

Up to now, researchers have developed a variety of technologies for the removal of TC from contaminated water. They include microbial degradation [10], chemical oxidation [11], physical adsorption [12], and membrane separation [13]. Among these, adsorption was a preferred method for the removal of various pollutants, which was due to its low cost and easy operation. Biochar is the material created by pyrolysis of available waste biomass under the condition of limiting oxygen, which usually has a high specific surface area and rich functional groups. Biochar or modified biochar, as an adsorbent by which it has been possible to remove TC from contaminated water, has received extensive attention. Chen et al. [14] reported on the adsorption capacity of TC, which increased from 150.2 mg g^{-1} to 166.3 mg g^{-1} when using the phosphoric acid-modified biochar. However, the main disadvantage of the adsorption technology was that TC was only transferred from one phase to the other, and it was not destroyed or disappeared [15].

As an environmentally friendly technology, photocatalysis has been used for the degradation of a variety of organic pollutants [16]. TiO_2 is a photocatalyst that is most widely studied for water purification [17,18], which has here been used due to its low cost, high stability, minimum photo-corrosion, high photo-activity, and low toxicity. Zhang et al. [19] found that TiO_2 can be removed by approximately 55% of TC in an aqueous solution. However, TiO_2 also has some disadvantages, such as small surface area, low light utilization, quick recombination of electron–hole pairs, easy agglomeration, low reactivity for low concentrations of contaminants, and difficulties in recovering in aqueous solutions. All these disadvantages limit the use of TiO_2 for contaminated water treatments. Researchers have used various methods to modify TiO_2 for the improvement of its photocatalytic removal of pollutants and its wide applicability. Xu et al. [20] reported on P-doped TiO_2, which had a smaller band gap value and could use more light energy than pure TiO_2 for a red shift of the absorption spectrum. Some studies have also loaded TiO_2 onto various materials, such as carbon nanotubes [21] and graphene [22]. Although the photocatalytic effect has been improved, the cost was high, and there was a lack of practical applications. Due to its high adsorption capacity and economic performance, biochar has received large attention as an excellent support material for TiO_2. Lu et al. [23] prepared a TiO_2–biochar composite catalyst and used it for the photocatalytic degradation of methyl orange. The modification by phosphoric acid may increase the specific surface area of biochar [24], which is beneficial for the adsorption of pollutants and the loading of TiO_2. Also, the combination of biochar and TiO_2 can reduce the band gap value and enhance the use of visible light [25], which is due to the formation of new states above the VB of the composite [26]. In fact, the adsorption of pollutants from the liquid phase to the solid surface of biochar is more conducive to photocatalytic degradation [27].

In the present study, phosphoric acid-treated peanut shell biochar (p-BC) has been used as a support material in the preparation of the TiO_2/p-BC composite, which has been characterized by SEM, FTIR, XRD, XPS, PL, and UV–vis DRS. The adsorption and photocatalytic performance have been evaluated by the removal efficiency of TC. The reusability of the composite was also studied by a recyclability experiment, and the photocatalytic degradation mechanism and the degradation pathways of TC were finally proposed.

2. Results and Discussion

2.1. Characterization

2.1.1. SEM

The surface morphologies of p-BC, TiO_2, and TiO_2/p-BC were examined by SEM analyses. The results showed a smooth and porous structure of p-BC (Figure 1a). As can be seen in Figure 1b, the TiO_2 particles showed irregular agglomerated morphologies with random shapes. As compared with the p-BC, the surface morphology of TiO_2/p-BC

(Figure 1c) was much rougher, with an accumulation of many TiO_2 particles. Thus, the pore structure of TiO_2/p-BC was severely blocked.

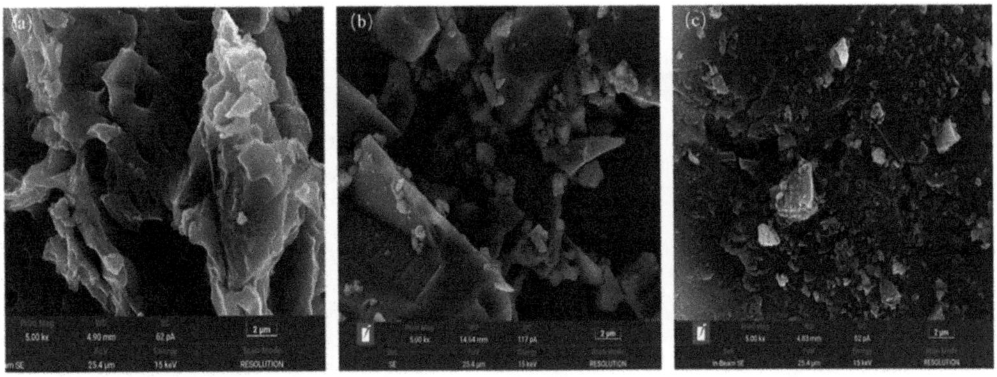

Figure 1. SEM images of (**a**) p-BC, (**b**) TiO_2, and (**c**) TiO_2/p-BC.

2.1.2. XPS Analyses

The survey spectrum of TiO_2/p-BC is presented in Figure 2a. According to the five peaks that are related to C1s, N1s, O1s, P2p, and Ti2p core levels, this spectrum clearly shows that the composite material mainly contains C, N, O, P, and Ti elements, respectively. As can be seen in Figure 2b, the C1s spectrum was deconvoluted into three peaks at 284.68 eV, 286.08 eV, and 289.48 eV, which corresponded to C–C, C–O, and C=O bonds, respectively. The high-resolution N1s spectrum in Figure 2c shows two peaks at 400.01 eV and 401.38 eV, which correspond to C=N–C and C–NHx bonds, respectively. Furthermore, the O1s spectrum was divided into two peaks at 530.68 eV and 531.98 eV, which corresponded to the O–Ti and O–H bonds (Figure 2d). Also, Figure 2e shows the high-resolution P2p spectrum with peaks at 134.20 and 135.15 eV, representing C–O–P and O=P–O, respectively. This spectrum indicated that the preparation of biochar by phosphoric acid-modified peanut shell was successful and that the phosphorus element was present in the prepared composite material. The fitting of the Ti2p spectrum is shown in Figure 2f, with characteristic peaks of Ti $2p^{2/3}$ and Ti $2p^{1/2}$ at 459.26 eV and 464.94 eV, respectively. This result indicated that the titanium element in the composite mainly existed in the form of Ti^{4+} [28]. Also, the binding energies of Ti $2p^{1/2}$ and Ti $2p^{3/2}$ were higher than the corresponding ones in pure TiO_2 (458.5 eV and 464.3 eV, respectively), which was due to the fact that the C atoms in the composite affected the electron cloud density around the Ti and O atoms, thereby forming Ti–O–C bonds [19].

2.1.3. XRD Analyses

The XRD patterns of p-BC, TiO_2, and TiO_2/p-BC are displayed in Figure 3a. For p-BC, the broad peak at about 25 degrees is the characteristic diffraction peak of p-BC. For TiO_2, the peaks at about 25.26, 28.61, 37.79, 48.04, 54.02, 55.01, 62.65, 68.73, 70.51, and 75.15 degrees corresponded to the 101, 110, 004, 200, 105, 211, 204, 116, 220, and 215 crystalline planes, respectively, of anatase TiO_2 (JCPDS No. 21-1272) [29]. Furthermore, the XRD pattern of TiO_2/p-BC lacked the characteristic diffraction peak of p-BC, which might be due to the overlap of characteristic peaks of TiO_2 at similar positions. Also, the characteristic diffraction peaks of anatase TiO_2 were present in the TiO_2/p-BC pattern, which indicated that the anatase phase of TiO_2 was successfully synthesized to p-BC, and the crystalline form of TiO_2 was not affected in this process. These results are consistent with previous reports [29,30].

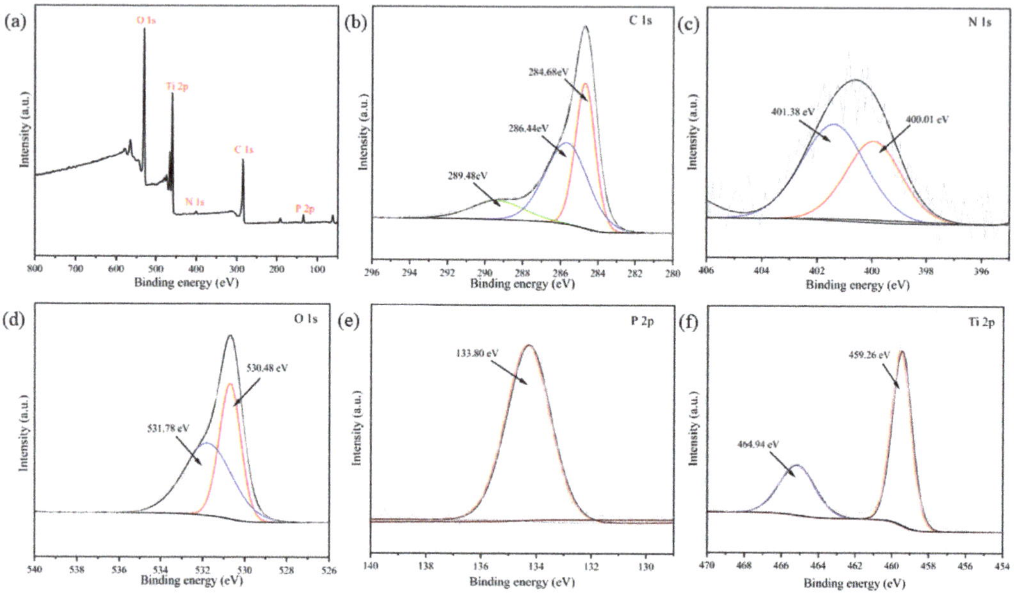

Figure 2. XPS spectra of TiO_2/p-BC: (**a**) survey, (**b**) C 1s, (**c**) N 1s, (**d**) O 1s, (**e**) P 2p, and (**f**) Ti 2p.

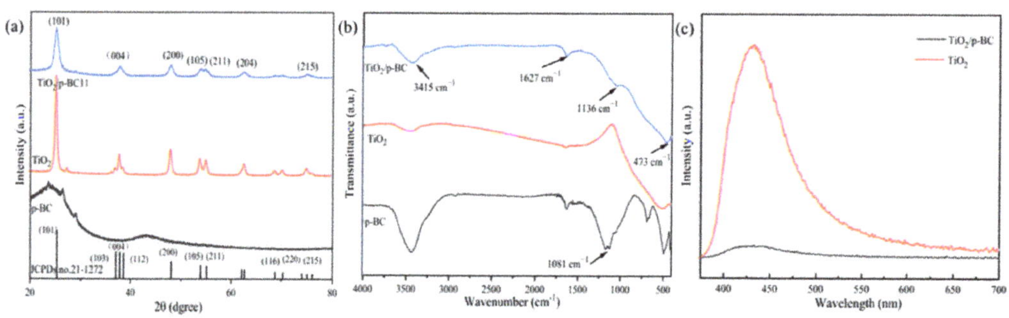

Figure 3. (**a**) XRD and (**b**) FTIR patterns of p-BC, TiO_2, and TiO_2/pBC. (**c**) PL emission spectra of TiO_2 and TiO_2/pBC.

2.1.4. FTIR Analyses

The FTIR results of p-BC, TiO_2, and TiO_2/p-BC are shown in Figure 3b. The characteristic peaks at about 3415 cm^{-1} were assigned to –OH stretching vibrations and the peaks at about 1627 cm^{-1} corresponded to stretching vibrations of C=O and C=C [31]. Furthermore, the characteristic Ti–O peak of TiO_2 (at about 473 cm^{-1}) was observed in the spectrum of TiO_2/p-BC [32]. Also, the peak at 1066 cm^{-1} was assigned to Ti–O–C stretching vibration, which suggested that new chemical bonds were formed between TiO_2 and p-BC [33]. These results indicated that TiO_2/p-BC was successfully prepared, and it was consistent with the SEM and XRD results.

2.1.5. PL Analyses

PL spectra have been used to evaluate the electron and hole recombination efficiency of different photocatalytic materials [34]. In general, the low and high PL intensities of the PL spectra indicated low and high recombination rates, respectively, of TiO_2 and TiO_2/p-BC (Figure 3c). However, the PL intensity of TiO_2/p-BC was much lower than that of TiO_2, which indicated that TiO_2/p-BC had low recombination efficiency of electrons and holes. In addition, this indicated a high photocatalytic activity of TiO_2/p-BC. As an explanation of these results, the conducting p-BC can assist in the transfer of photogenerated electrons, thereby preventing the recombination of electrons and holes [35].

2.1.6. UV-Vis DRS Analyses

The UV-vis DRS spectra of TiO_2 and TiO_2/p-BC are presented in Figure 4a. For TiO_2, the strong adsorption was obtained for a wavelength less than 400 nm, which indicates that they had no obvious absorption in the visible light. However, TiO_2/p-BC showed obvious absorption above the wavelength of 400 nm. This indicated that TiO_2/p-BC had improved light utilization and higher photocatalytic activity as compared with TiO_2. Furthermore, the bandgaps (Eg) of TiO_2 and TiO_2/p-BC were 3.21 eV and 2.73 eV, respectively (Figure 4b), which were calculated by using the Kubelka-Munk equation [36]. This result indicated that the addition of p-BC contributed to the reduction of the Eg of the composite. According to the Mott–Schottky equation [37], the flat band potential (E_{FB}) of TiO_2 and TiO_2/p-BC were determined as −0.50 eV and −0.83 eV, respectively (Figure 4c). Generally, the lower conduction band potential (E_{CB}) of n-type semiconductors was approximately equal to E_{FB} [38]. Based on the above analyses, the upper valence band potentials (E_{VB}) of TiO_2 (2.71 eV) and TiO_2/p-BC (1.90 eV) could be estimated.

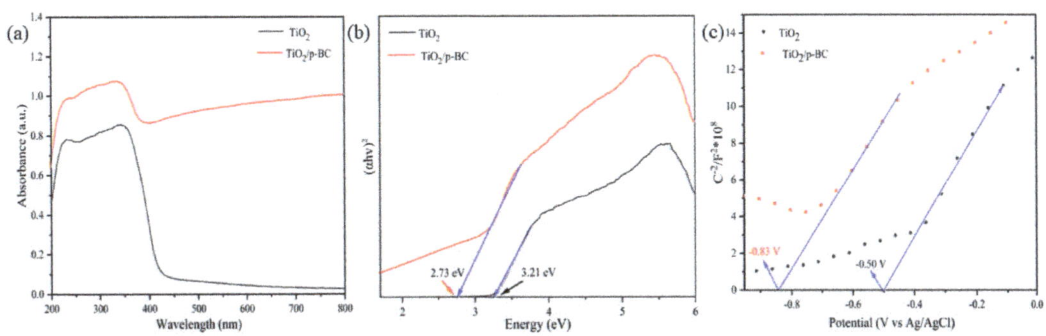

Figure 4. (**a**) UV–vis DRS spectra of TiO_2 and TiO_2/p-BC, (**b**) calculated corresponding bandgaps, and (**c**) corresponding Mott–Schottky curves.

2.2. Effect of Solution pH

2.2.1. Comparison of Removal Efficiency

Figure 5 shows the removal efficiency for three different samples on TC. The data show that p-BC can achieve a high removal efficiency of about 75.8% on TC. However, adsorption played a major role in this process, and light hardly contributed to the removal of TC. The adsorption only transfers TC from solution to p-BC, which may cause secondary pollution to the environment. The removal efficiency of TC by TiO_2 alone was 24.4 %. The adsorption of TiO_2 was weak, and photocatalytic was mainly responsible for the removal of TC. TiO_2/p-BC composite has excellent adsorption and photocatalytic effects on TC at the same time, and the removal efficiency was 94.5%. Compared with p-BC and TiO_2 alone, the removal efficiency of TC was increased by 18.7% and 70.1%, respectively. The recovered TiO_2/p-BC was sonicated in 30 ml deionized water for 30 min, and no TC was detected in the solution by the ultraviolet spectrophotometer, indicating that the adsorbed TC has been photocatalytically degraded. The synergistic effect of adsorption and photocatalysis was beneficial in improving the removal efficiency of TC by the TiO_2/p-BC composite.

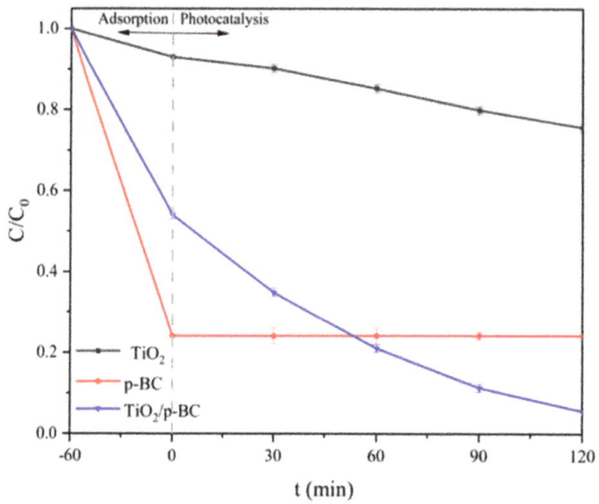

Figure 5. TC removal performance of different samples.

2.2.2. Effect of Solution pH

The initial pH value of the solution was a key parameter in the process of photocatalysis, which influenced the adsorption and photodegradation removal of pollutants. The zeta charge (pH_{PZC}) of TiO_2/p-BC was determined, as can be seen in Figure 6a. Also, at pH 4.23, the surface charge of TiO_2/p-BC was neutral. However, the surface charge was positive at pH values less than 4.23 and negative for pH values larger than 4.23. Also, the TC molecules had three different forms at different pH values: TC^+ ($pH < 3.3$), TC^0 ($3.3 < pH < 7.7$), and TC^- and TC^{2-} ($pH > 7.7$) [39]. The effects of pH on the removal of the TC are shown in Figure 6b. For the initial pH value of the solution, there were repulsive forces between the positive charges on the surface, which resulted in a low removal efficiency. Furthermore, the interactions between the charges on the surface were weak.

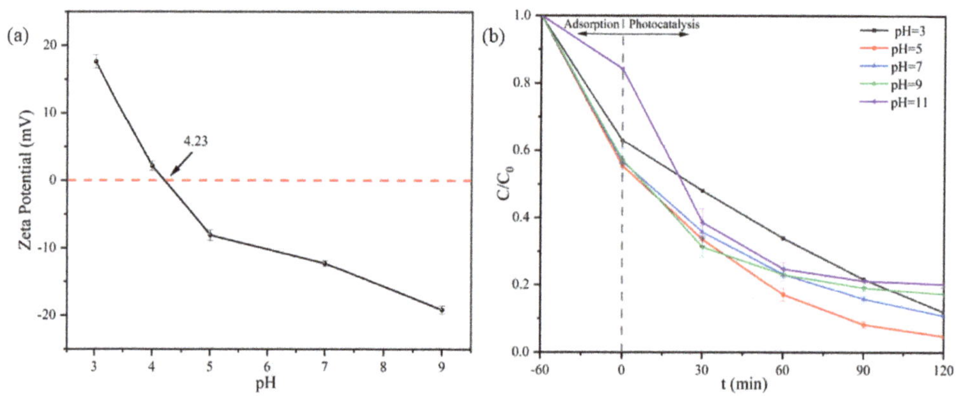

Figure 6. (**a**) Point of zeta charge of TiO_2/p-BC. (**b**) Removal efficiency of TC by TiO_2/p-BC at different pH values.

2.2.3. Effect of the TiO_2/p-BC Dosage

The effects of TiO_2/p-BC on the removal of TC are shown in Figure 7a. As can be seen in Figure 7a, with an increase in composite dosage from 0.025 g L^{-1} to 0.1 g L^{-1}, the removal efficiency increased from 47.4% to 96%. It thereafter increased with a further increase in the composite dosage (at a minor level). The removal of TC by the composite mainly included two processes: the adsorption of p-BC (dark stage) and the photocatalytic degradation of TiO_2 (light stage). Figure 7b shows the contribution of adsorption and photocatalytic degradation to the TC removal for different composite dosages. The results showed that the removal of TC by the removal of adsorption increased with an increase in composite dosage. With the dosage increase from 0.025 g L^{-1} to 0.1 g L^{-1}, the removal of TC by photocatalytic degradation increased from 27.3% to 50.7%. For a further increase in the composite dosage, the removal of photocatalytic degradation diminished further. The reason for this observation was that the high composite concentration blocked the penetration of light, thereby reducing the utilization of light [40]. Therefore, the composite dosage was selected as 0.1 g L^{-1}. At this dosage, the composite showed a high adsorption capacity and photocatalytic removal of TC and also ensured a high total removal efficiency.

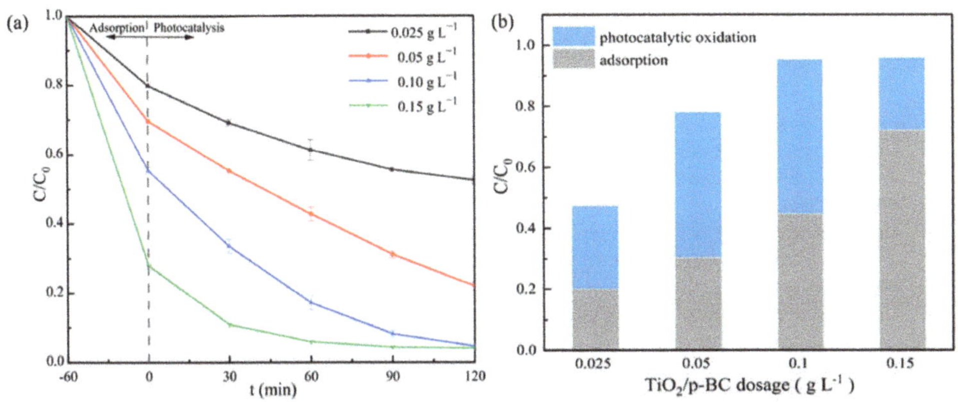

Figure 7. (**a**) Removal efficiency of TC at different dosages of TiO_2/p-BC. (**b**) Proportion of photocatalytic oxidation and adsorption at different dosages.

2.2.4. Effect of Initial TC Concentration

The effects of the initial concentration on the removal of TC are shown in Figure 8a. With the increase in initial TC concentration from 20 mg L^{-1} to 50 mg $^{-1}$, the removal efficiency decreased from 95.5% to 75.8%. Also, Figure 8b shows the contribution of adsorption and photocatalytic degradation to the TC removal under different initial concentrations. The results showed that adsorption played a key role in the removal of TC at the initial concentration of 20 mg L^{-1}, while the photocatalytic degradation contributed with a lesser degree to the removal. When the initial concentration was 30 mg L^{-1}, the composite had considerable adsorption and photocatalytic removal effect on TC, and the total removal efficiency was larger than 95%. Thereafter, with an increase in the initial concentration, the total removal efficiency of TC was less than 90%. One of the reasons was that the adsorption site of the composite was unchanged, and the increase in TC concentration led to a decrease in adsorption removal efficiency. In addition, the increase in TC concentration reduced the penetration of light and prevented light from reaching the surface of the composite, thereby reducing the utilization of light [41].

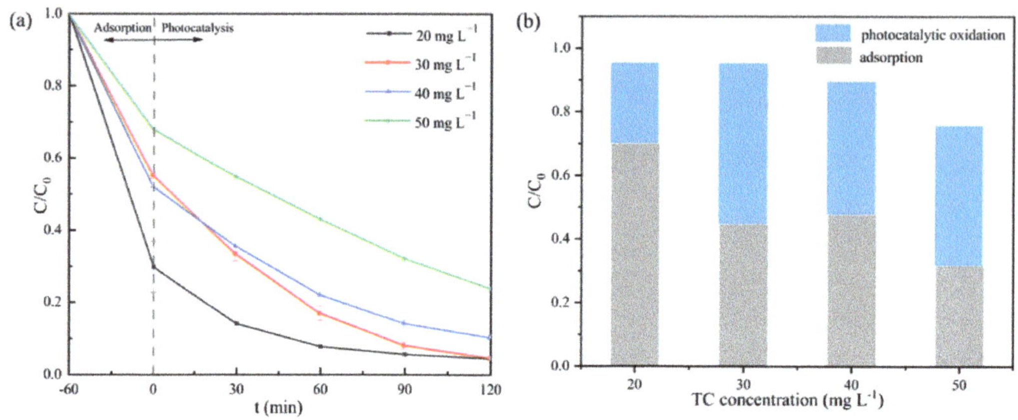

Figure 8. (**a**) Removal efficiency of TC at different initial concentrations of TiO_2/p-BC. (**b**) Proportion of photocatalytic oxidation and adsorption at different initial concentrations.

2.3. Reusability Test of TiO_2/p-BC

The reusability of the photocatalyst is related to the economic feasibility of the photocatalytic process, and it can also indicate the stability of the photocatalyst. In the present study, the reusability of TiO_2/p-BC has been evaluated by five successive photocatalysis experiments for the removal of TC. As can be seen in Figure 9, the removal efficiencies were 95.3%, 93.5%, 89.9%, 87.8%, and 86.5% after five cycles. Although the removal efficiency decreased slightly after each cycle, the removal efficiency of TC by TiO_2/p-BC remained above 86% after five cycles, indicating that it had excellent reusability and stability. The decrease in removal efficiency could be attributed to the accumulation of intermediates on the TiO_2/p-BC surface [30,42] and the detachment of a small amount of TiO_2 [43].

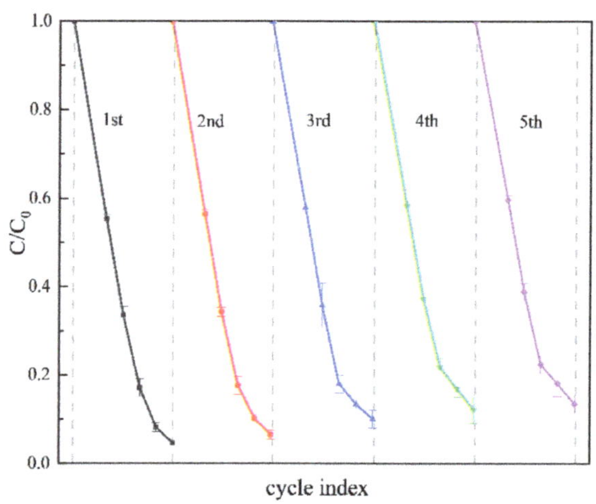

Figure 9. Cycles for the photodegradation of TC in the presence of TiO_2/p-BC.

2.4. Photocatalytic Removal Mechanism

Based on the above analysis results, a possible photocatalytic degradation mechanism of TC by TiO_2/p-BC has been proposed (Figure 10). It was found that p-BC, with an abundant porous structure and rich in functional groups, was an excellent carrier of TiO_2 and an adsorbent for the adsorption of TC. It, thereby, provided more contact opportunities for TiO_2 and TC. In addition, studies have confirmed that the combination with biochar was beneficial for the reduction of the bandgap energy of the photocatalyst. Compared with TiO_2 (3.21 eV), the bandgap of TiO_2/p-BC was 2.73 eV. In addition, the potentials of the conduction band (CB) and valence band (VB) were −0.83 eV and 1.90 eV, respectively. For sunlight excitation, the TiO_2/p-BC could generate electrons (e^-) from the valence band to the conduction band, with resulting holes (h^+) in the valence band. Acting as both an electron donor and acceptor, p-BC inhibited the recombination of electron-hole pairs through electron transport. The e^- could react with O_2 to generate $\cdot O_2^-$, and the h^+ could react with H_2O to generate ·OH [4]. Therefore, $\cdot O_2^-$, ·OH, and h^+ were all involved in the photocatalytic degradation of TC and eventually converted to CO_2 and H_2O.

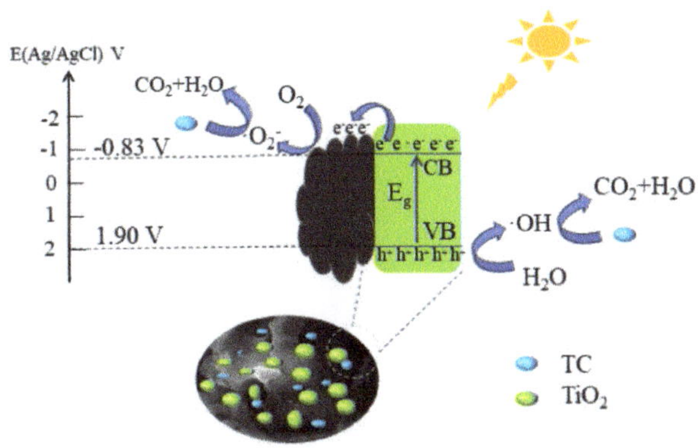

Figure 10. Schematic illustration of the TiO_2/p-BC photocatalytic enhancement mechanism in the TC degradation.

To explore the photocatalytic degradation mechanism of TC by TiO_2/p-BC, an active species capturing experiment was designed. BQ, IPA, and EDTA-2Na were then added to the reaction system as quenchers for $\cdot O_2^-$, ·OH, and h^+, respectively. The results (Figure 11) showed that the removal efficiency of TC decreased from 95.3% to 59.1%, 88.2%, and 74.7% in the presence of BQ, IPA, and EDTA-2Na, respectively. This indicated that $\cdot O_2^-$ made an important contribution to the photocatalytic degradation of TC, which was not the case for ·OH and h^+.

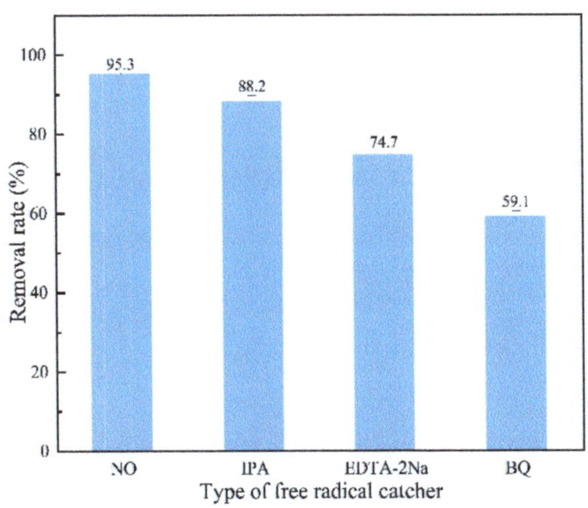

Figure 11. Quenching experiments of active species during the photocatalytic degradation of TC by TiO_2/p-BC (scavenger dosage: 0.1 mmol L^{-1}).

The intermediate products of TC photodegradation were analyzed by LCMS. The samples were detected after 15 min of illumination. The characteristic peak at *m/z* = 445 corresponded to the TC molecule. Also, the O_2^-, ·OH and h^+ that were generated in the reaction attacked the TC molecules, resulting in a C–C, C=C, C–N, and C–O cleavage in generating various intermediates. The became eventually degraded into CO_2, H_2O, and other small molecules. Based on the *m/z* value in the LCMS results (Figure S1 Supporting Information), four possible pathways for the photocatalytic degradation of TC have then been proposed (Figure 12).

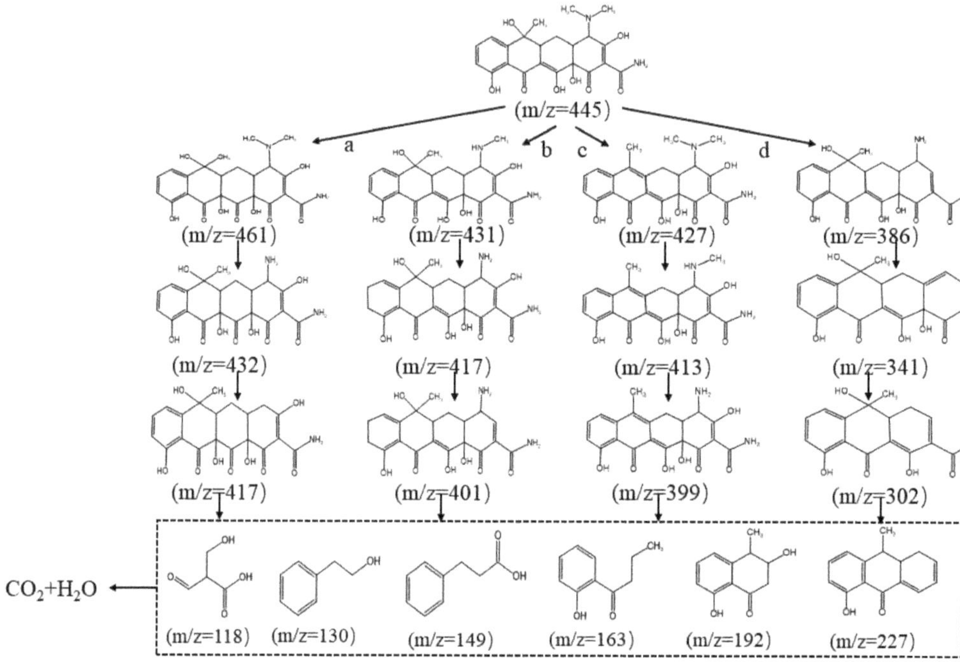

Figure 12. Possible degradation pathways of TC.

3. Materials and Methods

3.1. Materials

The peanut shell was obtained from a local market in Jinan, Shandong, China. Also, tetrabutyl titanate (99%), TC (99%), benzoquinone (BQ; 97%), ethylenediaminetetraacetic acid disodium salt (EDTA-2Na; 99%), and isopropyl alcohol (IPA; 99%) were purchased from Shanghai Macklin Biochemical Co. Ltd., Shanghai, China. In addition, anhydrous ethanol, glacial acetic acid, phosphoric acid, NaOH, and hydrochloric acid were purchased from Tianjin Fuyu Fine Chemical Co. Ltd., Tianjin, China. Furthermore, nitric acid was purchased from Shanghai Sinopharm Chemical Reagent Co. Ltd., Shanghai, China. All chemical reagents were analytically pure and without the need for further purification. Also, deionized water was used throughout all experiments.

3.2. Preparation of Modified Biochar (p-BC) and Composite Photocatalyst (TiO_2/p-BC)

The peanut shell was first cleaned with deionized water and then dried in an oven at 75 °C. Thereafter, it was crushed through a 100-mesh sieve. The method used for the following activation by the phosphoric acid was based on the previous work by Zhao et al. [44]. Briefly, 10 g of the peanut shell was impregnated with 30 mL of 50% H_3PO_4 for 24 h at room temperature. After filtration, the impregnated sample was dried in an oven at 110 °C for 5 h. Thereafter, the sample was placed in a crucible and pyrolyzed in a furnace at 500 °C for 2 h, with a slow rate of 5 °C min^{-1} in an N_2 atmosphere. As the next step, the sample was washed with deionized water until the filtrate had a neutral pH value. In the following step, the sample was dried in an oven at 110 °C and ground through a 100-mesh sieve to obtain a phosphoric acid-modified biochar (p-BC).

The TiO_2/p-BC composite was prepared through a modified sol–gel method [32]. The specific steps are as follows: (1) 6 mL of tetrabutyl titanate was added dropwise to 16 mL of anhydrous ethanol. Thereafter, the solution was stirred for 20 min to obtain a homogeneous solution, which was labeled as A. (2) Ten mL of anhydrous ethanol, 1.3 mL of deionized water, 0.7 mL of glacial acetic acid, and 0.7 mL of nitric acid (40 wt.%) were mixed into a homogeneous solution, and was labeled as B. (3) B was added dropwise to the A solution to obtain a mixed solution. Two grams of p-BC were added to this solution and stirred for 4 h before a static aging for 48 h. After 48 h, the gel was dried in an oven at 80 °C for 24 h and ground through a 100 mesh sieve. Finally, the solid was pyrolyzed in a furnace at 500 °C for 2 h, with a slow rate of 5 °C min^{-1} in an N_2 atmosphere. Thereby, the TiO_2/p-BC composite was obtained. The preparation process of the TiO_2/p-BC composite is displayed in Figure 13. Pure TiO_2 was prepared by the same method as above, except for the omission of p-BC.

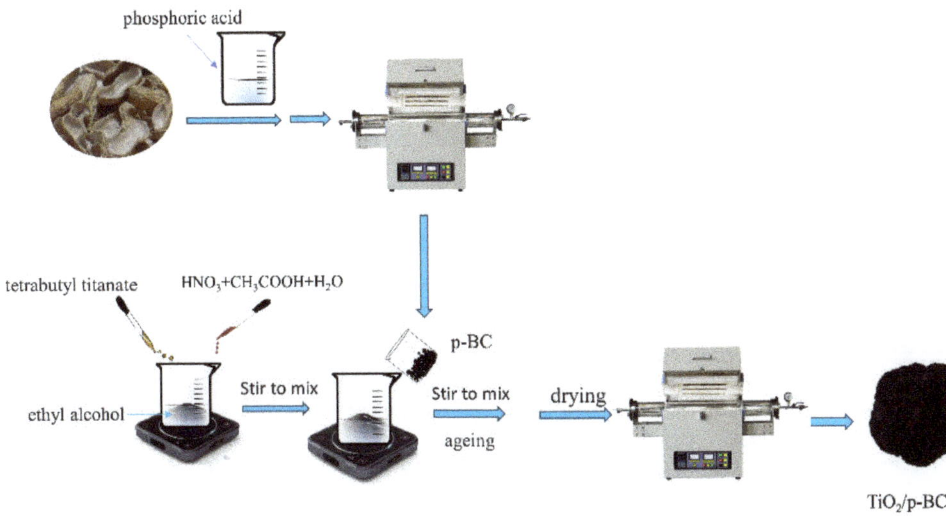

Figure 13. Schematic for the preparation of TiO_2/p-BC.

3.3. *Characterizations*

The surface morphologies and structures of the samples were characterized by scanning electron microscopy (SEM) (MIRA LMS, TESCAN, Brno, Czech Republic) with the working test parameters at 0.2–30 kV acceleration voltage. In addition, the surface chemical compositions and element-binding states of TiO_2/p-BC were analyzed by X-ray photoelectron spectroscopy (XPS) (K-Alpha, Thermo Scientific, Norristown, PA, USA) with Al Kα X-ray excitation source. Also, the crystalline phase of TiO_2/p-BC was examined by X-ray diffraction spectroscopy (XRD) (MiniFlex600, Rigaku, Tokyo, Japan) with Cu Kα radiation. Fourier-transform infrared spectroscopy (FTIR) (Nicolet iS20, Thermo Scientific, Norristown, PA, USA) was used to analyze the surface functional groups of the samples using KBr pellets in the range of 400–4000 cm^{-1}. Furthermore, the migration efficiency of the photoinduced electrons and holes was evaluated by photoluminescence spectroscopy (PL) at the excitation wavelength of 355 nm. (FLS1000, Edinburgh, UK). Ultraviolet–visible diffuse reflectance spectroscopy (UV-vis-DRS) (UV-3600, Shimadzu, Kyoto, Japan) was used to evaluate the optical absorption performance of TiO_2/p-BC. The zeta potential of the

TiO_2/p-BC was analyzed by nanoparticle size and zeta potential analyzer (DLS) (Zetasizer Nano ZS90, Malvern, UK). Finally, the TC concentration in the solutions was determined by an ultraviolet spectrophotometer at 356 nm, and a liquid chromatograph mass spectrometer (LCMS) (Ultimate 3000 UHPLC-Q, Thermo Scientific, Norristown, PA, USA) was used for the analyses of the intermediate products in the photodegradation process.

3.4. Photocatalytic Experiments

The photocatalytic experiments were conducted in a photochemical reactor (CME-PC4, microenerg, Beijing, China) with a 500 W xenon lamp as the sunlight source. For these experiments, 0.03 g of a photocatalyst and 30 mL of a 30 mg L^{-1} TC solution were evenly mixed in a quartz tube. Thereafter, this tube was placed in a dark environment for 1 h to ensure an adsorption/desorption equilibrium. As the next step, the photocatalytic reaction was conducted for a duration of 1 h. Three milliliters of the reaction solution was, thereafter, filtered through a 0.45 μm membrane, and the concentration of the residual TC in the solution was measured by a UV spectrophotometer. Magnetic stirrers were then used to ensure a uniform mixing of the solution during adsorption and photocatalytic degradation under dark conditions. Finally, the effective factors, with their ranges, for the removal of TC were especially investigated, including pH value (3–11), TC initial concentration (20 mg L^{-1}–50 mg L^{-1}), and photocatalyst dosage (0.025 g L^{-1}–0.15 g L^{-1}). The reusability of the photocatalyst was also evaluated by performing five repetitive cycles. After the photocatalytic experiment, the photocatalyst was recovered by filtration and washed several times. Thereafter, it was dried at 75 °C for the next coming use. The condition for each photocatalytic operation was identical to the one used for a fresh photocatalyst.

4. Conclusions

In the present study, the TiO_2/p-BC composite photocatalytic material has been prepared for the removal of TC. The characterization showed that the TiO_2 (with an anatase phase) was successfully synthesized to p-BC, and the crystalline form of TiO_2 was not affected in this process. Compared with TiO_2, the TiO_2/p-BC had a lower bandgap (2.73 eV) and recombination rates of electrons and holes. Thus, the TiO_2/p-BC demonstrated a higher light utilization and photocatalytic activity. Under the optimum conditions of an initial pH value of 5 in the solution, a composite dosage of 0.1 g L^{-1}, and an initial TC concentration of 30 mg $^{-1}$, the removal efficiency of TC by TiO_2/p-BC reached 95.3%. The removal efficiency remained above 86% after five consecutive cycles, which indicated that TiO_2/p-BC had excellent reusability and stability. Furthermore, the active species capturing experiment verified that $\cdot O_2^-$ made an important contribution to the photocatalytic degradation of TC, which was not the case with $\cdot$OH and h^+. At a low cost, the environmentally friendly, recyclable, and reusable photocatalytic material TiO_2/p-BC can be used for the efficient removal of TC and other organic pollutants.

Supplementary Materials: The following supporting information can be downloaded at https://www.mdpi.com/article/10.3390/catal14060357/s1. Figure S1: m/z of degrading TC over TiO_2/p-BC composite photocatalyst in different pathways (a–d) and m/z of the degraded small molecular fragments.

Author Contributions: Conceptualization, L.Z.; methodology, L.Z.; software, J.L., L.Z., H.W. and J.D.; validation.; Y.G.; formal analysis, M.S.; investigation, L.Z. and Z.Y.; resources, M.S.; data curation, L.J.; writing—original draft preparation, L.Z.; writing—review and editing, L.Z., Y.G. and X.Y.; funding acquisition, Y.G. All authors have read and agreed to the published version of the manuscript.

Funding: This research was funded by Shandong Provincial Key Research and Development Program, China (Major Scientific and Technological Innovation Project No. 2021CXGC011201; Rural Revitalization Project to Boost Science, Technology and Innovation, No. 2023TZXD003) and Shandong Provincial Natural Science Foundation, China (No. ZR2021MD126).

Data Availability Statement: The datasets used and analyzed in this study are available from the corresponding author upon reasonable request.

Conflicts of Interest: Author Jianhua Du was employed by the WSP Australia Pty Limited. The remaining authors declare that the research was conducted in the absence of any commercial or financial relationships that could be construed as a potential conflict of interest.

References

1. Ben, Y.; Fu, C.; Hu, M.; Liu, L.; Wong, M.H.; Zheng, C. Human health risk assessment of antibiotic resistance associated with antibiotic residues in the environment: A review. *Environ. Res.* **2019**, *169*, 483–493. [CrossRef] [PubMed]
2. Zhou, H.; Cui, J.; Pang, L.; Wangjin, Y.; Li, M.; Zhao, Z.; Huang, L. Removal of antibiotics and antibiotic resistance genes from urban rivers using a photocatalytic-and-bionic artificial ecosystem. *J. Clean. Prod.* **2022**, *348*, 131311. [CrossRef]
3. Zheng, D.; Yin, G.; Liu, M.; Chen, C.; Jiang, Y.; Hou, L.; Zheng, Y. A systematic review of antibiotics and antibiotic resistance genes in estuarine and coastal environments. *Sci. Total Environ.* **2021**, *777*, 146009. [CrossRef] [PubMed]
4. Wu, S.; Hu, H.; Lin, Y.; Zhang, J.; Hu, Y. Visible light photocatalytic degradation of tetracycline over TiO_2. *Chem. Eng. J.* **2020**, *382*, 122842. [CrossRef]
5. Yan, C.; Yang, Y.; Zhou, J.; Liu, M.; Nie, M.; Shi, H.; Gu, L. Antibiotics in the surface water of the Yangtze Estuary: Occurrence. distribution and risk assessment. *Environ. Pollut.* **2013**, *175*, 22–29. [CrossRef] [PubMed]
6. Yao, L.; Wang, Y.; Tong, L.; Deng, Y.; Li, Y.; Gan, Y.; Guo, W.; Dong, C.; Duan, Y.; Zhao, K. Occurrence and risk assessment of antibiotics in surface water and groundwater from different depths of aquifers: A case study at Jianghan Plain. central China. *Ecotoxicol. Environ. Saf.* **2017**, *135*, 236–242. [CrossRef] [PubMed]
7. Li, S.; Shi, W.; Liu, W.; Li, H.; Zhang, W.; Hu, J.; Ke, Y.; Sun, W.; Ni, J. A duodecennial national synthesis of antibiotics in China's major rivers and seas (2005–2016). *Sci. Total Environ.* **2018**, *615*, 906–917. [CrossRef] [PubMed]
8. Xiao, C.; Yuan, J.; Li, L.; Zhong, N.; Zhong, D.; Xie, Q.; Chang, H.; Xu, Y.; He, X.; Li, M. Photocatalytic synergistic biofilms enhance tetracycline degradation and conversion. *Environ. Sci. Ecotechnol.* **2023**, *14*, 100234. [CrossRef]
9. He, Y.; Huang, Z.; Ma, Z.; Yao, B.; Liu, H.; Hu, L.; Zhao, Q.; Yang, Q.; Liu, D.; Du, D. Highly efficient photocatalytic performance and mechanism of α-ZnTcPc/g-C_3N_4 composites for methylene blue and tetracycline degradation under visible light irradiation. *Appl. Surf. Sci.* **2019**, *498*, 143834. [CrossRef]
10. Chen, X.; Ke, Y.; Zhu, Y.; Xu, M.; Chen, C.; Xie, S. Enrichment of tetracycline-degrading bacterial consortia: Microbial community succession and degradation characteristics and mechanism. *J. Hazard. Mater.* **2023**, *448*, 130984. [CrossRef]
11. Weng, M.; Cai, M.; Xie, Z.; Dong, C.; Zhang, Y.; Song, Z.; Shi, Y.; Jin, M.; Wang, Q.; Wei, Z. Hydrodynamic cavitation-enhanced heterogeneous activation of persulfate for tetracycline degradation: Synergistic effects. degradation mechanism and pathways. *Chem. Eng. J.* **2022**, *431*, 134238. [CrossRef]
12. Rouhani, M.; Ashrafi, S.D.; Taghavi, K.; Joubani, M.N.; Jaafari, J. Evaluation of tetracycline removal by adsorption method using magnetic iron oxide nanoparticles (Fe_3O_4) and clinoptilolite from aqueous solutions. *J. Mol. Liq.* **2022**, *356*, 119040. [CrossRef]
13. Lin, Y.-C.; Zhuang, G.-L.; Tasi, P.-F.; Tseng, H.-H. Removal of protein. histological dye and tetracycline from simulated bioindustrial wastewater with a dual pore size PPSU membrane. *J. Hazard. Mater.* **2022**, *431*, 128525. [CrossRef] [PubMed]
14. Chen, T.; Luo, L.; Deng, S.; Shi, G.; Zhang, S.; Zhang, Y.; Deng, O.; Wang, L.; Zhang, J.; Wei, L. Sorption of tetracycline on H_3PO_4 modified biochar derived from rice straw and swine manure. *Bioresour. Technol.* **2018**, *267*, 431–437. [CrossRef]
15. Daghrir, R.; Drogui, P. Tetracycline antibiotics in the environment: A review. *Environ. Chem. Lett.* **2013**, *11*, 209–227. [CrossRef]
16. Li, L.; Yuan, X.; Zhou, Z.; Tang, R.; Deng, Y.; Huang, Y.; Xiong, S.; Su, L.; Zhao, J.; Gong, D. Research progress of photocatalytic activated persulfate removal of environmental organic pollutants by metal and nonmetal based photocatalysts. *J. Clean. Prod.* **2022**, *372*, 133420. [CrossRef]
17. Sendão, R.M.; da Silva JC, E.; da Silva, L.P. Photocatalytic removal of pharmaceutical water pollutants by TiO_2-Carbon dots nanocomposites: A review. *Chemosphere* **2022**, *301*, 134731. [CrossRef]
18. Zhang, Y.; Zhou, B.; Chen, H.; Yuan, R. Heterogeneous photocatalytic oxidation for the removal of organophosphorus pollutants from aqueous solutions: A review. *Sci. Total Environ.* **2023**, *856*, 159048. [CrossRef]
19. Zhang, X.; Chen, H.; Liu, S.; Zhang, B.; Zhu, H.; Chen, H.; Wen, B.; Chen, L. Preparation of TiO_2-graphitized carbon composite photocatalyst and their degradation properties for tetracycline antibiotics. *J. Mol. Struct.* **2022**, *1270*, 133897. [CrossRef]
20. Xu, L.; Tang, C.-Q.; Qian, J.; Huang, Z.-B. Theoretical and experimental study on the electronic structure and optical absorption properties of P-doped TiO_2. *Appl. Surf. Sci.* **2010**, *256*, 2668–2671. [CrossRef]
21. Ahmadi, M.; Motlagh, H.R.; Jaafarzadeh, N.; Mostoufi, A.; Saeedi, R.; Barzegar, G.; Jorfi, S. Enhanced photocatalytic degradation of tetracycline and real pharmaceutical wastewater using MWCNT/TiO_2 nano-composite. *J. Environ. Manag.* **2017**, *186*, 55–63. [CrossRef] [PubMed]
22. Trapalis, A.; Todorova, N.; Giannakopoulou, T.; Boukos, N.; Speliotis, T.; Dimotikali, D.; Yu, J. TiO_2/graphene composite photocatalysts for NOx removal: A comparison of surfactant-stabilized graphene and reduced graphene oxide. *Appl. Catal. B Environ.* **2016**, *180*, 637–647. [CrossRef]
23. Lu, L.; Shan, R.; Shi, Y.; Wang, S.; Yuan, H. A novel TiO_2/biochar composite catalysts for photocatalytic degradation of methyl orange. *Chemosphere* **2019**, *222*, 391–398. [CrossRef] [PubMed]
24. PePeng, H.; Gao, P.; Chu, G.; Pan, B.; Peng, J.; Xing, B. Enhanced adsorption of Cu (II) and Cd (II) by phosphoric acid-modified biochars. *Environ. Pollut.* **2017**, *229*, 846–853. [CrossRef]

25. Fazal, T.; Razzaq, A.; Javed, F.; Hafeez, A.; Rashid, N.; Amjad, U.S.; Rehman, M.S.U.; Faisal, A.; Rehman, F. Integrating adsorption and photocatalysis: A cost effective strategy for textile wastewater treatment using hybrid biochar-TiO_2 composite. *J. Hazard. Mater.* **2020**, *390*, 121623. [CrossRef] [PubMed]
26. Khan, H.; Shah, M.U.H. Modification strategies of TiO_2 based photocatalysts for enhanced visible light activity and energy storage ability: A review. *J. Environ. Chem. Eng.* **2023**, *11*, 111532. [CrossRef]
27. Kim, J.R.; Kan, E. Heterogeneous photocatalytic degradation of sulfamethoxazole in water using a biochar-supported TiO_2 photocatalyst. *J. Environ. Manag.* **2016**, *180*, 94–101. [CrossRef] [PubMed]
28. Yang, C.X.; Zhu, Q.; Dong, W.P.; Fan, Y.Q.; Wang, W.L. Preparation and characterization of phosphoric scid-modified biochar nanomaterials with highly efficient adsorption and photodegradation ability. *Langmuir* **2021**, *37*, 9253–9263. [CrossRef] [PubMed]
29. Yongheng, D.; Huayu, Y.; Jiang, L.; Qi, S.; Qianwen, Y.; Yuntao, Z. Direct Z-scheme P-TiO_2/g-C_3N_4 heterojunction for the photocatalytic degradation of sulfa antibiotics. *RSC Adv.* **2023**, *13*, 5957–5969. [CrossRef]
30. Wang, G.; Li, Y.; Dai, J.; Deng, N. Highly efficient photocatalytic oxidation of antibiotic ciprofloxacin using TiO_2@g-C_3N_4@biochar composite. *Environ. Sci. Pollut. Res.* **2022**, *29*, 48522–48538. [CrossRef]
31. Zheng, L.; Gao, Y.; Du, J.; Zhang, W.; Huang, Y.; Wang, L.; Zhao, Q.; Pan, X. A novel, recyclable magnetic biochar modified by chitosan–EDTA for the effective removal of Pb (II) from aqueous solution. *RSC Adv.* **2020**, *10*, 40196–40205. [CrossRef]
32. Zhang, H.; Wang, Z.; Li, R.; Guo, J.; Li, Y.; Zhu, J.; Xie, X. TiO_2 supported on reed straw biochar as an adsorptive and photocatalytic composite for the efficient degradation of sulfamethoxazole in aqueous matrices. *Chemosphere* **2017**, *185*, 351–360. [CrossRef] [PubMed]
33. Xie, Y.; Liu, A.; Bandala, E.R.; Goonetilleke, A. TiO_2-biochar composites as alternative photocatalyst for stormwater disinfection. *J. Water Process Eng.* **2022**, *48*, 102913. [CrossRef]
34. Khan, H.; Berk, D. Selenium modified oxalate chelated titania: Characterization. mechanistic and photocatalytic studies. *Appl. Catal. A Gen.* **2015**, *505*, 285–301. [CrossRef]
35. Mian, M.M.; Liu, G. Recent progress in biochar-supported photocatalysts: Synthesis. role of biochar, and applications. *RSC Adv.* **2018**, *8*, 14237–14248. [CrossRef] [PubMed]
36. Zhu, J.; Wang, J.; He, J.; Hu, L. Fabrication of CdS/ZnCr-LDH heterojunctions with enhanced of tetracycline hydrochloride photocatalytic degradation under visible light. *Opt. Mater.* **2023**, *136*, 113456. [CrossRef]
37. Zhang, Y.; Qi, F.; Li, Y.; Zhou, X.; Sun, H.; Zhang, W.; Liu, D.; Song, X.-M. Graphene oxide quantum dot-sensitized porous titanium dioxide microsphere: Visible-light-driven photocatalyst based on energy band engineering. *J. Colloid Interface Sci.* **2017**, *498*, 105–111. [CrossRef] [PubMed]
38. Geng, X.; Wang, L.; Zhang, L.; Wang, H.; Peng, Y.; Bian, Z. H_2O_2 production and in situ sterilization over a ZnO/g-C3N4 heterojunction photocatalyst. *Chem. Eng. J.* **2021**, *420*, 129722. [CrossRef]
39. Sun, S.; Yang, Z.; Cao, J.; Wang, Y.; Xiong, W. Copper-doped ZIF-8 with high adsorption performance for removal of tetracycline from aqueous solution. *J. Solid State Chem.* **2020**, *285*, 121219. [CrossRef]
40. Jiang, Y.; Liu, A. Cornstalk biochar-TiO_2 composites as alternative photocatalyst for degrading methyl orange. *Environ. Sci. Pollut. Res.* **2023**, *30*, 31923–31934. [CrossRef]
41. Zhang, L.; Song, X.; Liu, X.; Yang, L.; Pan, F.; Lv, J. Studies on the removal of tetracycline by multi-walled carbon nanotubes. *Chem. Eng. J.* **2011**, *178*, 26–33. [CrossRef]
42. Asgharzadeh, F.; Kalantary, R.R.; Gholami, M.; Jafari, A.J.; Kermani, M.; Asgharnia, H. TiO_2-decorated magnetic biochar mediated heterogeneous photocatalytic degradation of tetracycline and evaluation of antibacterial activity. *Biomass Convers. Biorefinery* **2023**, *13*, 8949–8959. [CrossRef]
43. Afzal, M.Z.; Zu, P.; Zhang, C.-M.; Guan, J.; Song, C.; Sun, X.-F.; Wang, S.-G. Sonocatalytic degradation of ciprofloxacin using hydrogel beads of TiO_2 incorporated biochar and chitosan. *J. Hazard. Mater.* **2022**, *434*, 128879. [CrossRef]
44. Zhao, N.; Zhao, C.; Lv, Y.; Zhang, W.; Du, Y.; Hao, Z.; Zhang, J. Adsorption and coadsorption mechanisms of Cr(VI) and organic contaminants on H_3PO_4 treated biochar. *Chemosphere* **2017**, *186*, 422–429. [CrossRef]

catalysts

MDPI

Article

Impact of Inorganic Anions on the Photodegradation of Herbicide Residues in Water by UV/Persulfate-Based Advanced Oxidation

Gabriel Pérez-Lucas, Aitor Campillo and Simón Navarro *

Department of Agricultural Chemistry, Geology and Pedology, School of Chemistry, University of Murcia, Campus Universitario de Espinardo, E-30100 Murcia, Spain; gpl2@um.es (G.P.-L.); aitor.campillol@um.es (A.C.)
* Correspondence: snavarro@um.es

Abstract: The removal of pesticides and other organic pollutants from water through advanced oxidation processes (AOPs) holds great promise. The main advantage of these technologies is that they remove, or at least reduce, pesticide levels by mineralization rather than transfer, as in conventional processes. This study first evaluated the effectiveness of $UV/S_2O_8^=$ compared to heterogeneous photocatalysis using UV/TiO_2 processes on the degradation of two commonly used herbicides (terbuthylazine and isoproturon) in aqueous solutions using a laboratory photoreactor. In addition, the effect of the UV wavelength on the degradation efficiency of both herbicides was investigated. Although the degradation rate was greater under $UV(254)/S_2O_8^=$ nm than under $UV(365)/S_2O_8^=$ nm, complete degradation of the herbicides (0.2 mg L^{-1}) was achieved within 30 min under UV-366 nm using a $Na_2S_2O_8$ dose of 250 mg L^{-1} in the absence of inorganic anions. To assess the impact of the water matrix, the individual and combined effects of sulfate ($SO_4^=$), bicarbonate (HCO_3^-), and chloride (Cl^-) were evaluated. These can react with hydroxyl ($HO^\bullet$) and sulfate ($SO_4^{\bullet-}$) radicals generated during AOPs to form new radicals with a lower redox potential. The results showed negligible effects of $SO_4^=$, while the combination of HCO_3^- and Cl^- seemed to be the key to the decrease in herbicide removal efficiency found when working with complex matrices. Finally, the main intermediates detected during the photodegradation process are identified, and the likely pathways involving dealkylation, dechlorination, and hydroxylation are proposed and discussed.

Keywords: matrix effect; herbicides; water treatment; persulfate; titanium dioxide; photooxidation

1. Introduction

Pesticides constitute an issue of considerable political interest in the fields of environment, agriculture, and human health under European legislation. The Directive on the Sustainable Use of Pesticides (2009/128/EC) [1] aims to reduce the risks and impacts of pesticide use on human health and the environment and to promote the use of integrated pest management and alternatives such as nonchemical approaches. In this context, water quality legislation is key to protecting public health and the environment in the European Union (EU). The Green Deal [2] and its associated strategies aim to reduce pesticide use and risks by 50% by 2030, with a focus on protecting ecosystems and enhancing biodiversity. For this purpose, the Water Framework Directive (WFD) 2000/60/EC [3] and its *"daughter directives"*, such as the Groundwater Directive (GWD) 2006/118/EC [4] and the Environmental Quality Standard Directive (EQSD) 2013/39/EU [5], help to protect European waters from pesticide and chemical pollution. In line with advances in science and technology over the past decades, on 26 October 2022, the EU Commission published its proposal to amend the WFD and EQSD [6]. These directives focus largely on so-called priority substances. In addition, the new Drinking Water Directive (DWD) 2020/2184 [7] sets quality objectives for pesticides at the tap and includes a *"risk-based approach"*, provisions for the protection of raw water, and the consideration of pesticide metabolites.

However, since the adoption of the EU Urban Wastewater Treatment Directive (UWWTD) 91/271/EEC [8] in 1991, the quality of Europe's rivers, lakes, and seas has improved dramatically. EU countries have used EU funding to build collection systems and wastewater treatment plants. However, there is still pollution that needs to be addressed and is not covered by current legislation. According to the European Environment Agency (EEA), a report based on data from Member States shows that only 38% of EU surface water bodies are in good chemical status, 46% are not in good chemical status, and 16% are in an unknown chemical status [9]. Another recent EEA technical report provides an overview of the occurrence of pesticides and their main metabolites (breakdown or reaction products) in surface water (lakes and rivers, 180 pesticides) and groundwater (159 pesticides), focusing on the European level (up to 39 European countries) from 2007 to 2017 [10]. This report shows exceedance rates of 5 to 15% for herbicides and 3 to 8% for insecticides in surface water, while in groundwater, the percentages were approximately 7% for herbicides and less than 1% for insecticides. Fungicides appeared to occur less frequently in both surface water and groundwater. To achieve maximum crop yields, herbicides are widely used. Agricultural activities, the cleaning of herbicide containers and equipment, and agricultural and herbicide manufacturing effluents are the main sources of herbicide water pollution. Two herbicides, isoproturon (ISP) and terbuthylazine (TBZ), are among the most frequently reported pesticides in surface water and groundwater in Europe. More recently, another study proposed a protocol for a systematic review and meta-analysis to determine the prevalence of pesticides in European water resources (watersheds, aquifers, rivers, the sea, and springs), wastewater (influent and effluent), and drinking water [11].

To address water pollution, the Commission has proposed an update to the Directive to enable a significant reduction in the discharge of pollutants into the EU's lakes, rivers, and seas [12]. In this sense, for some of the most persistent pollutants, such as recalcitrant pesticides, the polluted water (from urban, agricultural, and industrial sources) treated by conventional wastewater treatment plants (WWTPs) is in some cases insufficient to achieve the legally required level of purity [13]. This issue is of particular concern in areas where low rainfall does not provide sufficient water resources to meet the needs of agriculture, which requires increased reuse of effluents coming from WWTPs. Water scarcity and the uneven geographical distribution of rainfall are concerns in arid and semiarid areas where water management strategies advocate the reuse of treated wastewater in agriculture due to climate change. The EU has also addressed this issue by revising the minimum requirements for water reuse in the context of integrated management under Regulation 2020/741/EU on minimum requirements for water reuse [14]. The aim of this regulation is to ensure the safety of recycled water for agricultural irrigation, promote circularity, support resilience to climate change, and contribute to the objectives of the WFD by addressing water scarcity and related pressures on water resources. Growing public concern about the presence of pesticides in both wastewater and drinking water is leading to the development of new regulations that will undoubtedly have an impact on the design and operation of wastewater treatment plants in the coming years.

As stated in the EQSD, the sources of pollution must be identified, and the emissions of pollutants must be treated at the source in the most economical, environmental, and sustainable way. Concerns have increased, as has the need to identify them and apply new and effective techniques for their reduction and elimination. Conventional water treatment processes and technologies reduce pollution significantly but not as much as the current regulations require. The refractory nature of some pesticides, which can be toxic to the microorganisms used, often renders biological treatments ineffective [15]. In addition, approaches based on adsorption, flocculation, and ion exchange are not fully effective because they do not destroy contaminants but rather transfer contaminants to a solid phase that requires subsequent treatment. Membrane technology has been used extensively in recent years to remove micropollutants from wastewater, but during long-term processes, pollutants gradually accumulate in adsorbed materials until they reach saturation, after which they become inactive, and consequently, pollutant removal and membrane filtration efficiency decrease [16]; therefore, membrane technology is usually combined with other techniques, such as ozonation, activated carbon, or photooxidation [17].

Advanced oxidation processes (AOPs) have been the most studied in recent years for the removal of pesticides and other micropollutants from wastewater [18–20]. Pesticides are oxidized by highly reactive oxygen species (ROS), mainly hydroxyl radicals ($HO^{\bullet}/OH^-$, E^0 = 1.9–2.7 V vs. NHE) and others such as superoxide anion ($O_2^{\bullet-}$) and hydridodioxygen ($HO_2^{\bullet}$). The main advantage of these technologies is that they achieve the removal, or at least the reduction, of pesticide levels by mineralization rather than by transfer, as occurs in conventional processes [21]. These technologies are particularly useful in areas characterized by intensive agriculture and specific climatic patterns, where annual solar radiation is very high and water is scarce.

Among them, chemical reactions induced by a semiconductor photocatalyst that absorbs suitable radiation and remains unaffected during the process (heterogeneous photocatalysis) are of particular interest for water detoxification [22]. Semiconductors such as TiO_2, ZnO, Fe_2O_3, CdS, ZnS, and others can act as sensitizers for light-induced redox processes due to their electronic structure, which is characterized by a filled valence band (V_B) and an empty conduction band (C_B). The absorption of a photon with energy greater than the bandgap energy (E_g) leads to the formation of an electron (e^-)/hole (h^+) pair. In the absence of a suitable scavenger, the stored energy is dissipated by recombination within a few nanoseconds. If a suitable scavenger or surface defect is present to trap e^- or h+, recombination is prevented, and subsequent redox reactions can occur. The valence band h^+ is a strong oxidant (E^0 = 1.0–3.5 V vs. NHE) depending on the semiconductor and pH, while the conduction band e^- is a good reductant (E^0 = 0.5 to −1.5 V vs. NHE). Among the various semiconductor materials, TiO_2 (E_g = 3.1 eV) has been the most widely studied for environmental applications, mainly due to its nontoxicity, photostability, biological and chemical inertness, corrosion resistance, and chemical and thermal stability (over a wide pH range and at ambient pressure and temperature), among other properties. The photodegradation of pesticides in the presence of UV light and TiO_2 via redox reactions can be summarized according to the following mechanism [22]:

$$TiO_2 + h \rightarrow e^-_{CB} + h^+_{VB} \quad (1)$$

$$e^- + h^+ \rightarrow heat \quad (2)$$

$$h^+ + OH^- or\ H_2O \rightarrow HO^{\bullet} \quad (3)$$

$$e^- + O_2 \rightarrow O_2^{\bullet-} \quad (4)$$

$$O_2^{\bullet-} + H^+ \rightarrow HO_2^{\bullet} \quad (5)$$

$$HO_2^{\bullet} \rightarrow O_2 + H_2O_2 \quad (6)$$

$$O_2^{\bullet-} + H_2O_2 \rightarrow HO^{\bullet} + OH^- + O_2 \quad (7)$$

$$HO^{\bullet} + O_2^{\bullet-} + HO_2^{\bullet} + Pesticides \rightarrow Intermediates + CO_2 + H_2O + Mineral\ salts \quad (8)$$

More recently, AOPs based on sulfate radical anions ($SO_4^{\bullet-}$) have received increasing attention as alternatives to conventional $HO^{\bullet}$-based AOPs [23–25]. $SO_4^{\bullet-}$ has a similar redox potential ($SO_4^{\bullet-}/SO_4^{=}$, E^0 = 2.6–3.1 V vs. NHE) to that of $HO^{\bullet}$, higher selectivity, and, in some cases, longer half-lives ($t_{1/2}$ = 30–40 μs) than $HO^{\bullet}$ ($t_{1/2} \leq 1$ μs), allowing more contact and transfer of mass between radical and organic compounds [26]. $SO_4^{\bullet-}$ is typically generated from peroxydisulfate ($S_2O_8^{=}$, PDS), henceforth persulfate (PS) or peroxymonosulfate (HSO_5^-, PMS), which can be activated by energy (photochemical, sonochemical, and thermal), carbonaceous materials, electrochemical activation, alkaline conditions, transition metal ions, and various other oxidants such as ozone, hydrogen peroxide, and calcium peroxide [27,28]. PMS has a shorter bond length (1.460 Å) than PS

(1.497 Å), which explains its higher dissociation energy of the O-O bond (377 kJ mol^{-1}) compared to that of PS (92 kJ mol^{-1}) [23]. In most cases, UV/PS showed a better ability to oxidize organic pollutants than UV/PMS did. This can be attributed to the quantum yield of radical formation, as the quantum yield of PS_{254} is approximately 1.8 M $Einstein^{-1}$, which is much greater than that of PMS_{254} (0.5 M $Einstein^{-1}$) [23]. As a result, more energy is required for PMS to generate a radical during the homolytic cleavage of the peroxide bond [29]. In the thermal/photochemical decomposition pathway of PS, two moles of $SO_4^{\bullet-}$ are formed per mole of $S_2O_8^{=}$ as a result of the cleavage of the peroxide bond (Equation (9)). $SO_4^{\bullet-}$ reacts with H_2O at all pH values to form $HO^{\bullet}$ (Equation (10)), which is the primary reactive species under alkaline conditions (Equation (11)). At acidic (pH < 7) and alkaline (pH > 9) pH values, $SO_4^{\bullet-}$ and $HO^{\bullet}$ are the dominant reactive species, respectively, while both radicals participate similarly in reactions at circumneutral pH [23,30]. Finally, $SO_4^{\bullet-}$ promotes the mineralization of pesticides to CO_2 and H_2O (Equation (12)) according to the following reactions:

$$S_2O_8^{2-} + h\,(UV) \rightarrow 2SO_4^{\bullet-} \quad k = 5.8 \times 10^{-5}\ \mathrm{M^{-1}\,s^{-1}} \tag{9}$$

$$SO_4^{\bullet-} + H_2O \rightarrow HO^{\bullet} + SO_4^{2-} + H^{+} \quad k = 200\ \mathrm{M^{-1}\,s^{-1}} \tag{10}$$

$$SO_4^{\bullet-} + HO^{-} \rightarrow HO^{\bullet} + SO_4^{2-} \quad k = 6.5 \times 10^{7}\ \mathrm{M^{-1}\,s^{-1}} \tag{11}$$

$$SO_4^{\bullet-} + Pesticides \rightarrow Intermediates \rightarrow SO_4^{2-} + CO_2 + H_2O \tag{12}$$

However, both $SO_4^{\bullet-}$ and $HO^{\bullet}$ undergo rapid and unwanted reactions (Equations (13)–(15)) where they are consumed, which may limit their effectiveness [31]:

$$2SO_4^{\bullet-} \rightarrow S_2O_8^{2-} \quad k = 8.1 \times 10^{8}\ \mathrm{M^{-1}\,s^{-1}} \tag{13}$$

$$SO_4^{\bullet-} + HO^{\bullet} \rightarrow HSO_5^{-} \quad k = 1.0 \times 10^{10}\ \mathrm{M^{-1}\,s^{-1}} \tag{14}$$

$$SO_4^{\bullet-} + S_2O_8^{2-} \rightarrow SO_4^{2-} + S_2O_8^{\bullet-} \quad k = 6.1 \times 10^{9}\ \mathrm{M^{-1}\,s^{-1}} \tag{15}$$

In general, $SO_4^{\bullet-}$ is more susceptible to electron transfer reactions than is $HO^{\bullet}$. In contrast to $SO_4^{\bullet-}$, $HO^{\bullet}$ is more prone to hydrogen abstraction or addition [26,32]. Thus, $SO_4^{\bullet-}$-based oxidation is an alternative oxidative treatment to AOPs based on $HO^{\bullet}$ generation. $SO_4^{\bullet-}$ has several unique properties, such as being a very strong electron acceptor, which allows the degradation of persistent pesticides refractory to $HO^{\bullet}$. In addition, $S_2O_8^{=}$, an environmentally friendly nontoxic oxidizer, avoids the problem of transport limitations because it is relatively stable and can be produced in high quantities.

Overall, the elimination of pesticides and other micropollutants by $HO^{\bullet}$ and $SO_4^{\bullet-}$ based AOPs is largely influenced by the quality of the water matrix (dissolved constituents), which can be neutral, inhibitory, or promoting, depending on the process and the mechanism by which these water constituents react [23,33]. As with the experimental parameters, inorganic anions have an important influence on the performance of AOPs, although the influencing mechanisms of inorganic anions on their performance remain unclear. In general, inorganic anions can react with free radicals (usually called quenching effects), which affects the type and concentration of radicals, resulting in variations in the removal efficiency of the targeted organic pollutants. In addition to organic species, inorganic species such as Cl^{-}, HCO_3^{-}/CO_3^{2-}, NO_3^{-}/NO_2^{-}, PO_4H^{3-}, and/or SO_4^{2-}, among others, can also act as inhibitors either by scavenging and generating radicals such as $ClOH^{\bullet-}$, $Cl^{\bullet}$, $Cl_2^{\bullet-}$, $CO_3^{\bullet-}$, $HCO_3^{\bullet}$, $Br^{\bullet}$, $Br_2^{\bullet-}$, $NO^{\bullet}$, and/or $H_2PO_4^{\bullet}$, which have a lower E^0 than $HO^{\bullet}$ and $SO_4^{\bullet-}$ or by promoting the formation of ROS, as in the case of NO_3^{-}, which is capable of generating $HO^{\bullet}$ and $NO_2^{\bullet}$ radical species that promote the photodegradation of pesticides, especially those for which indirect photolysis is the main reactive pathway. In summary, the effect of inorganic anions on the performance of AOPs is complex. It is related to the type of inorganic anion, the type of oxidant, the types of organic pollutants targeted, and the surface physicochemical properties of the catalysts in heterogeneous catalysis. To accurately analyze the effect of inorganic anions on the performance of AOPs, it is necessary to understand their underlying influence mechanisms. Overall, the aim of this work was to evaluate the impact of inorganic ions on the removal of two herbicides (isoproturon and terbuthylazine) commonly found in wastewater using UV/persulfate-based advanced oxidation under laboratory conditions.

2. Results and Discussion

2.1. Photodegradation of Herbicides by UV, $Na_2S_2O_8$ and UV/$Na_2S_2O_8$

The degradation efficiency of the herbicides was defined according to the following equation (Equation (16)):

$$R_e(\%) = \left(\frac{H_0 - H_t}{H_0}\right) \times 100 \quad (16)$$

where R_e is the removal efficiency, H_0 is the initial concentration of herbicide and H_t is the concentration of herbicide at time t.

UV radiation has been proposed as an effective method for removing herbicides from water [34]. Most herbicides are photoactive because their structure generally contains aromatic rings, heteroatoms, and other functional groups that make them susceptible to absorbing UV–Vis radiation (direct photolysis) or reacting with photosensitive species capable of inducing the photodegradation of herbicides in water (indirect photolysis). The results of herbicide degradation using UV (254/366 nm), UV (366 nm), PS alone, UV (254/366 nm)/PS, and UV (366 nm)/PS combined (trials 1–5) are shown in Figure 1.

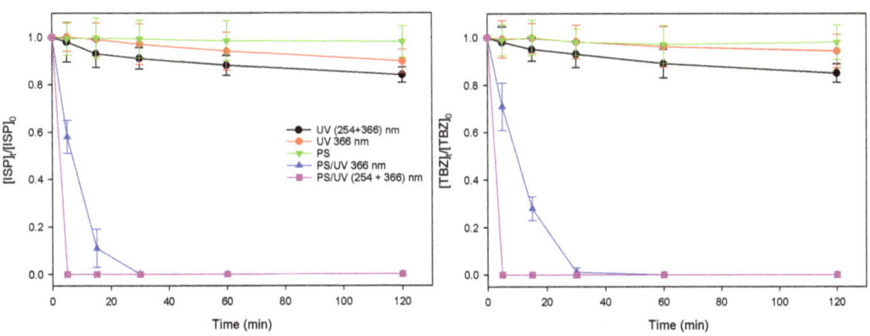

Figure 1. Evolution of herbicide residues over time in photolytic, PS, and UV/PS tests.

Minimal degradation (approximately 2% R_e) was observed for both herbicides when PS was used in the dark. When UV irradiation (254 or 366 nm) was used, the R_e of ISP increased to 16% (254 nm + 366 nm) and 10% (366 nm only), while TBZ removal was 15% (254 nm + 366 nm) and 7% (366 nm only) after a 120 min photoperiod. Some studies have shown that TBZ undergoes limited photodegradation and is considered photolytically and hydrolytically stable under environmentally relevant temperature and pH conditions [35]. However, photodegradation drastically increased in both cases when PS was activated with UV, either at a combined wavelength (254 + 366 nm) or at 366 nm only. After 5 min of irradiation (254 + 366 nm), the levels of both herbicides in the absence of inorganic anions were below LOD, whereas after 30 min of irradiation with 366 nm only, both herbicides practically disappeared. Similar results were obtained by Lin and Wu [36], who compared the effectiveness of PS activation at two wavelengths (254 nm and 365 nm) on the photodegradation of polyvinyl alcohol. PS activation by UV light might not be a promising method for industrial applications because of its high cost and complex equipment requirements. Although the activation of PS by UV/254 nm light improves the efficiency of the process, it is important to note that of the total radiant energy emitted by the Sun, only the part with a wavelength between 280–3000 nm, known as shortwave radiation, reaches the surface of our planet. This includes UVB (280–315 nm) and UVA

(315–400 nm) but not UVC (100–280 nm). Therefore, considering that the UVC fraction does not reach the Earth's surface, this fraction cannot be used in solar treatments for water purification because the spectrum of sunlight at the Earth's surface begins at approximately 300 nm. UV-366 nm/PS irradiation can therefore be considered an economical and effective process for the degradation of herbicides using natural sunlight. For this reason, the UVA fraction (366 nm) was used in subsequent tests because of the satisfactory results obtained.

2.2. Comparing UV/TiO_2 and UV/$Na_2S_2O_8$ Efficiencies

To compare the effectiveness of the two herbicides in degradation, UV/PS (test 5) and heterogeneous photocatalysis using UV/TiO_2 as a catalyst (test 6) were compared. Figure 2 shows the kinetic plots of both herbicides under these conditions. As seen, in both cases, a greater efficacy was observed for UV/PS than for UV/TiO_2, which is more notorious in the case of TBZ.

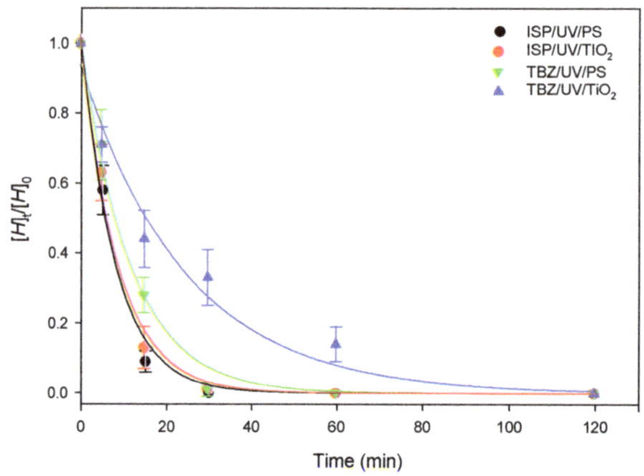

Figure 2. Degradation kinetics of herbicides in UV/PS and UV/TiO_2 systems.

According to the law of mass action, the photodegradation of herbicides can be modeled by assuming pseudo-first-order kinetics [37]. A kinetic study of the experimental results showed that the degradation process of both herbicides under UV basically conformed to the single first-order (SFO) kinetic model according to the following equations (Equations (17)–(19)):

$$-\frac{d[H]}{dt} = kt \tag{17}$$

$$[H]_t = [H]_0\, e^{-kt} \tag{18}$$

$$\ln[H]_t = \ln[H]_o - kt \tag{19}$$

where $[H]_t$ is the concentration of herbicide in solution at time t, $[H]_0$ is the initial herbicide concentration in solution, t is the time (min), and k is the apparent reaction rate constant (min^{-1}). The slope of $\ln(H_0/H_t)$ plotted against time gives the value of k. From the above equation, the time required for X% disappearance of herbicides (disappearance time) from the water can be calculated according to Equation (20):

$$DT_x = ln\left(\frac{100}{100-x}\right)/k \tag{20}$$

Consistent with the data obtained, the SFO model realistically fits the exponential decay curve with $R^2 \geq 0.97$ in all cases and a standard error of estimation ($S_{y/x}$) < 0.05. The ratio between the apparent rate constants (k_{PS}/k_{TiO2}) was > 1 in both cases, 1.1 for ISP and 3.3 for TBZ, indicating a higher reaction rate for both herbicides using UV/PS compared to

UV/TiO_2. For this reason, UV/PS was used to study the impact of inorganic anions on the photodegradation of the herbicides studied.

2.3. *Effect of Inorganic Anion Content on Herbicide Photodegradation Using UV/$Na_2S_2O_8$*

The effect of the composition of water on the photodegradation process is critical to assessing its suitability for real wastewater treatment, and its impact can be complex in some cases. In addition to dissolved organic matter, the unfavorable effects of some inorganic ions (mainly anions), such as Cl^-, HCO_3^-/CO_3^{2-}, NO_3^-/NO_2^-, PO_4H^{3-}, and/or SO_4^{2-}, can be explained by the fact that they reduce the oxidizing power of the solution. The scavenging of $HO^\bullet$ and $SO_4^{\bullet-}$ by various anions generates corresponding radicals such as $ClOH^{\bullet-}$, $Cl^\bullet$, $Cl_2^{\bullet-}$, $CO_3^{\bullet-}$, $HCO_3^\bullet$, $Br^\bullet$, $Br_2^{\bullet-}$, $NO^\bullet$, and/or $H_2PO_4^\bullet$, which have a lower E^0 than $HO^\bullet$ and $SO_4^{\bullet-}$ or promote the formation of ROS, as in the case of NO_3^-, which is capable of generating $HO^\bullet$ and $NO_2^\bullet$ radical species, which promote the photodegradation of pesticides, especially those for which indirect photolysis is the main reaction pathway [23,33,38]. Some anions and cations present in water are transparent to solar radiation, while nitrate (λ = 303 nm) and nitrite (λ = 355 nm) can absorb some of the radiation. Both absorb light and undergo homolysis to form $HO^\bullet$ and reactive nitrogen species such as $NO^\bullet$, $NO_2^\bullet$, N_2O_3, and/or N_2O_4, leading to herbicide degradation, although $HO^\bullet$ can be further scavenged by NO_2^- to form $NO_2^\bullet$ [39]. However, the concentrations of NO_3^-/NO_2 in real wastewater are usually low (<5 mg L^{-1}). For this reason, we assessed the influence of SO_4^{2-}, Cl^-, and HCO_3^-, which are present at much higher concentrations.

As shown in Figure 3, the removal efficiency of the herbicides was not affected by the presence of sulfate at the concentration studied (250 mg L^{-1}). Even the presence of SO_4^{2-} in the solution (test 7) had a positive effect on the photodegradation of both herbicides, especially for ISP, because higher values of k were obtained according to the results shown in Table 1.

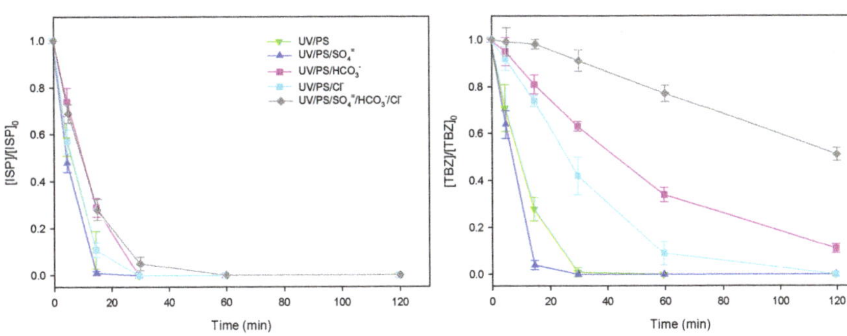

Figure 3. Anion-influenced UV/PS herbicide degradation.

Table 1. Kinetic parameters obtained following the SFO model for the photocatalytic degradation of herbicides in saline water.

Anion	Terbuthylazine (TBZ)				Isoproturon (ISP)			
	R	[1] k	[2] $S_{y/x}$	[3] DT_{50}	R	[1] k	[2] $S_{y/x}$	[3] DT_{50}
UV/PS	0.9952	0.1257	0.03	5.5	0.9953	0.1267	0.03	5.5
UV/PS/$SO_4^=$	0.9755	0.1287	0.08	5.4	0.9917	0.1672	0.04	4.1
UV/PS/HCO_3^-	0.9963	0.0175	0.02	40	0.9875	0.0844	0.05	8.2
UV/PS/Cl^-	0.9829	0.0307	0.06	23	0.9961	0.1284	0.03	5.4
UV/PS/$SO_4^=$ + HCO_3^- + Cl^-	0.9755	0.0054	0.03	128	0.9979	0.0849	0.03	8.2

[1] (min^{-1}); [2] Standard Error of Estimation; [3] (min).

$SO_4^=$ is not a strong scavenger of $HO^•$, such as Cl^-, HCO_3^-, and CO_3^{2-}. In contrast, it has been shown to promote the oxidative degradation of the antibiotics chloramphenicol [40] and ciprofloxacin [41] or polyvinyl alcohol [36], as occurs in our case. According to Cabrera-Reina et al. [42], the removal efficiency of acetamiprid, carbamazepine, and caffeine was also unaffected by the sulfate concentration in the range of 0–550 mg L^{-1}. Therefore, the impact of $SO_4^=$ on activated PS processes, as summarized in Equations (21) and (22), is very low.

$$SO_4^{2-} + HO^• \rightarrow SO_4^{•-} + OH^- \quad (21)$$

$$SO_4^{2-} + SO_4^{•-} \rightarrow S_2O_8^{2-} + e^- \quad (22)$$

Chloride ions (Cl^-) are major inorganic anions in natural water. The adverse effect of Cl^-, especially at high concentrations, on the degradation of many organic micropollutants has been reported by some authors [23]. In the case of ISP, a minimal influence was observed when chloride ions (150 mg L^{-1}) were present in the solution because $k_{PS}/K_{PS/Cl}$ is 0.98 (Table 1). However, a significant influence was observed for TBZ, where the $k_{PS}/K_{PS/Cl}$ ratio was 4.1, as shown in Figure 3. The adverse effects of high Cl^- concentrations could favor the scavenging of $HO^•$ and $SO_4^{•-}$ according to the following reactions (Equations (23)–(26)):

$$Cl^- + SO_4^{•-} \rightarrow SO_4^{2-} + Cl^• \quad (23)$$

$$Cl^- + HO^• \rightarrow ClOH^{•-} \quad (24)$$

$$ClOH^{•-} + H^+ \rightarrow H_2O + Cl^• \quad (25)$$

$$Cl^- + Cl^• \rightarrow Cl_2^{•-} \quad (26)$$

Compared to that of $HO^•$-based AOPs, radical scavenging by Cl^- in $SO_4^{•-}$-based AOPs is a major challenge. The reaction between Cl^- and $HO^•$ gives $ClOH^{•-}$ (Equation (24)), but this is a reversible reaction that mostly returns to $HO^•$ [43]. In contrast, $SO_4^{•-}$ produces $Cl^•$ (E^0 = 2.4 V) by one-electron abstraction from Cl^- (Equation (23)), which is often reflected in lower efficiencies in organic pollutant removal as well as greater generation of reactive chlorine species such as $Cl_2^{•-}$ (E^0 = 2.1 V), $ClO_2^•$ (E^0 = 0.9 V), ($ClO^•$ E^0 = 1.4 V), and $ClOH^{•-}$ (E^0 = 1.9) V, all of which have lower E^0 than $HO^•$ and $SO_4^{•-}$ [44].

An overall negative effect of HCO_3^- and CO_3^{2-} anions on $SO_4^{•-}$-based AOPs has traditionally been assumed [45]. The alkalinity of a water body is mainly influenced by CO_3^{2-} and HCO_3^- ions, which are generally present in natural waters at concentrations ranging from 50–250 mg L^{-1}. Both CO_3^{2-} and HCO_3^- are known to be radical scavengers in AOPs [23]. Among them, the radical scavenging ability of CO_3^{2-} is stronger than that of HCO_3^- due to its higher reaction rate constant with $HO^•$ (3.9×10^8 M^{-1} s^{-1} vs. 8.5×10^6 M^{-1} s^{-1}) [46].

However, depending on the chemical structure of the pesticide, the effect of HCO_3^-/CO_3^2 on the degradation rate can be neutral, positive, or negative, which is very difficult to predict because complex mixtures of pollutants are always present in wastewater [47,48]. Although $CO_3^{•-}$ has a lower redox potential than $HO^•$, it could exhibit better removal performance in degrading organic micropollutants, which could be due to its high selectivity and longer survival time in solution [49].

When HCO_3^- was added, the initial pH of the aqueous solution changed to approximately 8.3 rather than 6.7 without the addition of any anions, and the alkaline conditions reduced the degradation efficiency of both herbicides, as discussed by some authors [50]. CO_2, CO_3^{2-}, and HCO_3^- are present in aqueous media at pH > 4. CO_3^{2-} and HCO_3^-, which are responsible for water alkalinity, can compete with herbicides for $HO^•$ and $SO_4^{•-}$ to generate other weaker radicals, such as $CO_3^{•-}$ and/or $HCO_3^•$. Above pH = 10.3, CO_3^{2-} is the prevalent species, but at pH values below 8.3, all CO_3^{2-} is converted to HCO_3^- [51]. As the pH decreases, the HCO_3^- concentration also decreases, and the dissolved CO_2

concentration increases (Figure 4). The relationship between CO_3^{2-}/HCO_3^- and solution pH is represented as follows (Equation (27)):

$$pH = pK_a - log\frac{[HCO_3^-]}{[CO_3^{2-}]}\ pK_a = 10.3 \quad (27)$$

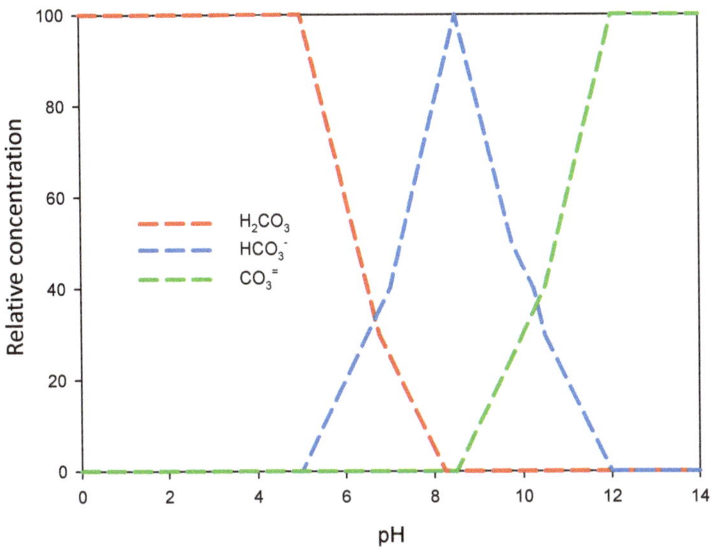

Figure 4. Speciation of carbonate species in water as a function of pH. Adapted from Manaham [51].

A high concentration (125 mg L^{-1}) of HCO_3^- strongly scavenges $HO^•$ and $SO_4^{•-}$, generating electrophilic $CO_3^{•-}$, a more selective and less reactive reaction radical with a lower electrode potential (E^0 = 1.6 V) than $HO^•$ and $SO_4^{•-}$ [52]. The detailed reaction mechanisms between $CO_3^{•-}$ and micropollutants are still largely unclear. In previous studies, the single electron transfer pathway was determined to be the dominant pathway [46]. The nature of $CO_3^{•-}$ makes it more selective toward electron-rich moieties such as -NH_2, -OH, and aromatic rings. $HCO_3^•$ has a lower reaction rate with organic pollutants than $HO^•$ does. Compared with $HO^•$, $SO_4^{•-}$ has a lower reaction rate with CO_3^{2-} ($6.1 \times 10^6\ M^{-1}\ s^{-1}$) and a similar reaction rate to CO_3^{2-} ($9.1 \times 10^6\ M^{-1}\ s^{-1}$) [53]. Nevertheless, the presence of CO_3^{2-}/HCO_3^- can cause the transformation of $HO^•$ and $SO_4^{•-}$, which can further affect the performance of AOPs. In radical-based treatment processes, $CO_3^{•-}$ and $HCO_3^•$ can be formed by the oxidation of CO_3^{2-} and HCO_3^- with highly reactive radical species, such as $HO^•$ and $SO_4^{•-}$ (Equations (28)–(31)) [54]:

$$CO_3^{2-} + HO^• \rightarrow CO_3^{•-} + OH^- \quad (28)$$

$$CO_3^{2-} + SO_4^{•-} \rightarrow CO_3^{•-} + SO_4^{2-} \quad (29)$$

$$HCO_3^- + HO^• \rightarrow H_2O + CO_3^{•-} \quad (30)$$

$$HCO_3^- + SO_4^{•-} \rightarrow SO_4^{2-} + HCO_3^• \rightarrow SO_4^{2-} + H^+ + CO_3^{•-} \quad (31)$$

In addition, carbonate precipitation can result in fouling of the membrane surface, which further affects the removal efficiency of micropollutants [55].

After studying the individual effects of $SO_4^=$, Cl^-, and HCO_3^- on the UV/PS process, their combined effect was then assessed (test 10). The R_e of both herbicides is presented in Figure 3. Cl^- (150 mg L^{-1}) and HCO_3^- (125 mg L^{-1}) generally had a negative effect on the herbicide R_e, although the magnitude depended on the chemical structure of each herbicide. As shown in Table 1, the R values ranged from 0.976 to 0.996 for TBZ and from 0.988 to 0.998 for ISP, with standard errors of estimation less than 0.08 and 0.05, respectively, indicating a good fit. The time to 50% disappearance (DT_{50}) for ISP was slightly greater in the presence of anions (8.2 min) than in the corresponding experiment carried out in the absence of anions (5.5 min). However, a substantial effect was observed for TBZ, as the DT_{50} was approximately 23 times greater than that in the absence of anions, mainly due to the impact of HCO_3^-.

The initial pH of the aqueous solution of herbicides in the absence/presence of inorganic anions decreased in all cases during the photoperiod (Table 2). This decrease is less pronounced in the presence of HCO_3^-, the predominant species at pH < 8.3. Acidification may be caused by the formation of acidic products as a consequence of herbicide degradation and acidic photoproducts derived from $S_2O_8^=$, such as HSO_4^-, with the release of H^+ (Equations (32) and (33)), as reported by Yang et al. [23]:

$$SO_4^{\bullet -} + H_2O \rightarrow HO^{\bullet} + HSO_4^- \quad (32)$$

$$HSO_4^- \rightarrow SO_4^{2-} + H^+ \quad (33)$$

Table 2. Changes in pH during the different tests.

Time (min)	pH									
	Trial									
	1	2	3	4	5	6	7	8	9	10
0	6.70	6.74	5.09	5.11	5.04	5.33	4.69	8.29	5.20	8.39
5	6.53	6.56	4.97	5.02	5.00	5.15	4.60	8.20	4.79	8.36
15	6.42	6.41	4.81	4.88	4.84	4.93	4.46	8.23	4.46	8.39
30	6.29	6.33	4.58	4.60	4.64	4.84	4.29	8.24	4.20	8.40
60	6.15	6.19	4.23	4.34	4.38	4.75	4.01	8.21	3.93	8.35
120	6.07	6.14	3.86	3.88	3.95	4.61	3.77	8.07	3.66	8.25

In addition, the concentration of $SO_4^=$ increased significantly throughout the process (Equation (25)), as seen from the EC values shown in Table 3, due to the transformation of the initially added $S_2O_8^{2-}$ into $SO_4^=$. Various methods have been proposed to remove $SO_4^=$ from water in recent decades, such as adsorption on activated carbon, neutralization with calcium carbonate, biological treatment, reverse osmosis and dialysis, and ion exchange. The choice of wastewater treatment method is usually based on the type of wastewater, removal rate, waste concentration, and cost of treatment. Adsorption on an ion exchange resin is the most popular method for removing $SO_4^=$ from water and wastewater and shows good potential for industrial wastewater treatment [56].

Table 3. Evolution of electrical conductivity (EC) in the different tests.

Time (min)	EC ($\mu S\ cm^{-1}$)									
	Trial									
	1	2	3	4	5	6	7	8	9	10
0	<5	<5	242	238	242	<5	760	377	610	1200
5	<5	<5	248	245	249	7	764	380	620	1234
15	<5	<5	255	252	254	10	769	383	644	1253
30	<5	<5	261	263	260	12	773	386	662	1259
60	<5	<5	273	270	266	18	795	389	674	1262
120	<5	<5	285	287	280	23	825	395	700	1265

Figure 5 shows the evolution of the DOC during the different tests. It is important to monitor the process using this tool because only DOC values close to zero guarantee that recalcitrant pollutants or intermediates with greater persistence and toxicity than the initial ones do not persist. In the absence of any ions (UV/PS) and in the presence of $SO_4^{=}$ (UV/PS/$SO_4^{=}$), a high mineralization rate was observed because the remaining DOC after 120 min was less than 5%. However, in the presence of HCO_3^- (UV/PS/HCO_3^-) and Cl^- (UV/PS/Cl^-), the residual DOC concentrations after 120 min were 20% and 10%, respectively, while in the presence of all anions in the reaction solution (UV/PS/$SO_4^{=}$ + HCO_3^- + Cl^-), the residual DOC concentration at the end of the experiment significantly increased (37%). This residual DOC fraction may be due to the partial removal of herbicides in an aqueous solution and the formation of nondegradable organic intermediates during irradiation.

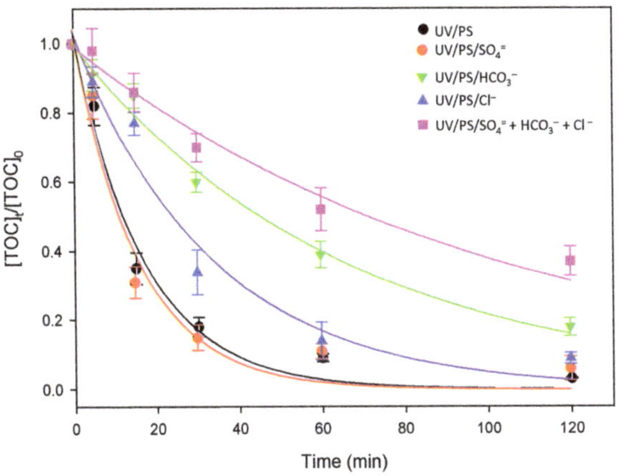

Figure 5. Evolution of DOC in the UV/PS system as a function of the anionic content.

2.4. Degradation Pathway of Herbicides with UV/$Na_2S_2O_8$

To gain a better understanding of the reaction mechanisms involved in the photodegradation of both herbicides, the evolution of key intermediates during the irradiation experiment was also compared with authentic analytical standards using quantitative HPLC-MS^2 analysis. Figure 6 shows the proposed degradation pathways for both herbicides.

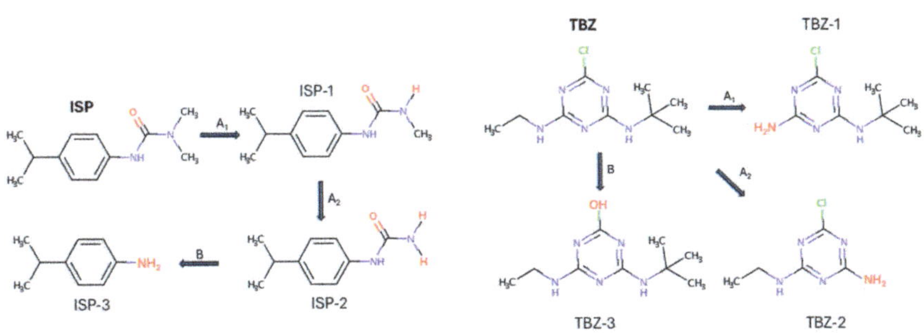

Figure 6. Proposed degradation pathways for ISP, including *N*-dealquilations (A1 and A2) and hydrolysis to aniline derivatives (B), and TBZ, involving dealkylations (A1 and A2), dechlorination, and subsequent hydroxylation (B).

The metabolic pathway involved in the degradation of ISP mainly involves two successive *N*-demethylations followed by hydrolysis to aniline-based metabolites, which can then be further degraded. Photooxidation of TBZ leads to the dealkylation of the amine groups, dechlorination, and subsequent hydroxylation [57].

Three transformation products of ISP, isoproturon-monodemethyl (ISP-1), isoproturon-desmethyl (ISP-2), and 4-isopropylaniline or cumidine (ISP-3), were identified during experiment 10, although their concentrations were below their LOQs (<2 μg L^{-1}) at the end of the photoperiod (120 min). The maximum concentration peak was found 30 min from the start, reaching levels between 4 and 8 μg L^{-1}. These intermediates were also isolated by other authors in aqueous suspensions of TiO_2 and ZnO under UV light [58,59]. ISP is classified as moderately hazardous (class II) by the World Health Organization (WHO) and cataloged as an endocrine disruptor [60]. In the case of TBZ, photooxidation of the parent compound was accompanied by the presence of desethyl-terbuthylazine (TBZ-1), desisopropyl-atrazine (TBZ-2), and hydroxy-terbuthylazine (TBZ-3), whose residual levels after 120 min were in the range of 1.2, 0.6, and 0.5 μg L^{-1}, respectively. TBZ, whose acceptable daily intake (ADI) is 0.004 mg kg^{-1} bw day^{-1}, is classified by the WHO as slightly hazardous (class III), while metabolites are not listed. Fenoll et al. [61] found TBZ-1 and TBZ-3 intermediates using TiO_2 and ZnO as photocatalysts. The presence of these degradants confirms the behavior observed in the evolution of DOC, indicating that there is no total mineralization of the herbicides in the presence of anions. It should be remembered that for mineralization of the pollutant to occur, not only must it disappear, but all the organic carbon must also be converted into inorganic carbon in the form of CO_2 since the DOC value is independent of the oxidation state of the compounds present in the aqueous solution.

3. Materials and Methods

3.1. *Herbicides, Solvents, and Reagents*

Analytical standards of the herbicides isoproturon (ISP), terbuthylazine (TBZ), and their main metabolites were purchased from Dr. Ehrenstorfer GmBh (Augsburg, Germany), all with purities ranging from 95–99.5%. The main physicochemical characteristics of the herbicides studied are shown in Table 4. Titanium dioxide (70 anatase/30 rutile, 99.5%, BET 50 m^2 g^{-1}, size < 21 nm) Aeroxide™ P25 was obtained from Nippon Aerosil Co., Ltd. (Osaka, Japan). All HPLC-grade solvents (H_2O, CH_3CN, CH_3OH, and CH_2O_2) and reagents ($Na_2S_2O_8$, NaCl, Na_2SO_4, and $NaHCO_3$) with purities > 98.5% were obtained from Scharlab (Barcelona, Spain).

Table 4. Main physicochemical properties of the herbicides studied.

Herbicide [1]	Structure	Formula	MW [2]	S_{H_2O} [3]	log K_{ow} [4]	*H* [5]
Isoproturon[PU]	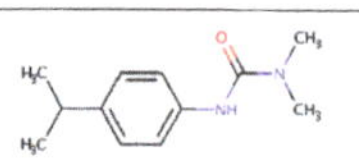	$C_{12}H_{18}N_2O$	206.3	70	2.5	1.5×10^{-5}
Terbuthylazine[TZ]	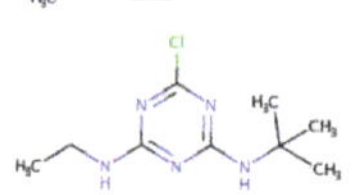	$C_9H_{16}ClN_5$	229.7	7	3.4	2.3×10^{-3}

[1] PU: Phenulurea; TZ: s-triazine; [2] Molecular weight; [3] Water solubility (mg L^{-1}); [4] Partition coefficient n-octanol/water; [5] Henry Law constant (Pa m^3 mol^{-1}).

3.2. *Experimental Setup*

The degradation of the herbicides was carried out by the UV/$S_2O_8^{=}$ or UV/TiO_2 process in a 2 L (10 cm × 25 cm long) Pyrex (jacketed borosilicate glass) batch cylindrical photoreactor (SBS, Barcelona, Spain) equipped with a magnetic stirrer, as previously described [62]. The study was carried out in batch circulation mode at a flow rate of 600 mL min^{-1}. The UV light source used was two Philips LIT 8 W (Amsterdam, The Netherlands) low-pressure mercury lamps with peak emissions at 254 and 366 nm. A typical irradiance of 10 mW cm^{-2} was measured using a Delta Ohm HD 2102.2 radiometer

(Caselle di Selvazzano, Italy). Nitrogen was continuously fed into the photoreactor at a flow rate of 0.5 L min^{-1} to ensure that oxygen did not interfere with the process. The system was thermostatically controlled by circulating water to maintain a constant temperature of 23 ± 1 °C. The photochemical reactor was filled with 2000 mL of Type II analytical grade water (pH 6.7, ORP 220 mV, conductivity < 20 μS cm^{-1}, TOC < 30 μg L^{-1}) and spiked with 0.2 mg L^{-1} of each herbicide. After the addition of 250 mg L^{-1} $Na_2S_2O_8$ or 250 mg L^{-1} TiO_2, the two UV lamps (1 × 254 nm + 1 × 366 nm or 2 × 366 nm) were switched on, and degradation was studied for 120 min. All experiments were replicated three times (n = 3). The initial pH of the water was 6.7. However, after the addition of $Na_2S_2O_8$, the pH decreased to approximately 5, while the addition of HCO_3^- increased the pH to 8.3. Table 5 summarizes the tests carried out.

Table 5. Summary of the trials carried out.

	Trials	pH	[a] EC	[b] $SO_4^=$	[b] Cl^-	[b] HCO_3^-
1	UV (1 × 254 nm + 1 × 366 nm)	6.7	<5	-	-	-
2	UV (2 × 366 nm)	6.7	<5	-	-	-
3	$Na_2S_2O_8$	5.1	240	-	-	-
4	$Na_2S_2O_8$/UV (254/366 nm)	5.1	242	-	-	-
5	$Na_2S_2O_8$/UV (366 nm)	5.0	246	-	-	-
6	TiO_2/UV (366 nm)	5.3	<5	-	-	-
7	$Na_2S_2O_8$/UV (366 nm)	4.7	760	250	-	-
8	$Na_2S_2O_8$/UV (366 nm)	8.3	377	-	-	125
9	$Na_2S_2O_8$/UV (366 nm)	5.2	610	-	150	-
10	$Na_2S_2O_8$/UV (366 nm)	8.4	1200	250	150	125

[a] μS cm^{-1}; [b] mg L^{-1}.

3.3. Sample Preparation and Analytical Determinations

The sample (25 mL) was passed through an Oasis® HLB 60 μm (500 mg) extraction cartridge purchased from Waters (Milford, MA, USA) using a Visiprep™ SPE vacuum manifold supplied by Supelco (Madrid, Spain) at a flow rate of approximately 3 mL min^{-1}. The extraction cartridge was previously conditioned with 3 mL of CH_3OH and equilibrated with 3 mL of acidified ultrapure water (pH = 3). After passing through the cartridge, the sample was washed with 5 mL of Milli-Q water (18.2 MΩ·cm), the eluate was discarded, and the column was air-dried. The analytes were then eluted with 5 mL of CH_3OH and the extracts were evaporated to dryness. Finally, the residues were dissolved in 500 μL of CH_3CN and filtered through 0.22 μm polytetrafluoroethylene (PTFE) syringe filters prior to chromatographic analysis.

The determination of herbicide residues was carried out using an HPLC system consisting of a Waters e2695 separation module (Waters, Milford, CT, USA) equipped with a quaternary pump and an autosampler and coupled to a Waters 2998 photodiode array detector (PDA). The data were processed using Water Empower software (version 3). Separation was performed on a 100 mm × 4.6 mm, 5 μm, Phenomenex Kinetex XB-C_{18} analytical column (Madrid, Spain) with a mobile phase consisting of CH_3CN (solvent A) and an aqueous solution of HCOOH (0.1%) (solvent B) under gradient mode, with the temperature of the column oven set to 25 °C. The following gradients were used: 30% of A for 1 min, linearly increased to 90% of A for 9 min, held for 1 min, and decreased to 30% of A for 2 min to allow equilibration before the next injection (5 min). The flow rate was maintained (0.5 mL min^{-1}), and the volume injected was 50 μL. The confirmation criteria were retention times and recovery wavelengths (190–400 nm). The detection wavelengths were 222 nm and 241 nm for TBZ and ISP, respectively. Standard solutions containing the herbicides were used to construct calibration curves (0.1–1000 μg L^{-1}). The limits of detection (LODs) and limits of quantitation (LOQs) were obtained by dividing the signal-to-noise ratios 3 and 10 times, respectively, by the angular coefficients of the calibration curves. In addition, HPLC-MS^2 analysis of the intermediates generated during the UV/PS process was performed on an Agilent (Santa Clara, CA, USA) 1200 HPLC system equipped with the same analytical column as previously described and coupled to an Agilent G6410A triple quadrupole (QqQ) mass spectrometer (MS) operating in electrospray ionization (ESI) positive ion mode. The most abundant fragment ion was selected for quantification, and the second was selected for identity confirmation. Data acquisition was performed using MassHunter software (V. 7.0).

The dissolved organic carbon (DOC) content was determined using a Multi N/C 3100 TOC Analyzer (Analytic Jena AG, Jena, Germany) after passing the samples through a nylon syringe filter (0.45 mm) prior to analysis to remove particulate OC from the sample. Both a GLP 21 pH meter and a GLP 31 conductivity meter (both from Crison Instruments, Barcelona, Spain) were used for pH and conductivity measurements, respectively. The statistical software SigmaPlot (Systat, Software Inc., San Jose, CA, USA) v.15 was used to fit the experimental data.

4. Conclusions

The need to increase the supply of water is linked to both the scarcity of the quantity available and the deterioration of its quality. Many families of chemical compounds, including pesticides, can have a negative impact on natural ecosystems and adversely affect water quality, thus posing a risk to human health. Many of them are carcinogenic, endocrine-disrupting, or even teratogenic and are highly persistent in the environment due to their complex structures and resistance to degradation. As a result of the current water scarcity, more emphasis should be placed on the 3*R*s (recovery, recycle, and reuse) approach to wastewater treatment, both industrial and agricultural. For this reason, it is now essential to develop remediation techniques that favor the total elimination of all traces of pollutants in water, both of natural and anthropogenic origin, to promote its subsequent reuse, all with the aim of achieving the zero-pollution goal announced in the European Green Deal in relation to the Chemicals Sustainability Strategy.

Increasingly stringent environmental regulations regarding the presence of emerging pollutants in wastewater and natural systems have driven the need for treatment technologies that minimize environmental hazards at a reasonable cost. Conventional water treatment processes and technologies reduce water pollution significantly but not to the extent required by current legislation. The use of artificial UV light in photochemical processes may not be a promising method for environmental applications due to its high cost and the need for complex equipment. However, in the case of solar-powered applications, the process is not expensive because it uses sunlight radiation.

AOPs, which are usually applied after biological processes, have recently emerged as effective tertiary treatments for the removal of pesticides and other micropollutants, but the oxidation rates of individual compounds can be strongly influenced by the constituents of the water matrix. Overall, the removal of the studied herbicides (isoproturon and terbuthylazine) by $HO^{\bullet -}$ and $SO4^{\bullet -}$-based AOPs is largely influenced by the quality of the water matrix (dissolved constituents), which can be neutral, inhibitory, or promoting, depending on the process and the mechanism by which these water constituents react. In addition to organic species, inorganic species, mainly chloride and carbonate/bicarbonate ions, can also act as inhibitors, either by scavenging or by generating new radicals with a lower E^0 than $HO^{\bullet}$ and $SO_4^{\bullet -}$. Herbicide photodegradation by UV/PS treatment is faster in pure water, largely due to the absence of organic and inorganic compounds that can absorb UV radiation and generate other ROS with lower oxidation potential.

Author Contributions: Conceptualization, S.N.; methodology, G.P.-L., A.C. and S.N.; investigation, G.P.-L., A.C. and S.N.; data curation, S.N.; writing–original draft preparation, G.P.-L. and S.N.; writing–review and editing, G.P.-L. and S.N.; supervision, S.N.; project administration, S.N.; funding acquisition, S.N. All authors have read and agreed to the published version of the manuscript.

Funding: This work was supported by the University of Murcia, Spain (Project 4711).

Data Availability Statement: The data will be made available upon request.

Conflicts of Interest: The authors declare no conflicts of interest.

References

1. EU. Directive 2009/128/EC of the European Parliament and of the Council of 21 October 2009 Establishing a Framework for Community action to achieve the sustainable use of pesticides. *Off. J. Eur. Union* **2009**, *L309*, 71–86. Available online: https://eur-lex.europa.eu/legal-content/EN/TXT/PDF/?uri=CELEX:32009L0128 (accessed on 15 February 2024).
2. EC. Communication from the Commission to the European Parliament, the Council, the European Economic and Social Committee and the Committee of the Regions—The European Green Deal, COM/2019/640. 2019. Available online: https://eur-lex.europa.eu/legal-content/EN/TXT/?uri=COM:2019:640:FIN (accessed on 15 February 2024).
3. EU. Directive 2000/60/EC of the European Parliament and of the Council of 23 October 2000 establishing a framework for Community action in the field of water policy. *Off. J. Eur. Union* **2000**, *L327*, 1–73. Available online: https://eur-lex.europa.eu/eli/dir/2000/60/oj (accessed on 15 February 2024).
4. EU. Directive 2006/118/EC of the European Parliament and of the Council of 12 December 2006 on the protection of groundwater against pollution and deterioration. *Off. J. Eur. Union* **2006**, *L372*, 19–31. Available online: https://eur-lex.europa.eu/legal-content/EN/TXT/PDF/?uri=CELEX:32006L0118 (accessed on 15 February 2024).
5. EU. Directive 2013/39/EU of the European Parliament and of the Council of 12 August 2013 amending Directives 2000/60/EC and 2008/105/EC as regards priority substances in the field of water policy. *Off. J. Eur. Union* **2013**, *L226*, 1–17. Available online: https://eur-lex.europa.eu/LexUriServ/LexUriServ.do?uri=OJ:L:2013:226:0001:0017:en:PDF (accessed on 15 February 2024).
6. EU. Proposal for a Directive of the European Parliament and of the Council amending Directive 2000/60/EC Establishing a Framework for Community Action in the Field of Water Policy, Directive 2006/118/EC on the Protection of Groundwater against Pollution and Deterioration and Directive 2008/105/EC on Environmental Quality Standards in the Field of Water Policy. COM/2022/540. 2022. Available online: https://eur-lex.europa.eu/legal-content/EN/TXT/?uri=CELEX:52022PC0540 (accessed on 15 February 2024).
7. EU. EU Parliament Council Directive (EU) 2020/2184 of the European Parliament and of the Council of 16 December 2020 on the quality of water intended for human consumption (recast). *Off. J. Eur. Union* **2020**, *L435*, 1–62. Available online: https://eur-lex.europa.eu/legal-content/EN/TXT/PDF/?uri=CELEX:32020L2184 (accessed on 15 February 2024).
8. EU. Council Directive of 21 May 1991 concerning urban wastewater treatment (91/271/EEC). *Off. J. Eur. Union* **1991**, *L135*, 40–91. Available online: https://eur-lex.europa.eu/legal-content/EN/TXT/PDF/?uri=CELEX:31991L0271 (accessed on 15 February 2024).
9. EEA. *European Waters—Assessment of Status and Pressures 2018*; EEA Report No 7/2018; European Environment Agency: Copenhagen, Denmark, 2018; Available online: https://www.eea.europa.eu/publications/state-of-water/ (accessed on 20 February 2024).
10. Mohaupt, V.; Völker, J.; Altenburger, R.; Birk, S.; Kirst, I.; Kühnel, D.; Küster, E.; Semeradova, S.; Šubelj, G.; Whalley, C. *Pesticides in European Rivers, Lakes and Groundwaters—Data Assessment. European Topic Centre on Inland, Coastal and Marine Waters*; Technical Report 1/2020; European Environment Agency: Magdeburg, Germany, 2020. [CrossRef]
11. Serrano-Valera, M.; Vela, N.; Piuvezam, G.; Mateo-Ramírez, F.; Santiago-Fernandes, I.D.; Martínez-Alcalá, I. Prevalence and concentration of pesticides in European waters: A protocol for systematic review and meta-analysis. *PLoS ONE* **2024**, *19*, e0282386. [CrossRef]
12. EC. Proposal for a Directive of the European Parliament and of the Council Concerning Urban Wastewater Treatment (Recast). SWD/2022/541. 2022. Available online: https://eur-lex.europa.eu/legal-content/EN/TXT/PDF/?uri=CELEX:52022SC0541 (accessed on 15 February 2024).
13. Parida, V.K.; Saidulu, D.; Majumder, A.; Srivastava, A.; Gupta, B.; Gupta, A.K. Emerging contaminants in wastewater: A critical review on occurrence, existing legislations, risk assessment, and sustainable treatment alternatives. *J. Environ. Chem. Eng.* **2021**, *9*, e105966. [CrossRef]
14. EU. Regulation (EU) 2020/741 of the European Parliament and of the Council of 25 May 2020 on minimum requirements for water reuse. *Off. J. Eur. Union* **2020**, *L177*, 32–55. Available online: https://eur-lex.europa.eu/legal-content/EN/TXT/PDF/?uri=CELEX:32020R0741 (accessed on 15 February 2024).
15. Alvarino, T.; Suarez, S.; Lema, J.; Omil, F. Understanding the sorption and biotransformation of organic micropollutants in innovative biological wastewater treatment technologies. *Sci. Total Environ.* **2018**, *615*, 297–306. [CrossRef]
16. Zhang, W.; Liang, W.; Zhang, Z.; Hao, T. Aerobic granular sludge (AGS) scouring to mitigate membrane fouling: Performance, hydrodynamic mechanism and contribution quantification model. *Water Res.* **2021**, *188*, e116518. [CrossRef] [PubMed]
17. Vasilachi, I.C.; Asiminicesei, D.M.; Fertu, D.I.; Gavrilescu, M. Occurrence and fate of emerging pollutants in water environment and options for their removal. *Water* **2021**, *13*, e181. [CrossRef]
18. Rout, P.R.; Zhang, T.C.; Bhunia, P.; Surampalli, R.Y. Treatment technologies for emerging contaminants in wastewater treatment plants: A review. *Sci. Total Environ.* **2021**, *753*, e141990. [CrossRef] [PubMed]
19. Saha, M.P. *Advanced Oxidation Processes for Effluent Treatment Plants*; Elsevier: Amsterdam, The Netherlands, 2021.
20. Soto-Verjel, J.; Maturana, A.Y.; Villamizar, S.E. Advanced catalytic oxidation coupled to biological systems to treat pesticide contaminated water: A review on technological trends and future challenges. *Water Sci. Technol.* **2022**, *85*, 1263–1294. [CrossRef] [PubMed]

21. Miklos, D.B.; Remy, C.; Jekel, M.; Linden, K.G.; Drewes, J.E.; Hübner, U. Evaluation of advanced oxidation processes for water and wastewater treatment—A critical review. *Water Res.* **2018**, *139*, 118–131. [CrossRef] [PubMed]
22. Augugliaro, V.; Palmisano, G.; Palmisano, L.; Soria, J. Heterogeneous photocatalysis and catalysis: An overview of their distinctive features. In *Heterogeneous Photocatalysis*; Marcì, G., Palmisano, L., Eds.; Elsevier: Amsterdam, The Netherlands, 2019; pp. 1–24. [CrossRef]
23. Yang, Q.; Ma, Y.; Chen, F.; Yao, F.; Sun, J.; Wang, S.; Yi, K.; Hou, L.; Li, X.; Wang, D. Recent advances in photoactivated sulfate radical-advanced oxidation process (SR-AOP) for refractory organic pollutants removal in water. *Chem. Eng. J.* **2019**, *378*, e122149. [CrossRef]
24. Lee, J.; Von Gunten, U.; Kim, J.H. Persulfate-based advanced oxidation: Critical assessment of opportunities and roadblocks. *Environ. Sci. Technol.* **2020**, *54*, 3064–3081. [CrossRef] [PubMed]
25. Brillas, E. Activation of persulfate and peroxymonosulfate for the removal of herbicides from synthetic and real waters and wastewaters. *J. Environ. Chem. Eng.* **2023**, *11*, e110380. [CrossRef]
26. Olmez-Hanci, T.; Arslan-Alaton, I. Comparison of sulfate and hydroxyl radical based advanced oxidation of phenol. *Chem. Eng. J.* **2013**, *224*, 10–16. [CrossRef]
27. Oh, W.D.; Dong, Z.; Lim, T.T. Generation of sulfate radical through heterogeneous catalysis for organic contaminants removal: Current development, challenges and prospects. *App. Catal. B Environ.* **2016**, *194*, 169–201. [CrossRef]
28. Wang, J.; Wang, S. Activation of persulfate (PS) and peroxymonosulfate (PMS) and application for the degradation of emerging contaminants. *Chem. Eng. J.* **2018**, *334*, 1502–1517. [CrossRef]
29. Wacławek, S.; Lutze, H.V.; Grübel, K.; Padil, V.V.T.; Černík, M.; Dionysiou, D.D. Chemistry of persulfates in water and wastewater treatment: A review. *Chem. Eng. J.* **2017**, *330*, 44–62. [CrossRef]
30. Matzek, L.W.; Carter, K.E. Activated persulfate for organic chemical degradation: A review. *Chemosphere* **2016**, *151*, 178–188. [CrossRef] [PubMed]
31. Ren, W.; Huang, X.; Wang, L.; Liu, X.; Zhou, Z.; Wang, Y.; Lin, C.; He, M.; Ouyang, W. Degradation of simazine by heat-activated peroxydisulfate process: A coherent study on kinetics, radicals and models. *Chem. Eng. J.* **2021**, *426*, e131876. [CrossRef]
32. Serrano, K.G. Indirect Electrochemical Oxidation Using Hydroxyl Radical, Active Chlorine, and Peroxodisulfate. In *Electrochemical Water and Wastewater Treatment*; Martínez-Huitle, C.A., Rodrigo, M.A., Scialdone, O., Eds.; Elsevier: Amsterdam, The Netherlands, 2018. [CrossRef]
33. Ribeiro, A.R.L.; Moreira, N.F.F.; Li Puma, G.; Silva, A.M.T. Impact of water matrix on the removal of micropollutants by advanced oxidation technologies. *Chem. Eng. J.* **2019**, *363*, 155–173. [CrossRef]
34. Orellana-Garcia, F.; Alvarez, M.A.; Lopez-Ramon, V.; Rivera-Utrilla, J.; Sanchez-Polo, M.; Mota, A.J. Photodegradation of herbicides with different chemical natures in aqueous solution by ultraviolet radiation. Effects of operational variables and solution chemistry. *Chem. Eng. J.* **2014**, *255*, 307–315. [CrossRef]
35. ECHA. *Proposal for Harmonized Classification and Labeling*; CLH-227-637-9 Report for Terbuthylazine; European Chemicals Agency: Merseyside, UK, 2014; Available online: https://echa.europa.eu/substance-information/-/substanceinfo/100.025.125 (accessed on 20 March 2024).
36. Lin, C.C.; Wu, M.S. UV/$S_2O_8{}^{2-}$ process for degrading polyvinyl alcohol in aqueous solutions. *Chem. Eng. Process.* **2014**, *85*, 209–215. [CrossRef]
37. Liu, B.; Zhao, X.; Terashima, C.; Fujishima, A.; Nakata, K. Thermodynamic and kinetics analysis of heterogeneous photocatalysis for semiconductor systems. *Phys. Chem. Chem. Phys.* **2014**, *16*, 8751–8760. [CrossRef]
38. Nafradi, M.; Alapi, T.; Bencsik, G.; Janaky, C. Impact of reaction parameters and water matrices on the removal of organic pollutants by TiO_2/LED and ZnO/LED heterogeneous photocatalysis using 365 and 398 nm radiation. *Nanomaterials* **2022**, *12*, e5. [CrossRef]
39. Mack, J.; Bolton, J.R. Photochemistry of nitrite and nitrate in aqueous solution: A review. *J. Photochem. Photobiol. A Chem.* **1999**, *128*, 1–13. [CrossRef]
40. Ghauch, A.; Baalbaki, A.; Amasha, M.; El Asmar, R.; Tantawi, O. Contribution of persulfate in UV-254 nm activated systems for complete degradation of chloramphenicol antibiotic in water. *Chem. Eng. J.* **2017**, *317*, 1012–1025. [CrossRef]
41. Lin, C.C.; Wu, M.S. Degradation of ciprofloxacin by UV/$S_2O_8{}^{2-}$ process in a large photoreactor. *J. Photochem. Photobiol. A Chem.* **2014**, *285*, 1–6. [CrossRef]
42. Cabrera-Reina, A.; Aliste, M.; Polo-López, M.I.; Malato, S.; Oller, I. Individual and combined effect of ions species and organic matter on the removal of microcontaminants by Fe^{3+}-EDDS/solar-light activated persulfate. *Water Res.* **2023**, *230*, e119566. [CrossRef]
43. Buxton, G.V.; Greenstock, C.L.; Helman, W.P.; Ross, A.B. Critical review of rate constants for reactions of hydrated electrons, hydrogen atoms and hydroxyl radicals in aqueous solution. *J. Phys. Chem. Ref. Data* **1988**, *17*, 513–886. [CrossRef]
44. Armstrong, D.A.; Huie, R.E.; Koppenol, W.H.; Lymar, S.V.; Merényi, G.; Neta, P.; Ruscic, B.; Stanbury, D.M.; Steenken, S.; Wardman, P. Standard electrode potentials involving radicals in aqueous solution: Inorganic radicals (IUPAC Technical Report). *Pure Appl. Chem.* **2015**, *87*, 1139–1150. [CrossRef]
45. Bennedsen, L.R.; Muff, J.; Søgaard, E.G. Influence of chloride and carbonates on the reactivity of activated persulfate. *Chemosphere* **2012**, *86*, 1092–1097. [CrossRef]

46. Canonica, S.; Kohn, T.; Mac, M.; Real, F.J.; Wirz, J.; von Gunten, U. Photosensitizer method to determine rate constants for the reaction of carbonate radical with organic compounds. *Environ. Sci. Technol.* **2005**, *39*, 9182–9188. [CrossRef]
47. Acero, J.L.; Benítez, F.J.; Real, F.J.; Rodríguez, E. Degradation of selected emerging contaminants by UV-activated persulfate: Kinetics and influence of matrix constituents. *Sep. Purif. Technol.* **2018**, *201*, 41–50. [CrossRef]
48. Lebik-Elhadi, H.; Frontistis, Z.; Ait-Amar, H.; Madjene, F.; Mantzavinos, D. Degradation of pesticide thiamethoxam by heat-activated and ultrasound-activated persulfate: Effect of key operating parameters and the water matrix. *Process Saf. Environ. Prot.* **2020**, *134*, 197–207. [CrossRef]
49. Sbardella, L.; Gala, I.V.; Comas, J.; Layret, R.R.; Gernjak, W. The impact of wastewater matrix on the degradation of pharmaceutically active compounds by oxidation processes including ultraviolet radiation and sulfate radicals. *J. Hazard. Mater.* **2019**, *380*, e120869. [CrossRef]
50. Lin, C.C.; Lee, L.T.; Hsu, L.J. Performance of UV/$S_2O_8^{2-}$ process in degrading polyvinyl alcohol in aqueous solutions. *J. Photochem. Photobiol. A Chem.* **2013**, *252*, 1–7. [CrossRef]
51. Manaham, S.E. *Environmental Chemistry*, 9th ed.; CRC Press: Boca Raton, FL, USA, 2010.
52. Zuo, Z.; Cai, Z.; Katsumura, Y.; Chitose, N.; Muroya, Y. Reinvestigation of the acid−base equilibrium of the (bi) carbonate radical and pH dependence of its reactivity with inorganic reactants. *Radiat. Phys. Chem.* **1999**, *55*, 15–23. [CrossRef]
53. Ma, J.; Yang, Y.; Jiang, X.; Xie, Z.; Li, X.; Chen, C.; Chen, H. Impacts of inorganic anions and natural organic matter on thermally activated persulfate oxidation of BTEX in water. *Chemosphere* **2018**, *190*, 296–306. [CrossRef]
54. Xiao, R.; Meng, Y.; Fu, Y.; Wacławek, S.; Wei, Z.; Spinney, R.; Dionysiou, D.; Hu, W.P. The overlooked carbonate radical in micropollutant degradation: An insight into hydration interaction. *Chem. Eng. J.* **2023**, *474*, e145245. [CrossRef]
55. Mikhaylin, S.; Bazinet, L. Fouling on ion-exchange membranes: Classification, characterization and strategies of prevention and control. *Adv. Colloid Interface Sci.* **2016**, *229*, 34–56. [CrossRef]
56. Haghsheno, R.; Mohebbi, A.; Hashemipour, H.; Sarraf, A. Study of kinetic and fixed bed operation of removal of sulfate anions from an industrial wastewater by an anion exchange resin. *J. Hazard. Mater.* **2009**, *166*, 961–966. [CrossRef]
57. Roberts, T.; Hutson, D. Metabolic Pathways of Agrochemicals. In *Part One: Herbicides and Plant Growth Regulators*; The Royal Society of Chemistry: Cambridge, UK, 1998.
58. Haque, M.M.; Muneer, M. Heterogeneous photocatalyzed degradation of an herbicide derivative, isoproturon in aqueous suspension of titanium dioxide. *J. Environ. Manag.* **2003**, *69*, 169–176. [CrossRef]
59. Fenoll, J.; Sabater, P.; Navarro, G.; Pérez-Lucas, G.; Navarro, S. Photocatalytic transformation of sixteen substituted phenylurea herbicides in aqueous semiconductor suspensions: Intermediates and degradation pathways. *J. Hazard. Mater.* **2013**, *244*, 370–379. [CrossRef]
60. Lewis, K.A.; Tzilivakis, J.; Warner, D.J.; Green, A. An international database for pesticide risk assessments and management. *Hum. Ecol. Risk Assess.* **2016**, *22*, 1050–1064. [CrossRef]
61. Fenoll, J.; Hellin, P.; Martinez, C.M.; Flores, P.; Navarro, S. Semiconductor-sensitized photodegradation of s-triazine and chloroacetanilide herbicides in leaching water using TiO_2 and ZnO as catalyst under natural sunlight. *J. Photochem. Photobiol. A Chem.* **2012**, *238*, 81–87. [CrossRef]
62. Vela, N.; Fenoll, J.; Garrido, I.; Navarro, G.; Gambín, M.; Navarro, S. Photocatalytic mitigation of triazinone herbicide residues using titanium dioxide in slurry photoreactor. *Catal. Today* **2015**, *252*, 70–77. [CrossRef]

catalysts

MDPI

Article

Interstitial N-Doped TiO_2 for Photocatalytic Methylene Blue Degradation under Visible Light Irradiation

Dezheng Li [1,†], Vilanculo Clesio Calebe [1,†], Yuqiao Li [1,†], Huimin Liu [1,*] and Yiming Lei [2,*]

1 School of Chemical and Environmental Engineering, Liaoning University of Technology, Jinzhou 121001, China; ldz221882002@163.com (D.L.); clesio.vilanculo@gmail.com (V.C.C.); yuqiao.li@njust.edu.cn (Y.L.)
2 Departament de Química, Facultat de Ciències, Universitat Autònoma de Barcelona (UAB), Cerdanyola del Valles, 08193 Barcelona, Spain
* Correspondence: liuhuimin08@tsinghua.org.cn (H.L.); yiming.lei@uab.cat (Y.L.)
† These authors contributed equally to the manuscript.

Abstract: Photocatalysis is a promising method for methylene blue (MB) degradation due to its effectiveness and environmental compatibility. Among the photocatalysts, titanium dioxide (TiO_2) has been widely used for MB degradation due to its exceptional photocatalytic activity. However, the wide bandgap limits the degradation efficiency of TiO_2 under visible light. Here, an interstitial nitrogen-doped TiO_2 (5%N_T/TiO_2) used thiourea as the N source was fabricated for visible light-derived MB degradation. The 5%N_T/TiO_2 exhibited an extended absorption range of visible light. Moreover, photoelectrochemical measurements showed an improvement in the photocurrent response and charge transfer behavior on N/TiO_2. Thus, 5%N_T/TiO_2 had enhanced photocatalytic activity compared with pristine TiO_2 and substitutive N-doped TiO_2 (5%N_{AB}/TiO_2). The accelerated photocatalytic MB degradation process on N/TiO_2 could be mainly attributed to the interstitial N doping, which caused the appearance of new energy states and extended optical properties. Through comparing the impact of interstitial and substitutive in TiO_2 activity, our work proposes a suitable form of element doping to enhance the optical properties and photocatalytic activity of TiO_2 and even other semiconductors, providing guidance for future work.

Keywords: TiO_2; nitrogen doping; photocatalytic activity; methylene bule degradation; visible light

1. Introduction

Methylene blue (MB) is a common dye used in various industries [1,2]. However, with its widespread utilization, MB leads to significant environmental challenges, including water pollution, toxicity to aquatic life, bioaccumulation, and potential human health risks [3,4]. Thus, effective removal of MB is crucial for ecosystems and human health. Several strategies have been developed for MB remediation, such as advanced oxidation processes (AOPs), adsorption techniques, bioremediation, and membrane filtration [5,6]. Among them, based on AOP technology, photocatalytic degradation of organic compounds is an ideal method for the removal of industrial pollutants with solar energy as the only input energy source. Improving photocatalytic degradation efficiency has been attracting a great deal of research interest.

The design of the photocatalyst is so important for photocatalytic degradation that it determines the whole reaction's efficiency [7–10]. Titanium dioxide (TiO_2) is a widely studied semiconductor material known for its excellent photocatalytic properties, chemical stability, and non-toxicity. It has garnered significant attention for applications in environmental remediation, particularly in the degradation of organic pollutants. However, the practical application of TiO_2 is limited by its wide bandgap (~3.2 eV for anatase), which restricts its photo-response to the ultraviolet light region, constituting only about 5% of the solar spectrum. In order to improve the photoconversion efficiency and photocatalytic degradation properties of TiO_2, it is necessary to develop a reliable strategy for extending the photo-response range of TiO_2 photocatalysts. Nitrogen doping is a promising approach to modify the electronic structure of TiO_2 and enhance its capabilities for visible light absorption [11,12]. By incorporating nitrogen atoms into the TiO_2 lattice, the bandgap can be narrowed, allowing it to absorb visible light and thus improving its photocatalytic efficiency under solar irradiation [13–15]. Although numerous studies have proved the positive influences of N element doping in the photocatalytic activity of TiO_2, the effects of different N precursors on achieving N doping should be further explored to elucidate the impact of N atoms on the structure and activity of TiO_2.

In this study, nitrogen-doped TiO_2 catalysts (denoted x% N_T/TiO_2 and x% N_{AB}/TiO_2, with x being the wt% of N element) were prepared from two different N sources (i.e., thiourea and ammonium bicarbonate) and used for photocatalytic MB degradation under visible light irradiation. Compared with TiO_2 and N_{AB}/TiO_2, N_T/TiO_2 showed better photocatalytic degradation performance. The content of N elements was optimized, showing that 5% N_T/TiO_2 possessed the highest MB degradation efficiency, with more than 60% of the MB removed within 150 min. In order to elucidate the effect of N element doping and clarify the advantages of thiourea as the N source, a series of characterizations were conducted, which confirmed the interstitial N doping in N_T/TiO_2. N atoms from thiourea were mainly incorporated into the lattice of N_T/TiO_2. On the other hand, N atoms from ammonium bicarbonate were mainly incorporated into N_{AB}/TiO_2 in the form of oxygen substitution. Moreover, N_T/TiO_2 possessed an improved capacity for visible light absorption. The photoelectrochemical experiments indicated that the interstitial N doping increased the number of photoexcited electrons from N_T/TiO_2 under visible light illumination. Therefore, with enhanced optical properties and photoelectric response, N_T/TiO_2 with interstitial N atoms displayed the optimal MB degradation efficiency. This work proved the advantages of interstitial N doping in improving the photocatalytic activity of semiconductors represented by TiO_2.

2. Results and Discussion

2.1. Optimization of the Amount of N Doping

The concentration of nitrogen doping was optimized by gradually increasing the amount of nitrogen elements in 5%N_T/TiO_2 in terms of the degradation efficiency of MB. Figure 1 shows the efficiencies of photocatalytic MB degradation with different amounts of N doping. Under visible light irradiation, undoped TiO_2 exhibited ~15% MB degradation after 150 min (Figure 1). The 5%N_T/TiO_2 had the highest MB photodegradation efficiency. After 150 min, approximately 56% of the MB was degraded. The decline in the activity of 10%N_T/TiO_2 and 15%N_T/TiO_2 could be attributed to the excessive N atoms, which may be recombination sites of photoexcited charges, as found by previous reports [16–18]. After determining the optimal amount of N doping, namely 5%, in the subsequent research and analysis, the amount of doping remained consistent, and only the N source was changed.

2.2. XRD Patterns of TiO_2, 5%N_T/TiO_2, and 5%N_{AB}/TiO_2 Catalysts

X-ray diffraction (XRD) analysis was performed to determine the crystalline phases and crystallographic structure of 5%N_T/TiO_2, 5%N_T/TiO_2, and TiO_2 catalysts (Figure 2a). The undoped TiO_2 mainly consisted of the rutile crystalline phase. The peaks centered at 27.5°, 36.1°, 41.3°, and 54.3° corresponded to the (110), (101), (111), and (211) crystal planes of the rutile phase, respectively [19]. When TiO_2 was doped with thiourea as the nitrogen source, 5%N_T/TiO_2 mainly showed the anatase phase. The peaks centered at 25.2°, 37.8°, 48.0°, and 55.0° corresponded to the (101), (004), (200), and (211) crystal planes of the anatase phase, respectively [19]. However, the 5%N_{AB}/TiO_2 with ammonium bicarbonate as the nitrogen source mainly showed a mixed crystal phase of rutile and anatase. This transformation in the crystal form may be attributed to the fact that when N is doped into the TiO_2 lattice, it will replace some of the O atoms or enter the lattice's interstitial positions. This will cause a distortion in the lattice structure of TiO_2 and generate stress. When the stress accumulates to a certain extent, in order to reduce the system's energy, the crystal structure will change, thus leading to a change in the crystal form. Moreover, the (101) crystal plane of 5%N_T/TiO_2 shifted from 25.2° to 26.1° (Figure 2b). It indicated that the interstitial doping, which can cause lattice distortion, had been achieved in 5%N_T/TiO_2. However, 5%N_{AB}/TiO_2 maintained two TiO_2 phases without a peak shift, meaning that the doping mode of N in 5%N_{AB}/TiO_2 may be substitutive doping.

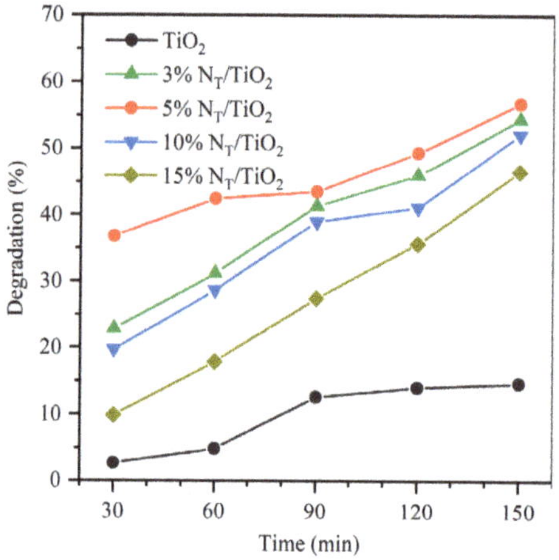

Figure 1. The degradation rates of an MB solution by N_T/TiO_2 photocatalysts with different amounts of N doping.

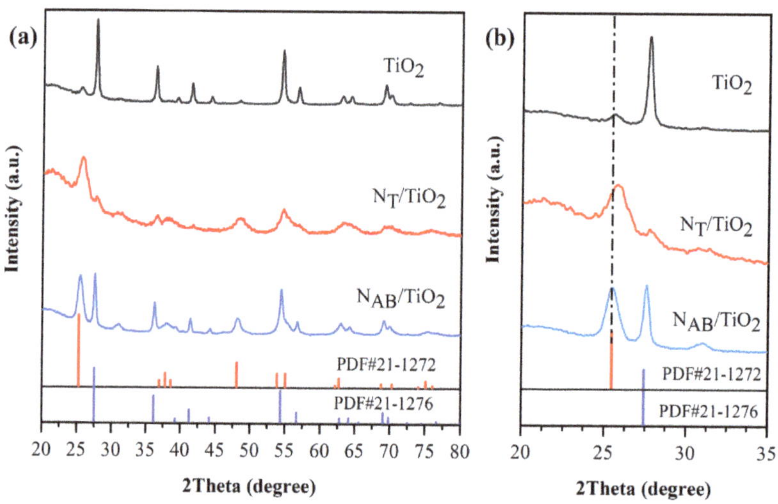

Figure 2. XRD patterns of (**a**) TiO_2, 5%N_T/TiO_2, and 5%N_{AB}/TiO_2, and (**b**) local magnification image of XRD.

2.3. *SEM Images and Corresponding Element Mappings of 5%N_T/TiO_2*

The elemental distribution of the 5%N_T/TiO_2 catalyst was observed by scanning electron microscopy (SEM). As can be seen from Figure 3b–d, the N element showed a very uniform distribution state on the sample and had the same distribution as the Ti and O elements. This phenomenon strongly indicated that the N element had been successfully doped into the lattice of TiO_2.

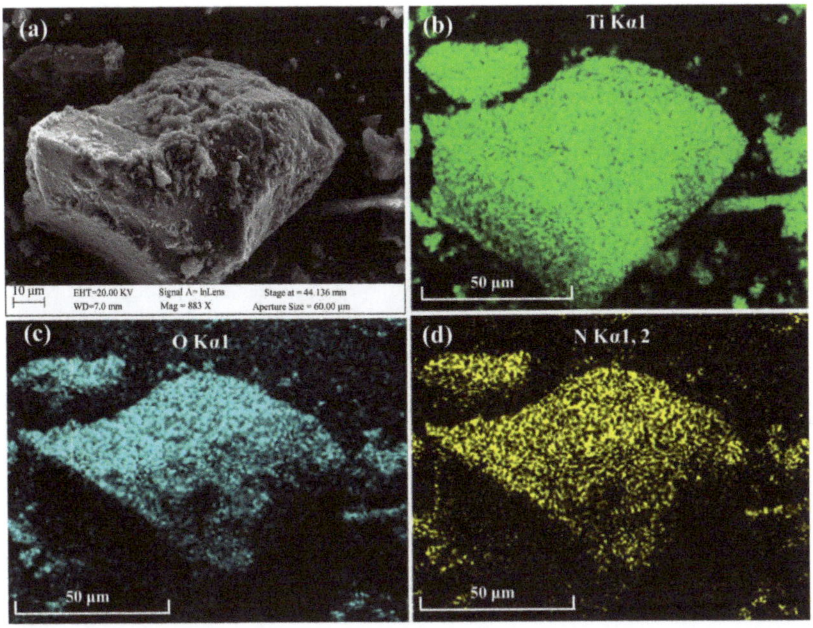

Figure 3. (**a**) SEM image and (**b**) the corresponding Ti element, (**c**) O element and (**d**) N element mappings of 5%N_T/TiO_2.

2.4. XPS Spectra of TiO_2, 5%N_T/TiO_2, and 5%N_{AB}/TiO_2 Catalysts

X-ray photoelectron spectroscopy (XPS) analysis was conducted to investigate the surface elemental composition of the TiO_2, 5%N_T/TiO_2, and 5%N_{AB}/TiO_2 catalysts. The results revealed that the XPS results of the catalysts showed the presence of Ti 2p, O 1s, and N 1s peaks (Figure 4). The XPS spectra of 5%N_T/TiO_2 and 5%N_{AB}/TiO_2 revealed the presence of nitrogen, indicating that nitrogen was doped into TiO_2.

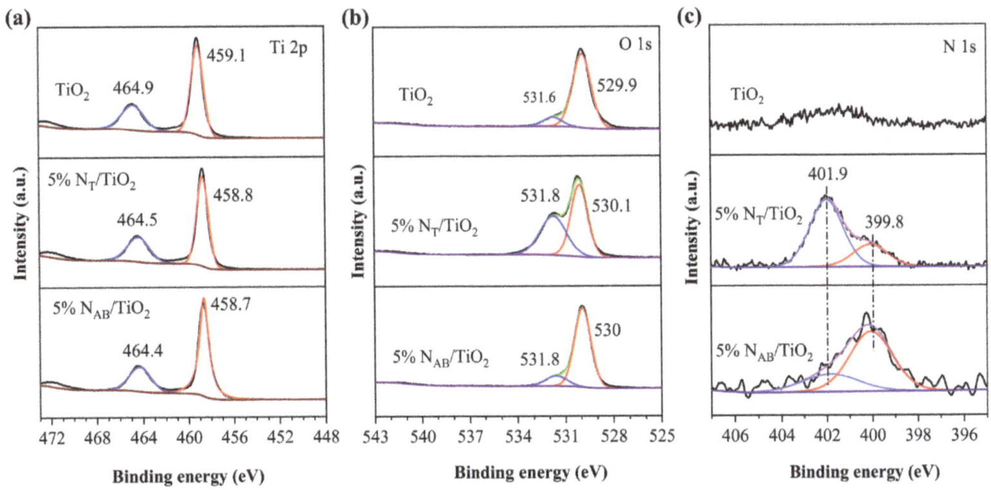

Figure 4. XPS survey spectra. (**a**) Ti 2p XPS spectra, (**b**) O 1s spectra, and (**c**) N 1s spectra of the catalysts.

Figure 4a illustrates the HR-XPS spectra of Ti 2p in TiO_2, 5%N_T/TiO_2, and 5%N_{AB}/TiO_2. For TiO_2, the Ti 2p3/2 and 2p1/2 core energy level peaks appeared at 459.1 and 464.9 eV, respectively, which were contributed by the O–Ti–O in TiO_2 [20–22]. Compared with TiO_2, in 5%N_T/TiO_2, the Ti 2p3/2 and 2p1/2 core energy level peaks appeared at 458.8 and 464.5 eV, respectively, which were the Ti 2p peaks of the N–Ti–N or O–Ti–N in N–TiO_2 [21–23]. For 5%N_{AB}/TiO_2, the Ti 2p peaks of 5%N_{AB}/TiO_2 shifted to the lower binding energy with changes of ~0.4 eV. The peak shifts of Ti 2p indicated the successful incorporation of N in both 5%N_T/TiO_2 and 5%N_{AB}/TiO_2 [20,21].

In Figure 4b, TiO_2 exhibits an O 1s peak at 529.9 eV, which was attributed to the lattice oxygen (Ti–O–Ti). And the peak at 531.8 eV was surface-adsorbed oxygen. In 5%N_{AB}/TiO_2 and 5%N_T/TiO_2, the peaks had an obvious shift owing to the interstitial doping of N into the TiO_2 lattice to form hyponitrite $(N_2O_2)^{2-}$, which was further confirmed in the FTIR analyses, as discussed below.

Figure 4c clearly shows the binding energies of N 1s at 399.8 and 401.9 eV for 5%N_T/TiO_2 and 5%N_{AB}/TiO_2, respectively. For the 5%N_T/TiO_2, according to previous literature [24], the peak at 401.9 eV indicated the formation of Ti–O–N bonds based on the form of interstitial N in the lattice gap of 5%N_T/TiO_2. In the XPS spectrum of 5%N_{AB}/TiO_2, the peak of N 1s at 399.8 eV could be attributed to the Ti–N–Ti bonding, which was due to the substitution of oxygen in the 5%N_T/TiO_2 lattice by N and thus was consistent with the XRD results.

Through a comparison of the XPS spectra, the difference between using thiourea and ammonium bicarbonate as N source could be attributed to the different chemical bonds between N and TiO_2. In N_T/TiO_2, more N atoms existed in the form of interstitial N in the lattice gap of TiO_2 and formed Ti–O–N bonds, indicating that the O atoms had not been replaced. The following Fourier transform infrared spectroscopy (FTIR) spectra also confirmed this idea due to the existence of N^+–O types such as $(N_2O_2)^{2-}$. However, in N_{AB}/TiO_2, more N atoms tended to form Ti–N–Ti bonds, suggesting that many O atoms were replaced by N atoms.

2.5. *TEM Images of TiO_2, 5%N_T/TiO_2, and 5%N_{AB}/TiO_2 Catalysts*

Transmission electron microscopy (TEM) characterization was used to observe the morphology of the TiO_2 and N/TiO_2 catalysts. Undoped TiO_2 typically showed mainly rutile phases with particle sizes ranging from 25 to 50 nm (Figure 5a). After N element doping, the crystal phase of 5%N_T/TiO_2 transformed into the anatase phase, while 5%N_{AB}/TiO_2 became mainly a mixture of rutile and anatase phases (Figure 5b,c). The particle sizes of 5%N_T/TiO_2 and 5%N_{AB}/TiO_2 were slightly smaller than that of TiO_2. Among them, 5%N_T/TiO_2 had the smallest particle size (5.5–8 nm). High-resolution transmission eectron microscopy (HRTEM) images showed clear lattice fringes with an interplanar spacing of 3.20 Å and 2.48 Å (Figure 5d,f), corresponding to the (110) and (101) facets of rutile TiO_2, respectively. The lattice spacings of 3.48 Å and 3.51 Å (Figure 5e,f), respectively, corresponded to those of the (101) plane of anatase TiO_2 in 5%N_T/TiO_2 and 5%N_{AB}/TiO_2. It can be observed that due to the difference in the doping methods, significant differences in the lattice spacings of 5%N_T/TiO_2 and 5%N_{AB}/TiO_2 were induced. This was because 5%N_T/TiO_2 involved interstitial doping, and the N doping was in the lattice gap of TiO_2, which led to lattice distortion.

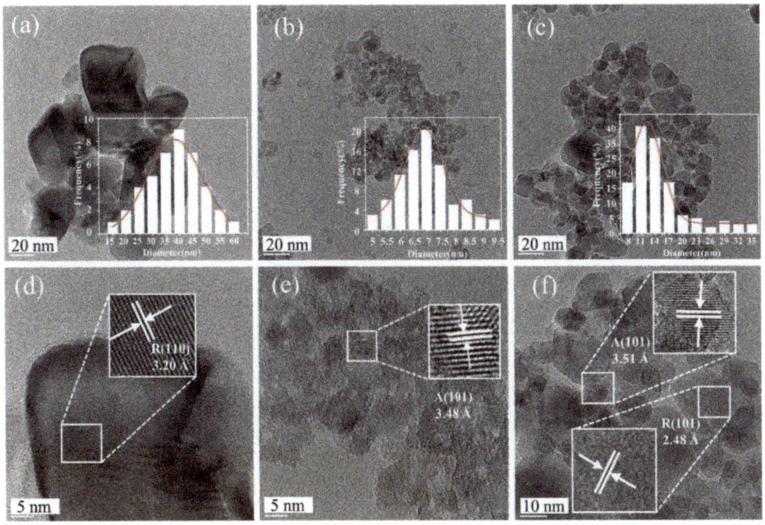

Figure 5. (**a**,**d**) TEM and HRTEM images for TiO_2, (**b**,**e**) 5%N_T/TiO_2, and (**c**,**f**) 5%N_{AB}/TiO_2. R, rutile; A, anatase.

2.6. N_2 Adsorption–Desorption Isotherm Analysis of TiO_2, 5%N_T/TiO_2, and 5%N_{AB}/TiO_2 Catalysts

N_2 adsorption–desorption isotherms analysis was used to determine the specific surface area and porosity of the TiO_2, 5%N_T/TiO_2, and 5%N_{AB}/TiO_2 catalysts. The N_2 isotherms of TiO_2, 5%N_T/TiO_2, and 5%N_{AB}/TiO_2 displayed IV-type isotherms (Figure 6a). This indicated that all the catalysts had mesoporous structures with pore size distributions of 2–40 nm (Figure 6b). The 5%N_T/TiO_2 had a lower pore size distribution. Figure 6b and Table 1 show that the N doping, which can create additional surface defects and porosity, led to higher specific surface area and pore size. The larger specific surface area is beneficial for mass transfer during photocatalytic reactions. According to previous reports, the N atoms can affect the properties of porosity, including the average pore size and distribution, and the specific surface area [25,26]. In detail, under high-temperature processing, NH_3 gas was produced via in-situ breakdown of ammonia precursors (e.g., thiourea) so that the porous structure was formed during synthesis [27,28]. On the other hand, the decomposition of ammonium bicarbonate produced CO_2, further leading to the formation of interconnected mesopores in the TiO_2 lattice [29].

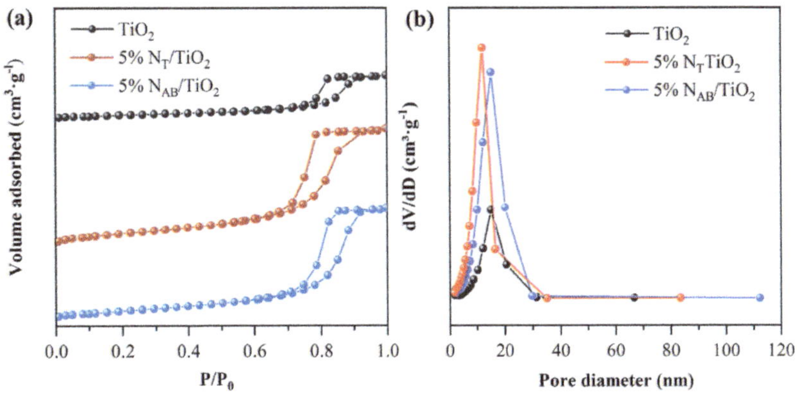

Figure 6. (**a**) N_2 adsorption–desorption isotherms and (**b**) pore size distribution of the catalysts.

Table 1. BET surface area and average pore size of the catalysts.

Catalysts	Surface Area ($m^2.g^{-1}$)	Average Pore Size (nm)
TiO_2	37.0	9.5
5%N_T/TiO_2	107.7	3.5
5%N_{AB}/TiO_2	90.0	9.6

2.7. FTIR of TiO_2, 5%N_T/TiO_2, and 5%N_{AB}/TiO_2 Catalysts

FTIR spectroscopy was employed to identify the presence of nitrogen-related species. (Figure 7). The characteristic peaks in the low wavenumber range (500 cm^{-1}) were attributed to TiO_2. The spectra showed characteristic peaks of TiO_2 at 3400 cm^{-1}, 1630 cm^{-1}, and 700–500 cm^{-1}, which were attributed to the Ti–OH bond, the OH bending vibration of water molecules, and Ti–O–Ti bond stretching vibrations, respectively. Notably, the 5%N_T/TiO_2 presented an additional peak at 1388 cm^{-1}, which was also attributed to the stretching vibration of N–O. Several bands in the low-frequency region observed at 1050 and 1250 cm^{-1} belonged to the N^+–O type substances embedded in the TiO_2 network. The peak at 1150 cm^{-1} was the vibration peak of $(N_2O_2)^{2-}$ formed by interstitial N and O.

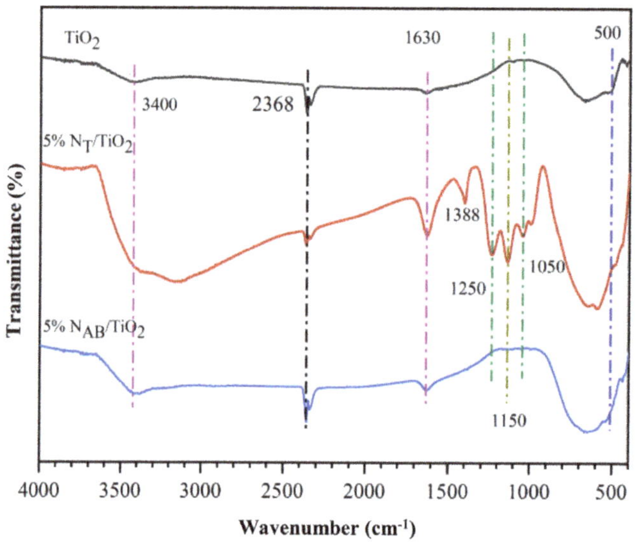

Figure 7. FTIR spectra of pure TiO_2, 5%N_T/TiO_2, and 5%N_{AB}/TiO_2 catalysts.

2.8. Optical Properties of TiO_2, 5%N_T/TiO_2, and 5%N_{AB}/TiO_2 Catalysts

The light-harvesting capacities of the TiO_2, 5%N_T/TiO_2, and 5%N_{AB}/TiO_2 catalysts were evaluated via Ultraviolet-visible (UV–vis) spectroscopy, as shown in Figure 8a. TiO_2 showed a typical absorption edge at 420 nm (corresponding to the 3.05 eV bandgaps in Figure 8b), suggesting the lack of capacity for a visible light response in TiO_2. On the contrary, N doping enhanced the visible light absorption of 5%N_T/TiO_2 and 5%N_{AB}/TiO_2, for which the light absorption edge was ~520 nm. Therefore, the incorporation of nitrogen into the TiO_2 lattice led to the formation of new energy states, as the N 2p position was more negative than the O 2p state, leading to a decrease in the energy bandgap and a shift in the optical absorption towards the visible light region.

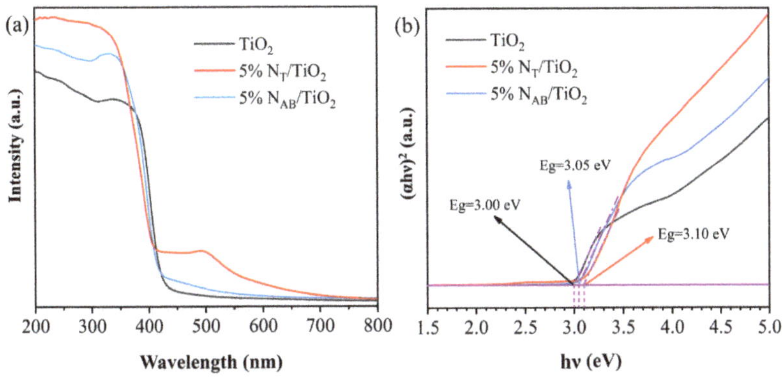

Figure 8. (**a**) UV–vis spectra of the TiO_2, $5\%N_T/TiO_2$, and $5\%N_{AB}/TiO_2$ catalysts. (**b**) Bandgaps of the TiO_2, $5\%N_T/TiO_2$, and $5\%N_{AB}/TiO_2$ catalysts.

2.9. Photoelectrochemical Measurements of TiO_2, $5\%N_T/TiO_2$, and $5\%N_{AB}/TiO_2$ Catalysts

The photoelectrochemical (PEC) measurement was conducted to study the photocurrent responses and charge transfer behavior (Figure 9a). TiO_2 exhibited a lower photocurrent density under visible light irradiation due to its narrow light absorption range. However, $5\%N_T/TiO_2$ and $5\%N_{AB}/TiO_2$ showed an enhanced photocurrent density under visible light, indicating that nitrogen doping allowed the $5\%N_T/TiO_2$ and $5\%N_{AB}/TiO_2$ to convert visible light into photoexcited charges. In the results of electrochemical impedance spectroscopy (EIS) (Figure 9b), the $5\%N_T/TiO_2$ and $5\%N_{AB}/TiO_2$ catalysts showed lower semicircles compared with TiO_2, indicating lower charge transfer resistance and better charge separation [30,31]. Among them, $5\%N_T/TiO_2$ had the lowest resistance and the strongest photocurrent because of the smaller lattice spacing of $5\%N_{AB}/TiO_2$, as shown in the HRTEM image. In general, a smaller lattice spacing means that the atoms or ions are closer to each other and the electrons move more easily in the lattice, resulting in lower resistance and stronger photogenerated carriers [32,33].

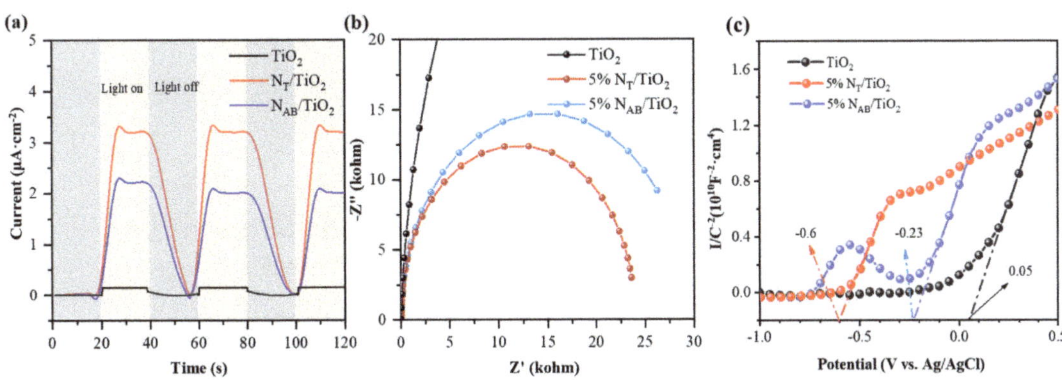

Figure 9. (**a**) Photoinduced i–t curves, (**b**) EIS curves, and (**c**) Mott–Schottky plots for the TiO_2, $5\%N_T/TiO_2$, and $5\%N_{AB}/TiO_2$ catalysts.

Mott–Schottky (MS) plots were used for estimating the band structure features of the catalysts. TiO_2, $5\%N_T/TiO_2$ and $5\%N_T/TiO_2$ can be considered as an n-type semiconductors according to the positive slope of the MS curve. The estimated flat band potential for $5\%N_T/TiO_2$ was –0.6 V vs. an Ag/AgCl electrode, which can be converted to –0.36 V vs. a normal hydrogen electrode (NHE). In general, the conduction band (E_{CB}) edge of typical n-type semiconductors is 0.20 V below its flat band potential. Then, the E_{CB} value of 5%N_T/TiO_2 was determined to be −0.56 V vs. NHE. The valence band (E_{VB}) potential of the 5%N_T/TiO_2 can be expressed by using the equation

$$E_{CB} = E_{VB} - E_g \quad (1)$$

In this way, the E_{VB} position of 5%N_T/TiO_2 was calculated to be +2.54 V vs. NHE, while the E_{VB} position of TiO_2 and 5%N_{AB}/TiO_2 were calculated to be +3.09 and +2.86 V vs. NHE, respectively. The 5%N_T/TiO_2 showed the most negative E_{CB} position compared with the potential of TiO_2 (+0.09 V vs. NHE) and 5%N_{AB}/TiO_2 (−0.19 V vs. NHE). Moreover, the 5%N_T/TiO_2 showed a more negative E_{CB} position (−0.56 V vs. NHE) compared with the potential of $O_2/{\cdot}O_2^-$ (−0.33 V vs. NHE), suggesting that the photogenerated electrons of 5%N_T/TiO_2 could capture dissolved O_2 to generate ${\cdot}O_2^-$. Moreover, the E_{VB} value of 5%N_T/TiO_2 was +2.54 V vs. NHE, which was higher than the potentials of OH/·OH (+2.40 V vs. NHE) and H_2O/·OH (+2.38 V vs. NHE), indicating that photogenerated holes of 5%N_T/TiO_2 could directly yield ·OH radicals.

2.10. Photocatalytic MB Degradation Activity

It should be noted that the activity of 5%N_{AB}/TiO_2 was lower than that of 5%N_T/TiO_2 (Figure 10a), suggesting that interstitial doping was more effective than substitutive doping. This is because the interstitial doped N atoms were located in the lattice gaps of TiO_2. XRD and HRTEM showed that this doping method reduced the lattice spacing of TiO_2, meaning that the atoms were arranged more closely. This is beneficial for electron transfer, and this fact could be proved by the EIS curves. TEM and BET showed that interstitial doping made the grain size of TiO_2 smaller and the specific surface area larger. In addition, the UV–vis and Mott–Schottky curves indicated that interstitial doped TiO_2 had a stronger visible light absorption ability and adjusted its energy band structure to obtain stronger photogenerated carriers. This was confirmed by the photoinduced i–t curves.

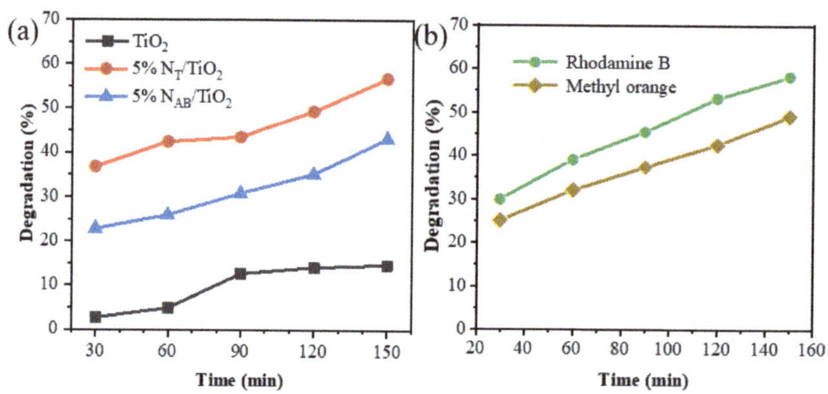

Figure 10. (**a**) Comparison of the rates of MB degradation by TiO_2, 5%N_T/TiO_2, and 5%N_{AB}/TiO_2 photocatalysts. (**b**) The degradation rates of rhodamine B and methyl orange by 5%N_T/TiO_2 photocatalysts.

At present, many studies have carried out various modifications of TiO_2 for the degradation of methylene blue dye. For comparison, a summary of very recent studies on the photodegradation of MB by different TiO_2-based photocatalysts is listed in Table 2. The percentages of degradation obtained by doped TiO_2 prepared under our conditions were very near to those reported in the literature for TiO_2 catalysts. Moreover, in our work, 5%N_T/TiO_2 still had good degradation performance for other organic dyes. As shown in Figure 10b, under the same reaction conditions, 5%N_T/TiO_2 could decolorize 59.0% of rhodamine B in 150 min or 50.0% of methyl orange. It was demonstrated that 5%N_T/TiO_2 has good general applicability. In addition, to explore the stability of the catalyst, we carried out SEM and XRD characterization analyses on the catalyst after 150 min of reaction with methylene blue. As shown in Figure 11a–e, the morphology, element distribution, and crystal form of the catalyst after the reaction were almost the same as those before the reaction. This indicated that the catalyst had good stability.

Table 2. Photocatalytic activity of recently studied TiO_2-based photocatalysts for MB degradation.

	Irradiation Time (min)	Degradation Efficiency (%)	Ref.
$5\%N_T/TiO_2$	150	56.5	This work
TiO_2–Fe_2O_3	60	62.9	[34]
P25	120	20.0	[35]
TiO_2/CS	180	33.9	[36]
TiO_2-GO_x	60	50.0	[37]
AlHF-TiO_2	60	30.0	[38]
Ag-TiO_2	50	40.0	[39]
Hg-doped TiO_2	120	56.7	[40]

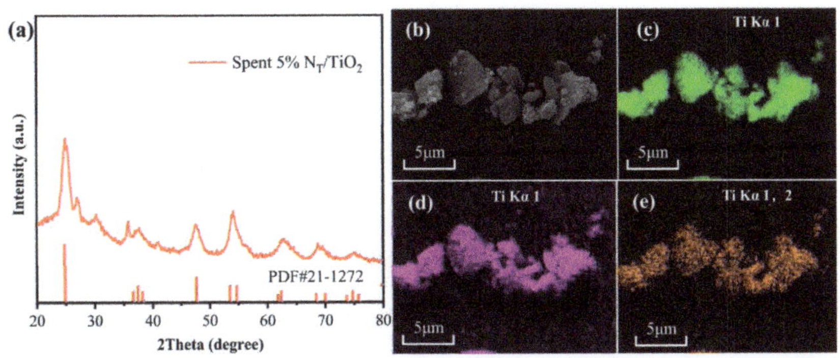

Figure 11. (**a**) XRD patterns of spent $5\%N_T/TiO_2$. (**b**) SEM image and (**c**) the corresponding Ti element, (**d**) O element, and (**e**) N element mappings of spent $5\%N_T/TiO_2$.

Previous studies demonstrated that the photocatalytic reaction pathway is believed to involve the reaction of MB with the generated ·OH radicals, producing a range of intermediate products to reach complete mineralization with the formation of CO_2 and H_2O [41]. To study the main active components in the degradation process of MB on the surface of TiO_2 and to understand the degradation mechanism in more detail, trapping experiments were carried out. Silver nitrate, methanol, and isopropanol were used as scavengers to capture photogenerated electrons (e^-), photogenerated holes (h^+), and hydroxyl radicals (·OH), respectively. To demonstrate the involvement of these radicals, a mixture of MB and silver nitrate (2% v/v) or methanol (2% v/v) or isopropanol (2% v/v) was irradiated under the same conditions. The results thus obtained are shown in Figure 12a. As can be observed, the addition of a photogenerated electron scavenger inhibited the degradation of MB. These results indicated that holes were the primary active species in the degradation of MB, while ·OH and photogenerated holes (h^+) radicals were likely of secondary importance in photodegradation.

According to the results above and literature reports [42], the possible photocatalytic mechanism of the $5\%N_T/TiO_2$ photocatalyst was plotted as shown in Figure 12b. The photodegradation process depends on the generation and separation of carriers; under light conditions, once the semiconductor absorbs energy that is higher than the energy of its energy band, electrons will be excited from the valence band across the forbidden band to the conduction band. The doping of interstitial N reduced the width of the forbidden band of TiO_2, which greatly increased the density of photogenerated carriers. O_2 adsorbed on the surface of TiO_2 formed active substances ($·O_2^-$) after accepting electrons. At the same time, holes remained on the former, and the formed holes reacted directly with MB molecules, and H_2O was adsorbed on the TiO_2 surface to form the active species (·OH).

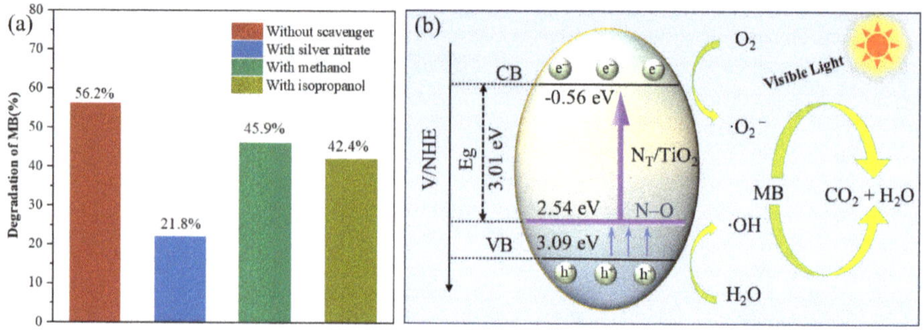

Figure 12. (**a**) Effects of the addition of radical scavengers on the photodegradation of MB and (**b**) schematic diagram of the photocatalytic mechanism of the N_T/TiO_2 photocatalyst.

3. Materials and Methods

3.1. Preparation of the Catalysts

3.1.1. Preparation of 5%N_T/TiO_2

First, 40 mL of tetrabutyl titanate (Aladdin, Shanghai, China) was uniformly dispersed in 50 mL of an ethanol solution. It was vigorously stirred at 60 °C for 2 h and was named Solution A. Subsequently, 3 mL of nitric acid was added to 30 mL of distilled water and vigorously stirred for 5 min, and this was named Solution B. Immediately after that, thiourea (Aladdin, Shanghai, China) was added to 20 mL of ethanol and thoroughly stirred for 5 min. Subsequently, 3 mL of nitric acid was added and stirred at 60 °C for 1 h, and this was named Solution C. Then, under stirring, Solution B and Solution C were poured into Solution A at intervals of 5 min successively. Finally, the obtained mixture was poured into the polytetrafluoroethylene liner in the autoclave and maintained at 100 °C for 24 h. After centrifugal washing with ethanol, the sample was heated to 150 °C at a rate of 2 °C/min and maintained at this temperature for 1 h to remove the residual ethanol. After this stage was complete, the sample continued to be heated at the same rate until it reached 450 °C and was maintained at this temperature for 2 h.

3.1.2. Preparation TiO_2 and 5%N_{AB}/TiO_2

The preparation methods of 5%N_{AB}/TiO_2 and TiO_2 were the same as that for 5%N_T/TiO_2. The only difference was that the nitrogen precursor doped in 5%N_{AB}/TiO_2 was ammonium bicarbonate (Aladdin, Shanghai, China), and for TiO_2, no nitrogen precursor was added.

3.2. Characterization of the Catalysts

The crystalline structure and phase composition of the TiO_2, 5%N_T/TiO_2, and 5%N_T/TiO_2 catalysts were determined by X-ray diffraction (XRD) on an X-Pert diffractometer (Rigaku, Kyoto, Japan) equipped with graphite monochromatized Cu-Kα radiation. The elemental composition, chemical states, and surface chemistry of TiO_2, 5%N_T/TiO_2, and 5%N_T/TiO_2 were analyzed using Agilent 5100 X-ray Photoelectron Spectroscopy (XPS, Agilent Technologies, Dallastown, PA, USA), where the tube's voltage was 15 kV, the tube's current was 10 mA, and an ultra-high vacuum chamber with a mu-metal magnetic shield was used. The specific surface areas were determined using a surface area analyzer (Micromeritics, Norcross, GA, USA) by the Brunauer–Emmett–Teller (BET) method. The morphology and size of the TiO_2, 5%N_T/TiO_2, and 5%N_T/TiO_2 catalysts were observed using a transmission electron microscope (TEM-16-TS-008, Thermo Fisher Scientific, Waltham, MA, USA). The acceleration voltage was 200 kV, the point resolution was 0.248 nm, and the maximum magnification was 1.05 million times to visualize the morphology, size, and surface features of the TiO_2, 5%N_T/TiO_2, and 5%N_T/TiO_2 catalysts. Fourier transform infrared spectroscopy

(FTIR, Bruker Vertex 70 infrared spectrometer, Bruker Corporation, Billerica, MA, USA) was used to analyze the chemical bonds and functional groups present on the surface of the TiO_2, 5%N_T/TiO_2 and 5%N_T/TiO_2 catalysts. FTIR spectra in transmission mode were recorded at a resolution of 4 cm^{-1} in the range of 400 to 4000 cm^{-1} using the KBr tabbing technique. The light absorption properties of the TiO_2, 5%N_T/TiO_2 and 5%N_T/TiO_2 catalysts were determined by UV–vis spectroscopy with a wavelength range from 200 to 800 nm on a Japanese Shimadzu UV-3600i Plus Spectrometer. The photocurrent and electrochemical impedance spectroscopy were measured on an electrochemical workstation (Shanghai Chenhua Instrument Co., Ltd., Shanghai, China).

3.3. *Evaluation of the Catalysts*

The performance of different catalysts in the photocatalytic degradation of MB was evaluated in a photocatalytic reactor under visible light irradiation (300 W xenon lamp). Firstly, 0.02 g of each catalyst was added to 20 mL MB. Adsorption and desorption equilibrium were attained by agitating the suspensions with a magnetic stirrer for 1 h. Afterward, the condensate water was turned on to keep the reaction temperature at room temperature. Then the solution was irradiated with a 300 W xenon lamp without a light intensity meter as the source of visible light. An aliquot was collected periodically with a syringe from the reactor every 30 min during a 150 min interval. Finally, the catalyst was filtered and recycled. Changes in the concentration of MB were observed from its characteristic absorption at 664 nm using a UV–vis spectrometer. The degradation efficiency of MB at each time point was calculated using the following formula [43,44]

$$\text{Degradation efficiency } (\%) = \frac{C_o - C_t}{C_o} \times 100\% \quad (2)$$

where C_o is the initial concentration of MB, and C_t is the concentration of MB at a time.

4. Conclusions

In conclusion, this work provides a straightforward method to fabricate nitrogen-doped TiO_2 photocatalysts based on thiourea as the N source. The systematic characterizations confirmed the presence of nitrogen in TiO_2 crystals. UV–vis spectra showed the extended light absorption range on 5%N_T/TiO_2. Thus, 5%N_T/TiO_2 had a better photocurrent response during PEC measurement. Through optimizing the nitrogen concentration, 5%N_T/TiO_2 showed higher MB degradation activity than TiO_2 and 5%N_{AB}/TiO_2 under visible light illumination. Our study clarified the advantages of interstitial N doping in improving photocatalytic activity under visible light conditions. More importantly, this work demonstrated that the introduction of nitrogen facilitates the utilization of solar light energy and makes TiO_2 a promising material for environmental remediation and beyond.

Author Contributions: D.L.: software, data curation, and writing—review and editing. V.C.C.: validation and writing—original draft preparation. Y.L. (Yuqiao Li): validation and visualization. H.L.: supervision, writing—review, and funding. Y.L. (Yiming Lei): conceptualization, supervision, and editing. All authors have read and agreed to the published version of the manuscript.

Funding: This work received financial support from the Young Talent Plan of Liaoning Province (XLYC2203068), the Scientific Research Foundation of Technology Department of Liaoning Province of China (2022-MS-379), and the National Natural Science Foundation of China (21902116).

Data Availability Statement: The data presented in this study are available on request from the corresponding author.

Acknowledgments: The program of China Scholarships Council (No. 202206250016) is acknowledged.

Conflicts of Interest: The authors declare no conflicts of interest.

Abbreviations

$X\%N_T/TiO_2$: x% represents the amount of N doping, and T indicates that the selected N source was thiourea. $X\%\ N_{AB}/TiO_2$: x% represents the amount of N doping, and AB indicates that the selected N source was ammonium bicarbonate.

References

1. Sadek, O.; Touhtouh, S.; Dahbi, A.; Hajjaji, A. Photocatalytic degradation of methylene blue on multilayer TiO_2 coatings elaborated by the sol-gel spin-coating method. *Water. Air. Soil. Pollut.* **2023**, *234*, 698. [CrossRef]
2. Ma, J.; Tian, Z.; Li, L.; Lu, Y.; Xu, X.; Hou, J. Loading nano-CuO on TiO_2 nanomeshes towards efficient photodegradation of methylene blue. *Catalysts* **2022**, *12*, 383. [CrossRef]
3. Ma, Y.; Tao, L.; Bai, S.; Hu, A. Green synthesis of Ag nanoparticles for plasmon-assisted photocatalytic degradation of methylene blue. *Catalysts* **2021**, *11*, 1499. [CrossRef]
4. Hariharalakshmanan, R.K.; Watanabe, F.; Karabacak, T. In situ growth and UV photocatalytic effect of ZnO nanostructures on a Zn plate immersed in methylene blue. *Catalysts* **2022**, *12*, 1657. [CrossRef]
5. Al-Ghouti, M.A.; Dib, S.S. Utilization of nano-olive stones in environmental remediation of methylene blue from water 03 chemical sciences 0306 physical chemistry (incl. structural). *J. Environ. Heal. Sci. Eng.* **2020**, *18*, 63–77. [CrossRef]
6. Haq, F.; Kiran, M.; Chinnam, S.; Farid, A.; Khan, R.U.; Ullah, G.; Aljuwayid, A.M.; Habila, M.A.; Mubashir, M. Synthesis of bioinspired sorbent and their exploitation for methylene blue remediation. *Chemosphere* **2023**, *321*, 138000. [CrossRef]
7. Xu, Y.; Shi, X.; Hua, R.; Zhang, R.; Yao, Y.; Zhao, B.; Liu, T.; Zheng, J.; Lu, G. Remarkably catalytic activity in reduction of 4-nitrophenol and methylene blue by Fe_3O_4@COF supported noble metal nanoparticles. *Appl. Catal. B Environ.* **2020**, *260*, 118142. [CrossRef]
8. Malato, S.; Fernández-Ibáñez, P.; Maldonado, M.I.; Blanco, J.; Gernjak, W. Decontamination and disinfection of water by solar photocatalysis: Recent overview and trends. *Catal. Today* **2009**, *147*, 1–59. [CrossRef]
9. Bairamis, F.; Konstantinou, I.; Petrakis, D.; Vaimakis, T. Enhanced performance of electrospun nanofibrous $TiO_2/g\text{-}C_3N_4$ photocatalyst in photocatalytic degradation of methylene blue. *Catalysts* **2019**, *9*, 880. [CrossRef]
10. Babyszko, A.; Wanag, A.; Sadłowski, M.; Kusiak-Nejman, E.; Morawski, A.W. Synthesis and characterization of SiO_2/TiO_2 as photocatalyst on methylene blue degradation. *Catalysts* **2022**, *12*, 1372. [CrossRef]
11. Zhang, C.; Zhou, Y.; Bao, J.; Sheng, X.; Fang, J.; Zhao, S.; Zhang, Y.; Chen, W. Hierarchical honeycomb Br-, N-codoped TiO_2 with enhanced visible-light photocatalytic H_2 production. *ACS Appl. Mater. Interfaces* **2018**, *10*, 18796–18804. [CrossRef] [PubMed]
12. Assayehegn, E.; Solaiappan, A.; Chebude, Y.; Alemayehu, E. Fabrication of tunable anatase/rutile heterojunction N/TiO_2 nanophotocatalyst for enhanced visible light degradation activity. *Appl. Surf. Sci.* **2020**, *515*, 145966. [CrossRef]
13. Chakraborty, A.K.; Ganguli, S.; Sabur, M.A. Nitrogen doped titanium dioxide (N-TiO_2): Electronic band structure, visible light harvesting and photocatalytic applications. *J. Water Process Eng.* **2023**, *55*, 104183. [CrossRef]
14. Bissinger, D.; Honerkamp, J.H.; Roldan, J.; Bremes, J.; Kannen, K.; Lake, M.K.; Roppertz, A. Development of catalytically functionalized polyester-based filters produced by flame spray pyrolysis. *Top. Catal.* **2024**, *67*, 539–550. [CrossRef]
15. Khan, T.T.; Rafiqul Bari, G.A.K.M.; Kang, H.J.; Lee, T.G.; Park, J.W.; Hwang, H.J.; Hossain, S.M.; Mun, J.S.; Suzuki, N.; Fujishima, A.; et al. Synthesis of N-doped TiO_2 for efficient photocatalytic degradation of atmospheric NO_x. *Catalysts* **2021**, *11*, 109. [CrossRef]
16. Das, D.; Shyam, S. Reduced work function in anatase ⟨101⟩ TiO_2 films self-doped by O-vacancy-dependent Ti^{3+} bonds controlling the photocatalytic dye degradation performance. *Langmuir* **2024**, *40*, 10502–10517. [CrossRef]
17. Divyasri, Y.V.; Lakshmana Reddy, N.; Lee, K.; Sakar, M.; Navakoteswara Rao, V.; Venkatramu, V.; Shankar, M.V.; Gangi Reddy, N.C. Optimization of N doping in TiO_2 nanotubes for the enhanced solar light mediated photocatalytic H_2 production and dye degradation. *Environ. Pollut.* **2020**, *269*, 116170. [CrossRef]
18. Bhowmick, S.; Saini, C.P.; Santra, B.; Walczak, L.; Semisalova, A.; Gupta, M.; Kanjilal, A. Modulation of the work function of TiO_2 nanotubes by nitrogen doping: Implications for the photocatalytic degradation of dyes. *ACS Appl. Nano Mater.* **2023**, *6*, 50–60. [CrossRef]
19. Zhang, X.; Zuo, G.; Lu, X.; Tang, C.; Cao, S.; Yu, M. Anatase TiO_2 sheet-assisted synthesis of Ti^{3+} self-doped mixed phase TiO_2 sheet with superior visible-light photocatalytic performance: Roles of anatase TiO_2 sheet. *J. Colloid. Interface Sci.* **2017**, *490*, 774–782. [CrossRef]
20. Sathish, M.; Viswanathan, B.; Viswanath, R.P. Alternate synthetic strategy for the preparation of CdS nanoparticles and its exploitation for water splitting. *Int. J. Hydrog. Energy* **2006**, *31*, 891–898. [CrossRef]
21. Chen, X.; Burda, C. Photoelectron spectroscopic investigation of nitrogen-doped titania nanoparticles. *J. Phys. Chem. B* **2004**, *108*, 15446–15449. [CrossRef]
22. Pustovalova, A.A.; Pichugin, V.F.; Ivanova, N.M.; Bruns, M. Structural features of N-containing titanium dioxide thin films deposited by magnetron sputtering. *Thin Solid. Film.* **2017**, *627*, 9–16. [CrossRef]
23. Ren, W.; Ai, Z.; Jia, F.; Zhang, L.; Fan, X.; Zou, Z. Low temperature preparation and visible light photocatalytic activity of mesoporous Carbon-Doped Crystalline TiO_2. *Appl. Catal. B Environ. Energy* **2007**, *69*, 138–144. [CrossRef]

24. Di Valentin, C.; Pacchioni, G.; Selloni, A.; Livraghi, S.; Giamello, E. Characterization of paramagnetic species in N-doped TiO_2 powders by EPR spectroscopy and DFT calculations. *J. Phys. Chem. B* **2005**, *109*, 11414–11419. [CrossRef]
25. Li, B.; Dai, F.; Xiao, Q.; Yang, L.; Shen, J.; Zhang, C.; Cai, M. Nitrogen-doped activated carbon for a high energy hybrid supercapacitor. *Energy Environ. Sci.* **2015**, *9*, 102–106. [CrossRef]
26. Zhang, Z.; Lu, S.; Shen, G.; Zhao, Y.; Zhu, T.; Gao, Q.; Sun, N.; Wei, W. Controllable and rapid synthesis of nitrogen-doped ordered mesoporous carbon single crystals for CO_2 capture. *J. CO_2 Util.* **2021**, *56*, 101851. [CrossRef]
27. Rustamaji, H.; Prakoso, T.; Devianto, H.; Widiatmoko, P.; Saputera, W.H. Urea nitrogenated mesoporous activated carbon derived from oil palm empty fruit bunch for high-performance supercapacitor. *J. Energy Storage* **2022**, *52*, 104724. [CrossRef]
28. Wang, Y.; Yin, C.; Qin, H.; Wang, Y.; Li, Y.; Li, X.; Zuo, Y.; Kang, S.; Cui, L. A urea-assisted template method to synthesize mesoporous N-doped CeO_2 for CO_2 capturet. *Dalt. Trans.* **2015**, *44*, 18718–18722. [CrossRef]
29. Ghosh, A.; Ghosh, S.; Seshadhri, G.M.; Ramaprabhu, S. Green synthesis of nitrogen-doped self-assembled porous carbon-metal oxide composite towards energy and environmental applications. *Sci. Rep.* **2019**, *9*, 5187. [CrossRef]
30. Jiang, Z.; Ran, M.; Liu, K.; Huang, Y.; Li, Z.; Shen, T.; Li, W.; Khojiev, S.; Hu, Z.; Liu, J. Directing Ni/Al layered double hydroxides nanosheets on tubular graphite carbon nitride for promoted photocatalytic hydrogen production. *Mater. Today Chem.* **2024**, *39*, 102135. [CrossRef]
31. Jiang, L.; Yu, H.; Shi, L.; Zhao, Y.; Wang, Z.; Zhang, M.; Yuan, S. Optical band structure and photogenerated carriers transfer dynamics in FTO/TiO_2 heterojunction photocatalysts. *Appl. Catal. B Environ.* **2016**, *199*, 224–229. [CrossRef]
32. Li, Y.; Lai, R.; Luo, X.; Liu, X.; Ding, T.; Lu, X.; Wu, K. On the absence of a phonon bottleneck in strongly confined CsPbBr 3 perovskite nanocrystals. *Chem. Sci.* **2019**, *10*, 5983–5989. [CrossRef]
33. Zhu, Y.; Cui, Q.; Chen, J.; Chen, F.; Shi, Z.; Zhao, X.; Xu, C. Inhomogeneous trap-state-mediated ultrafast photocarrier dynamics in $CsPbBr_3$ microplates. *ACS Appl. Mater. Interfaces* **2021**, *13*, 6820–6829. [CrossRef]
34. Kumar, M.R.A.; Abebe, B.; Nagaswarupa, H.P.; Murthy, H.C.A.; Ravikumar, C.R.; Sabir, F.K. Enhanced photocatalytic and electrochemical performance of TiO_2-Fe_2O_3 nanocomposite: Its applications in dye decolorization and as supercapacitors. *Sci. Rep.* **2020**, *10*, 1249. [CrossRef]
35. Wan, J.; Wei, M.; Hu, Z.; Peng, Z.; Wang, B.; Feng, D.; Shen, Y. Ternary composites of TiO_2 nanotubes with reduced graphene oxide (RGO) and meso-tetra (4-carboxyphenyl) porphyrin for enhanced visible light photocatalysis. *Int. J. Hydrog. Energy* **2016**, *33*, 14692–14703. [CrossRef]
36. Ling, M.F.C.; Hui, K.C.; Sambudi, N.S. Modification of TiO_2 with clam-shell powder for photodegradation of methylene blue. *J. Sol. Gel Sci. Technol.* **2022**, *102*, 412–421. [CrossRef]
37. Kim, B.C.; Jeong, E.; Kim, E.; Hong, S.W. Bio-organic-inorganic hybrid photocatalyst, TiO_2 and glucose oxidase composite for enhancing antibacterial performance in aqueous environments. *Appl. Catal. B Environ. Energy* **2018**, *242*, 194–201. [CrossRef]
38. Magnone, E.; Kim, M.-K.; Lee, H.J.; Park, J.H. Testing and substantial improvement of TiO_2/UV photocatalysts in the degradation of methylene blue. *Ceram. Int.* **2019**, *10*, 3359–3367. [CrossRef]
39. Ridha, N.J.; Alosfur, F.K.M.; Kadhim, H.B.A.; Ahmed, L.M. Synthesis of Ag decorated TiO_2 nanoneedles for photocatalytic degradation of methylene blue dye. *Mater. Res. Express* **2021**, *8*, 125013. [CrossRef]
40. Abbas, F.; Bensaha, R. Effect of annealing time on structural and optical proprieties of mercury (Hg^{+2}) doped TiO_2 thin films elaborated by sol-gel method for future photo-catalytic application. *Opt.* **2021**, *247*, 167846. [CrossRef]
41. da Silva, C.G.; Faria, J.L. Photochemical and photocatalytic degradation of an azo dye in aqueous solution by UV irradiation. *J. Photochem. Photobiol. A Chem.* **2003**, *155*, 133–143. [CrossRef]
42. Zhou, Q.; Zhang, L.; Zuo, P.; Wang, Y.; Yu, Z. Enhanced photocatalytic performance of spherical BiOI/MnO_2 composite and mechanism investigation. *RSC Adv.* **2018**, *8*, 36161–36166. [CrossRef] [PubMed]
43. Trandafilović, L.V.; Jovanović, D.J.; Zhang, X.; Ptasińska, S.; Dramićanin, M.D. Enhanced photocatalytic degradation of methylene blue and methyl orange by ZnO: Eu nanoparticles. *Appl. Catal. B Environ.* **2017**, *203*, 740–752. [CrossRef]
44. Girotto, G.Z.; Thill, A.S.; Matte, L.P.; Vogt, M.A.H.; Machado, T.V.; Dick, L.F.P.; Mesquita, F.; Bernardi, F. Ni/$SrTiO_3$ nanoparticles for photodegradation of methylene blue. *ACS Appl. Nano Mater.* **2022**, *5*, 13295–13307. [CrossRef]

catalysts

MDPI

Article

TiO_2 Catalysts Co-Modified with Bi, F, SnO_2, and SiO_2 for Photocatalytic Degradation of Rhodamine B Under Simulated Sunlight

Lu Qiu [1], Hanliang Li [2], Wenyi Xu [3], Rongshu Zhu [3,*] and Feng Ouyang [3,*]

1 College of Eco-Environmental Engineering, Guizhou Minzu University, Guiyang 550025, China; 202001089@gzmu.edu.cn
2 Environmental Science and Engineering Department, Liaoning Technical University, Fuxin 123000, China; lihanliang2012@126.com
3 Shenzhen Key Laboratory of Organic Pollution Prevention and Control, Harbin Institute of Technology, Shenzhen 518055, China
* Correspondence: rszhu@hit.edu.cn (R.Z.); ouyangfh@hit.edu.cn (F.O.)

Abstract: The organic pollutants discharged from industrial wastewater have caused serious harm to human health. The efficient photocatalytic degradation of organic pollutants under sunlight shows promise for industrial applications and energy utilization. In this study, a modified TiO_2 photocatalyst doped with bismuth (Bi) and fluorine (F) and composited with SnO_2 and SiO_2 was prepared, and its performance for the degradation of Rhodamine B (RhB) under simulated sunlight was evaluated. Through the optimization of the doping levels of Bi and F, as well as the ratio of SnO_2 and SiO_2 to TiO_2, the optimal catalyst reached degradation efficiency of 100% for RhB within 20 min under simulated sunlight, with a first-order reaction rate constant of 0.291 min^{-1}. This value was 15, 41, 6.5, and 3.3 times higher than those of TiO_2/SnO_2, Bi/TiO_2, Bi-TiO_2/SnO_2, and F/Bi-TiO_2/SnO_2, respectively. The active species detection showed that h^+ was the most crucial active species in the process. The role of Bi and F addition and SnO_2-SiO_2 compositing was investigated by characterization. Bi formed a chemical bonding with TiO_2 by doping into TiO_2. The absorbance intensity in the UV and visible light regions was improved by SnO_2 and F modification. Composite with SiO_2 led to a larger surface area that allowed for more RhB adsorption sites. These beneficial modifications greatly enhanced the photocatalytic activity of the catalyst.

Keywords: photocatalysis; TiO_2; bismuth; SnO_2; rhodamine B; organic pollutant removal

1. Introduction

Industrial wastewater discharge always introduces organic pollutants, such as dyes, herbicides, pesticides, analgesics, and antibiotics into the water environment, causing serious water problems and posing severe threats to the ecosystem and human health. Photocatalytic treatment is an advanced oxidation process (AOP) that is promising, economical, efficient, and sustainable for degrading organic pollutants in water using solar energy. It is well known that titanium dioxide (TiO_2) has long been one of the most widely researched and applied semiconductor materials in the field of photocatalysis due to its suitable band structure, low cost, non-toxicity, and high stability [1,2]. However, the wide bandgap of TiO_2 (3.0–3.2 eV) limits its spectral response only in the ultraviolet region. Another disadvantage is its rapid recombination of photoelectrons and holes [3,4]. These defects limit its practical efficiency in the photocatalytic degradation of organic pollutants under sunlight. To overcome this disadvantage, researchers have explored various strategies, such as doping with metals and non-metals [5–7], loading co-catalysts [8,9], designing heterojunctions [10,11], and adjusting morphology [12,13]. By modulating TiO_2 properties from different perspectives, these endeavors are intended to enhance its absorption and utilization of light, restrain photo-generated carrier recombination, and ultimately improve photocatalytic efficiency.

Doping TiO_2 with metals or non-metals is a well-established method for boosting its photocatalytic activity. Research on non-metal doping always includes C, S, N, B, and F dopants. Among these, fluorine doping has been demonstrated to enhance catalytic activity of TiO_2 in organic pollutant degradation, possibly by inducing the formation of surface oxygen vacancies [14,15] and Ti^{3+} defects [16] or increasing surface hydroxyl groups [17,18]. In addition, there are many kinds of doped metals. The addition of impurities in the pure semiconductor proves effective in generating electron–hole traps and narrowing the bandgap [19], thereby improving photocatalytic efficiency. Among these doped metals, bismuth has been used as a dopant effectively. Bi_2O_3 is a common Bi (III)-based compound with a suitable bandgap width (2.3–2.8 eV) [20,21]. Bi_2O_3 and TiO_2 exhibit matched band structures. It has been proved that doped Bi can reduce the bandgap and promote the separation of photo-generated carriers [22–24]. The bandgap reduction resulted from the up-shift valence band that was triggered by the hybridization of Bi (6 s) orbital with O (2p) orbital [25]. Research studies have pointed out that the modification of various forms of TiO_2 with Bi_2O_3 enhanced its visible light catalytic activity against certain organic dyes [26,27]. Additionally, coupling TiO_2 with other semiconductors to form photocatalytic heterojunctions is a strategy to facilitate the separation of photo-generated carriers. Tin dioxide (SnO_2), an ideal and typical oxide semiconductor material with a bandgap of 3.6 eV, shares certain similarities with TiO_2 in terms of the crystal structure [28], making it conducive to constructing heterojunction structures [29,30]. Tin modification of TiO_2 can reduce the binding of photo-generated electron–hole pairs, thus facilitating the degradation of complex pollutants [31,32]. It was reported that the forming of $Ti_{1-x}Sn_xO_2$ solid solutions generated type II heterojunctions [32], which achieved effective charge separation. Moreover, the formation of TiO_2-SnO_2 composite aids in the reduction in the bandgap, which shifts the optical absorption from UV to the visible region [33,34]. The effective charge separation was achieved by promoting the holes (h^+) to the lower valence band (VB) of TiO_2 and the electrons (e^-) to the lower conduction band (CB) of SnO_2 [35–38]. For these reasons, the SnO_2-TiO_2 composite showed improved activity in the photocatalytic degradation of organic pollutants under visible light [39–41].

Moreover, the diminutive specific surface area of TiO_2 leads to limited adsorption capacity and a small number of activity sites for the reactants [42,43]. Therefore, to improve the adsorption ability of TiO_2 towards organic pollutants, combining TiO_2 with other materials with a large surface area serves as a promising approach to achieve this target [44–47].

To sum up, numerous methods and diverse added chemical compounds have been explored for modifying TiO_2 to enhance its photocatalytic performance. Therefore, a combination of co-doping and compositing holds theoretical feasibility in comprehensively improving properties of TiO_2, such as absorption of ultraviolet and visible light, separation of photo-generated carrier, high surface area, etc., and in further bolstering the degradation efficiency of organic pollutants under simulated sunlight. Up to now, F, Bi, or Sn modification has been reported to enhance the performance of TiO_2. However, the co-doping of F and Bi, along with SnO_2 and SiO_2 compositing, to synthesized TiO_2 has not been reported to our knowledge. The TiO_2 catalyst modified by F-Bi-Sn-Si was synthesized for the first time, and its performance on RhB degradation was investigated in terms of optical properties.

Herein, this study employed homogeneous precipitation and sol–gel methods to synthesize F/Bi-TiO_2/SnO_2/SiO_2 photocatalysts. These materials were utilized for the photocatalytic degradation of RhB under simulated solar light. Various techniques were employed to characterize the optical and physicochemical properties, with an aim to elucidate the impact of different element doping or compositing on catalyst structure and properties. Given the enhanced performance of the prepared catalysts, the efficient degradation of RhB was achieved under simulated solar light.

2. Results

2.1. Impact of Bi Addition and Compositing with SnO_2 on Catalyst Properties

2.1.1. Impact on Crystal Phase

To analyze the influence of Bi and SnO_2 modification on the crystal phase, X-ray diffraction (XRD) patterns of Bi-TiO_2/SnO_2 catalysts with different Bi doping amounts and different SnO_2/TiO_2 molar ratios were assessed. Figure 1A shows the XRD spectra of Bi-TiO_2/SnO_2 catalysts with varying Bi amounts. All catalysts exhibited diffraction peaks at 25.3°, 37.9°, and 48.0°, corresponding to (101), (004), and (200) crystal faces of anatase TiO_2 (JCPDS 21–1272), while other weak diffraction peaks were assigned to rutile TiO_2. The peaks at 26.81°, 34.31°, and 54.18° corresponded to (110), (101), and (211) crystal faces of rutile SnO_2. The results indicated the existence of TiO_2 and SnO_2 primarily in the forms of anatase and rutile phases, respectively. No diffraction peaks of Bi oxides were observed in the XRD patterns, suggesting that no crystalline Bi oxides were formed. This could be attributed to Bi being doped into the TiO_2 lattice in the ionic form or being highly dispersed [24,48,49]. The crystal grain size was calculated and listed in Table S1. Although no Bi diffraction peaks were present, the crystal grain size showed varying degrees of decrease with the addition of Bi, indicating that Bi doping reduced the crystal grain size of TiO_2 [24,50]. Figure 1B illustrates the influence of the SnO_2/TiO_2 molar ratio on the XRD spectra of Bi-TiO_2/SnO_2 catalysts. All catalysts exhibited characteristic peaks of anatase TiO_2 and rutile SnO_2. Different Sn/Ti molar ratios did not cause a change in the crystal form, indicating that the addition of SnO_2 did not alter the TiO_2 crystal phase.

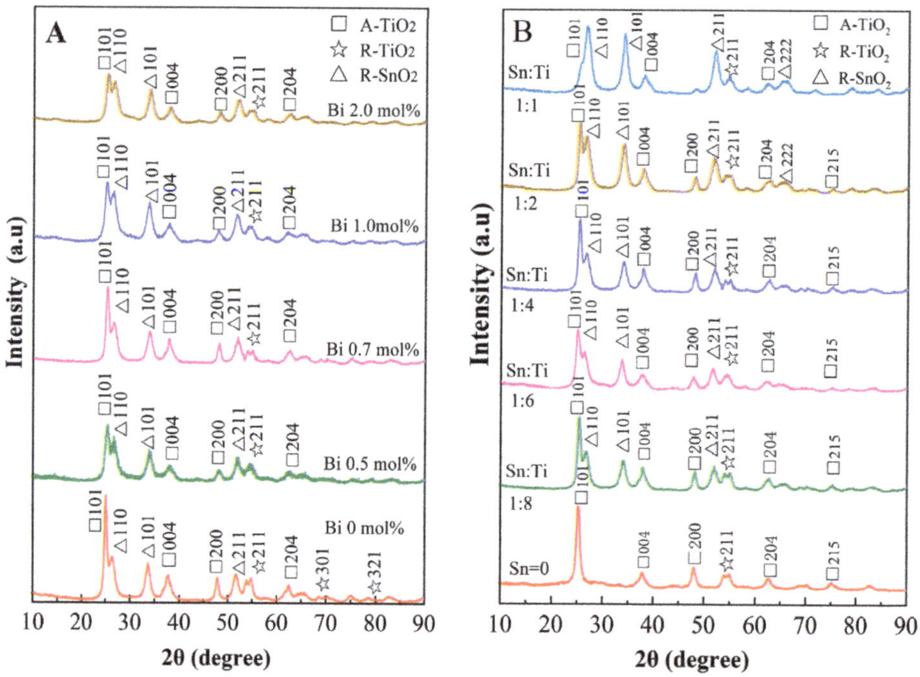

Figure 1. XRD patterns of Bi-TiO_2/SnO_2 samples with (**A**) different Bi amounts and (**B**) different SnO_2/TiO_2 molar ratios.

2.1.2. X-Ray Photoelectron Spectroscopy Analysis

The X-ray photoelectron spectroscopy (XPS) was introduced to evaluate the surface chemical and elemental states of the Bi-TiO_2/SnO_2 catalyst. Figure 2A presents the Ti 2p spectrum of Bi-TiO_2/SnO_2, with distinct Ti $2p_{3/2}$ and Ti $2p_{1/2}$ peaks appearing at 458.72 eV and 464.38 eV, respectively. These peaks exhibited a slight shift towards higher values compared to the standard features of TiO_2 (458.6 eV and 464.3 eV), indicating that Ti in the sample was in a combined state with Bi [49]. The O 1s spectrum of Bi-TiO_2/SnO_2 in Figure 2B could be decomposed into two peaks, with one binding energy at 530.28 eV, between the O 1s standard binding energies of TiO_2 and SnO_2 (530.08 eV and 530.94 eV), indicating lattice oxygen [51]. The other peak at 532.02 eV was assigned to surface hydroxyl oxygen [52]. Figure 2C exhibits the Bi 4f spectrum of the Bi-TiO_2/SnO_2 catalyst. The binding energy values of Bi $4f_{7/2}$ and Bi $4f_{5/2}$ were 159.23 eV and 164.18 eV, respectively. However, according to the binding energy handbook, the XPS peak positions for Bi^{3+} in Bi $4f_{7/2}$ and Bi $4f_{5/2}$ are 158.5 eV and 162.2 eV, respectively. Therefore, it indicated that, due to electron transfer interactions with TiO_2, Bi existed in a higher-valence $Bi^{3+\sigma}$ state, which formed a certain type of chemical bonding with TiO_2 [53]. Combined with the previous XRD result that Bi entered into the TiO_2 lattice, the $Bi^{3+\sigma}$ occupation in the catalyst lattice could cause lattice distortion. To compensate for the energy change triggered by lattice distortion, more oxygen vacancies were generated on the catalyst, acting as traps for photo-generated electrons and hindering their recombination with holes, ultimately elevating the efficiency of the photocatalytic reaction. On the other hand, the hybridization of Bi with O orbital could form hybrid energy [25]. Figure 2D shows the high-resolution spectrum of Sn 3d for Bi-TiO_2/SnO_2 catalysts. The Sn $3d_{5/2}$ peak appeared at 486.51 eV and the Sn $3d_{3/2}$ peak at 494.88 eV, with a separation of 8.37 eV, illustrating that tin existed as Sn (IV) [28,54]. The full survey spectrum for the Bi-TiO_2/SnO_2 sample is depicted in Figure 2E.

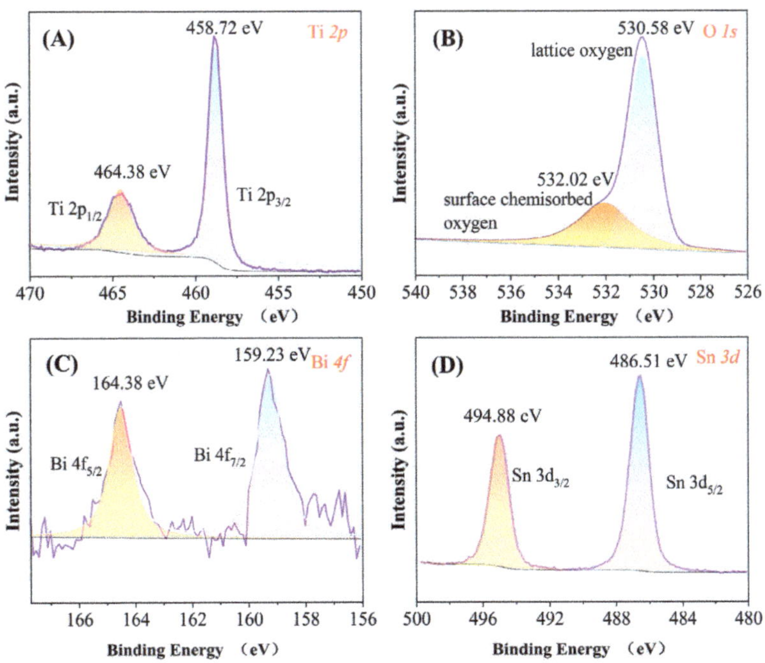

Figure 2. *Cont.*

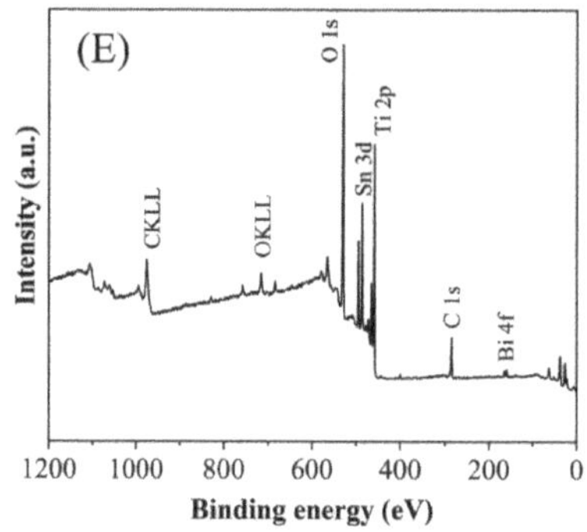

Figure 2. High-resolution XPS spectra of the as-prepared Bi-TiO_2/SnO_2 of (**A**) Ti 2p, (**B**) O 1s, (**C**) Bi 4f, (**D**) Sn 3d, and (**E**) the overall XPS spectrum.

2.1.3. Impact on Surface Morphology

To analyze the influence of Bi and SnO_2 addition on the surface morphology of the catalyst, scanning electron microscopy (SEM) was performed for TiO_2/SnO_2, Bi-TiO_2/SnO_2, and Bi-TiO_2. Figure 3A presents the SEM image of TiO_2/SnO_2. It can be seen that, without Bi addition, the TiO_2/SnO_2 exhibited a dense and smooth surface of the primary particles. Figure 3C presents that Bi doping induced a loose and porous structure on the catalyst's surface, indicating that the addition of Bi promoted the dispersion of particles, possibly due to the inhibiting effect of Bi doping during the formation of TiO_2 crystals, which is consistent with the XRD results. Comparing Figure 3B with Figure 3C, the results showed that the addition of SnO_2 had no obvious effect on the surface morphology from the SEM images.

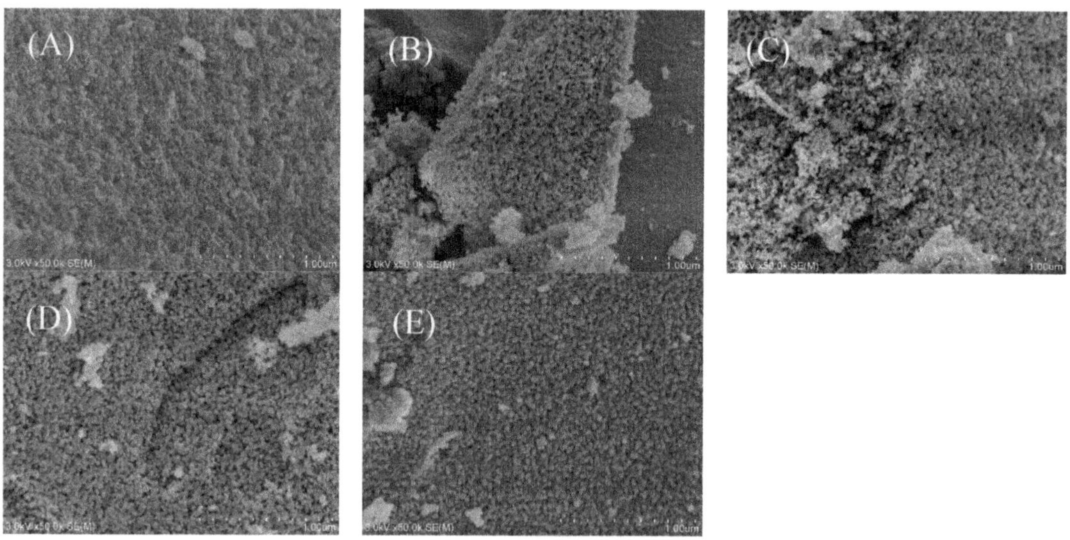

Figure 3. SEM images of (**A**) TiO_2/SnO_2; (**B**) Bi-TiO_2; (**C**) Bi-TiO_2/SnO_2; (**D**) F-Bi-TiO_2/SnO_2; and (**E**) F-Bi-$TiO_2/SnO_2/SiO_2$.

2.1.4. Impact on Optical Properties

Figure 4A displays the UV–vis diffuse reflectance spectra (DRSs) of TiO_2/SnO_2, Bi-TiO_2, and Bi-TiO_2/SnO_2. It can be seen that, compared to TiO_2/SnO_2, Bi-TiO_2/SnO_2 exhibited stronger absorbance in UV light regions but weaker absorbance in visible light regions, indicating improved UV light-harvesting properties due to Bi doping. Compared to Bi-TiO_2, Bi-TiO_2/SnO_2 showed enhanced absorbance in both the UV and visible light regions. It indicated that the combination with SnO_2 can improve the absorption of visible and ultraviolet light. The bandgaps were calculated by Tauc plot based on DRS results (Figure 4B), and the bandgap energies (E_g) of Bi-TiO_2, TiO_2/SnO_2, and Bi-TiO_2/SnO_2 were 3.00, 2.94, and 2.97 eV, respectively. This calculated result showed that the Bi and SnO_2 modifications have no significant effects on the bandgap.

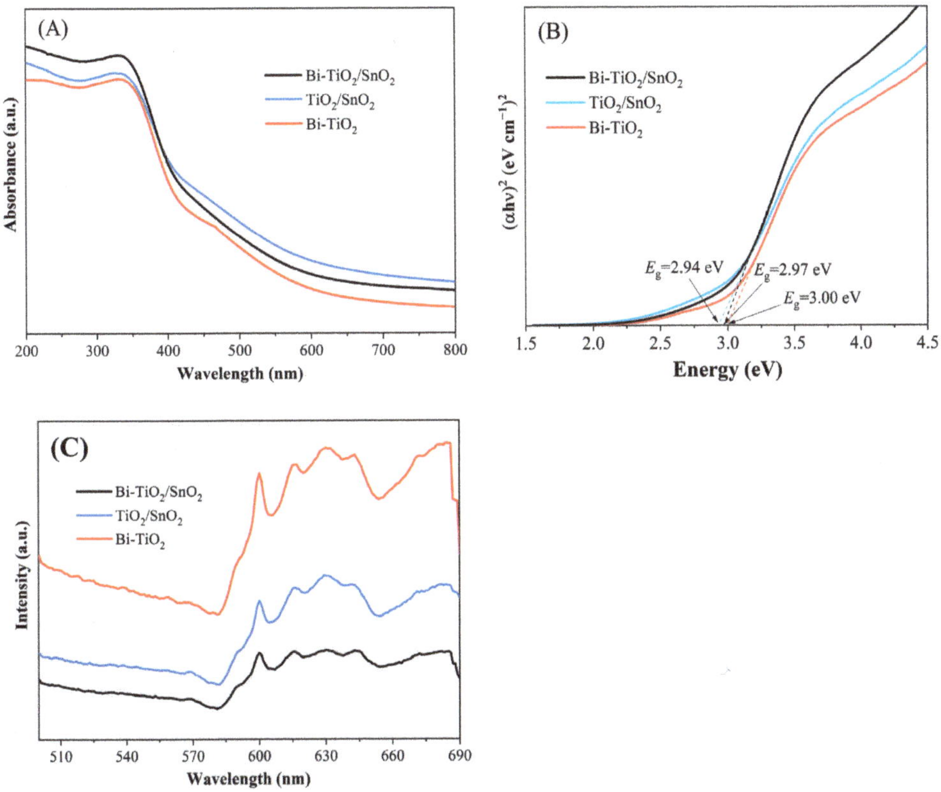

Figure 4. (**A**) UV–vis DRS of TiO_2/SnO_2, Bi-TiO_2, Bi-TiO_2/SnO_2; (**B**) the corresponding Tauc plot; and (**C**) the corresponding photoluminescence spectra.

Figure 4C shows the photoluminescence (PL) spectra of TiO_2/SnO_2, Bi-TiO_2, and Bi-TiO_2/SnO_2. The PL spectra are related to the transfer behavior of photo-generated carriers. A higher PL intensity indicates a higher recombination rate of photo-generated carriers [55]. As shown in Figure 4C, the PL spectrum of Bi-TiO_2/SnO_2 did not exhibit new emission peaks compared to TiO_2/SnO_2 and Bi-TiO_2, indicating that Bi doping and SnO_2 composite did not induce new emission phenomena. Compared to TiO_2/SnO_2, Bi-TiO_2/SnO_2 had relatively weak emission intensity, indicating that a certain amount of Bi doping inhibited the recombination of photo-generated carriers. Coupled with the XRD and XPS results, the reason could be that Bi doping triggers lattice distortion, generating impurity levels and more oxygen vacancies that serve as traps for photo-generated electrons. Compared to Bi-TiO_2, Bi-TiO_2/SnO_2 exhibited a significantly reduced PL intensity, indicating that the appropriate amounts of SnO_2 can enhance the separation of photo-generated carriers. This may be attributed to the transfer of photo-generated electrons from CB of TiO_2 to SnO_2 and the movement of holes from VB of SnO_2 to TiO_2, which improve the separation efficiency of photon-generated carriers [38].

2.2. Impact of Bi and SnO_2 Addition on RhB Photocatalytic Degradation

The influence of Bi amounts and the molar ratio of SnO_2 to TiO_2 on the photocatalytic activity of Bi-TiO_2/SnO_2 under simulated sunlight were investigated. Figure 5A shows the RhB degradation curves of Bi-TiO_2/SnO_2 prepared with varying Bi amounts. The degradation efficiency initially increased and then decreased with increasing Bi doping. The catalyst exhibited the highest activity when the Bi doping level was 0.7 mol%, achieving an RhB degradation rate of 96.56% after 60 min of reaction. Beyond the doping level of 0.7 mol%, the catalytic activity decreased. The reason may be that a moderate amount of Bi is an effective charge separator, but beyond a certain amount, it begins to act as a new recombination center [50].

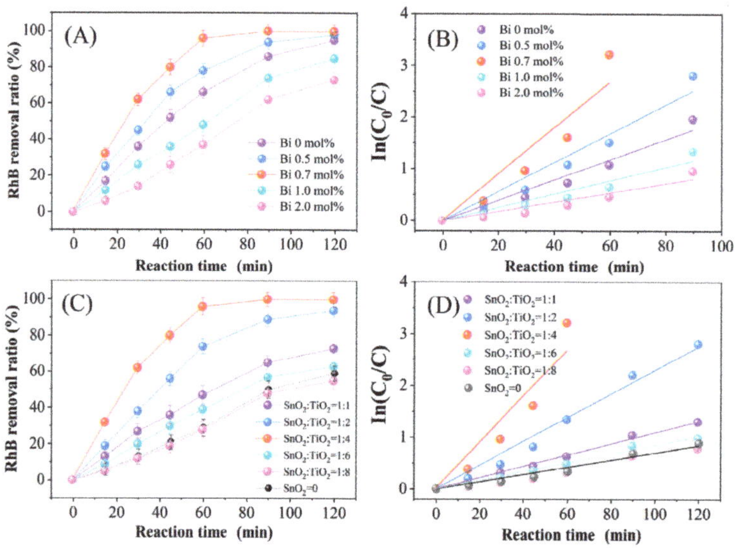

Figure 5. (**A**) Photocatalytic removal curves and (**B**) pseudo-first-order kinetics curves for the degradation of RhB by Bi-TiO_2/SnO_2 with different Bi amounts; (**C**) photocatalytic removal curves and (**D**) pseudo-first-order kinetics curves of Bi-TiO_2/SnO_2 with different SnO_2/TiO_2 molar ratios.

Figure 5C illustrates the degradation curves of Bi-TiO_2/SnO_2 catalysts with different SnO_2/TiO_2 molar ratios. With an increasing SnO_2 proportion, the degradation efficiency of RhB initially increased and then decreased. The optimum catalytic activity was observed at a SnO_2/TiO_2 molar ratio of 1:4, indicating that the composite of TiO_2 with SnO_2 in a specific ratio is advantageous for boosting the photocatalytic activity.

The activity test results suggested that the optimal Bi doping amount was 0.7 mol%, and the optimal SnO_2:TiO_2 molar ratio was 1:4. The RhB degradation processes were fitted by the pseudo-first-order model (Figure 5B,D). The results showed that the optimized Bi-TiO_2/SnO_2 had a reaction rate constant (k) of 0.0621 min^{-1}. This value was 6.4 times that of the Bi-TiO_2 catalyst and 2.3 times that of the TiO_2/SnO_2 catalyst (Figure S2). It indicated that the addition of Bi and the combination with SnO_2 significantly boosted the catalyst's activity, which was consistent with the improvements in light absorption performance observed through DRS and PL.

2.3. *Impact of F Addition and Compositing with SiO_2 on Catalyst Properties*

2.3.1. Surface Morphology

The effect of F addition and compositing with SiO_2 on the surface morphology of the catalyst was analyzed by SEM. Figure 3C–E show the surface morphology images of Bi-TiO_2/SnO_2, F/Bi-TiO_2/SnO_2, and F/Bi-TiO_2/SnO_2/SiO_2. It can be observed that all three catalysts consist of primary particles with relatively uniform sizes that aggregate into amorphous secondary particles. Comparing Figures 3C and 3D, the F addition made the catalyst particles more uniformly dispersed, with numerous tiny grains scattered on the surface, inhibiting particle aggregation. This was speculated to be caused by the etching effect of hydrofluoric acid on the sample surface [56]. Comparing Figures 5C and 3E, compositing with SiO_2 led to a more uniform distribution of the particles, with a smoother surface. It was inferred that, due to the porous structure of SiO_2 [57], catalytic active sites were evenly distributed on the catalyst surface while suppressing grain growth and aggregation.

2.3.2. Crystal Phase

To understand the impact of F and SiO_2 addition on the crystal phase of the catalyst, XRD characterization was performed on Bi-TiO_2/SnO_2, F/Bi-TiO_2/SnO_2, and F/Bi-TiO_2/SnO_2/SiO_2 (Figure 6). Comparing the spectra of Bi-TiO_2/SnO_2 and F/Bi-TiO_2/SnO_2 revealed that F addition resulted in finer and higher TiO_2 diffraction peaks. The crystal grain size calculated in Table S1 indicated that F modification restrained the growth of TiO_2 grains [58]. In fact, F was mostly doped into the TiO_2 lattice, which was demonstrated by the XPS characterization of F-Bi-TiO_2/SnO_2 (Figure S3).

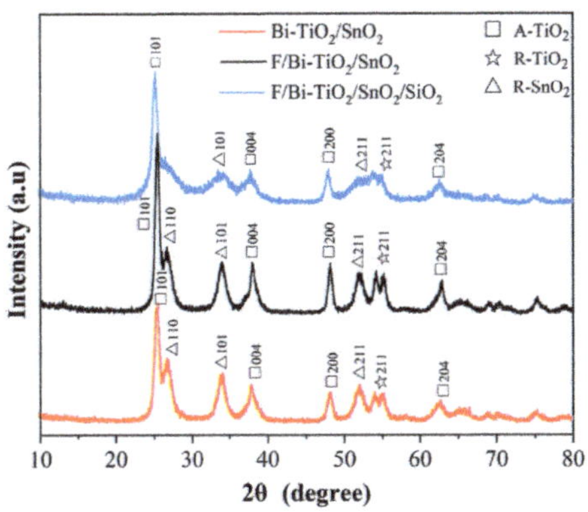

Figure 6. XRD patterns of Bi-TiO_2/SnO_2, F-Bi-TiO_2/SnO_2, and F-Bi-TiO_2/SnO_2/SiO_2.

Comparing the XRD spectra of F/Bi-TiO_2/SnO_2 and F/Bi-TiO_2/SnO_2/SiO_2, the introduction of SiO_2 reduced the intensity of the diffraction peaks of rutile SnO_2 and anatase TiO_2. This may be due to the hindrance of SnO_2 and TiO_2 grain growth when combined with SiO_2, resulting in smaller particle sizes (Table S1).

2.3.3. Nitrogen Adsorption Desorption Analysis

To obtain information about the pore structure and specific surface area of these catalysts, N_2 adsorption–desorption tests were conducted on Bi-TiO_2/SnO_2, F/Bi-TiO_2/SnO_2, and F/Bi-TiO_2/SnO_2/SiO_2 catalysts. The obtained adsorption–desorption isotherms are shown in Figure 7A. According to new classification by Donohue and Aranovich [59], the adsorption–desorption isotherms of the three catalysts are type (IV) isotherms with hysteresis loops. The BJH (Barrett–Joyner–Halenda) plot in Figure 7B indicated that the catalysts' pore size distribution was mainly between 2 and 25 nm, suggesting that these catalysts are mesoporous materials. Based on the Brunauer–Emmett–Teller equation, the specific surface areas of Bi-TiO_2/SnO_2, F/Bi-TiO_2/SnO_2, and F/Bi-$TiO_2/SnO_2/SiO_2$ were calculated as 73.0, 69.1, and 112.27 m^2/g, respectively. It can be seen that F addition did not increase the specific surface area of the catalyst, while the addition of SiO_2 led to a significant increase in the specific surface area, which was likely due to SiO_2 forming a mesoporous material [57]. The increase in the specific surface area is beneficial for providing more adsorption and reactive sites for the reactants, thus enhancing the reaction efficiency.

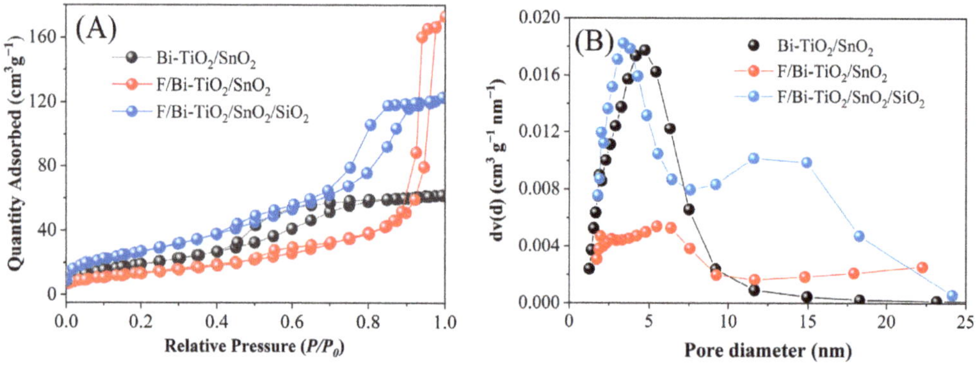

Figure 7. (**A**) N_2 adsorption–desorption isotherms and corresponding pore size distribution curves of Bi-TiO_2/SnO_2, F-Bi-TiO_2/SnO_2, and F-Bi-$TiO_2/SnO_2/SiO_2$. (**B**) The BJH plot.

2.3.4. Optical Properties

Figure 8A presents the UV-vis DRS for Bi-TiO_2/SnO_2, F/Bi-TiO_2/SnO_2, and F/Bi-$TiO_2/SnO_2/SiO_2$ catalysts. Compared with Bi-TiO_2/SnO_2, F/Bi-TiO_2/SnO_2 exhibited a significant red shift in the absorption edge, and the absorbance intensity in the UV and visible light regions was greatly enhanced. This indicated that fluoride ions, as dopants, alter the band structure of TiO_2, possibly due to an increase in the oxygen vacancy and Ti^{3+} defects introduced by fluorine doping into the TiO_2 lattice [47,60,61]. It reduced the bandgap and expanded the absorption range into the visible light region. Moreover, the absorption intensity was enhanced due to the formation of an intermediate level between the VB and CB of the metal oxide. The intermediate level can react as a trap to retard the charge recombination [28,62]. In comparison to F/Bi-TiO_2/SnO_2, further compositing with SiO_2 in F/Bi-$TiO_2/SnO_2/SiO_2$ did not cause a red shift in the absorption edge, and although there was an enhancement in UV absorption, there was no improvement in the visible light region. This is possibly due to no change in the structure of the metal oxides by composite with SiO_2. Instead, the role of SiO_2 was to inhibit the aggregation and growth of crystal grains and, more importantly, to provide a larger specific surface area, thereby increasing the adsorption and active site quantity for organic pollutants. Based on the Tauc plot (Figure 8B), the bandgaps of Bi-TiO_2/SnO_2, F/Bi-TiO_2/SnO_2, and F/Bi-$TiO_2/SnO_2/SiO_2$ were calculated as 2.97, 2.86, and 2.94 eV, respectively. It indicated that F doping reduced the bandgap of the photocatalyst and widened the visible light absorption range, while SiO_2 composite provided a larger specific surface area for the photocatalytic reaction, offering more reactive sites and dispersing the active sites.

To compare the separation efficiency of photo-generated carriers in the three catalysts, PL characterization was performed (Figure 8C). The PL spectra of the three catalysts show similar linear shapes. F and SiO_2 modifications, therefore, did not induce new fluorescence phenomena, indicating that fluorescence effects are primarily related to the microcrystalline surface structure of the catalyst. Compared to Bi-TiO_2/SnO_2 and F/Bi-TiO_2/SnO_2, the fluorescence intensity of F/Bi-$TiO_2/SnO_2/SiO_2$ significantly decreased, suggesting that both F modification and compositing with SiO_2 suppressed the recombination of photo-generated carriers. The reason probably lies in that F doping into TiO_2 and promotes the formation of lattice oxygen vacancies, which trap electrons and cause a decreased PL intensity [47,63]. In addition, the role of SiO_2 is to disperse TiO_2 and SnO_2 grains on its surface, and then form a more favorable structure for the separation of photo-generated carriers, thereby elevating the photocatalytic efficiency.

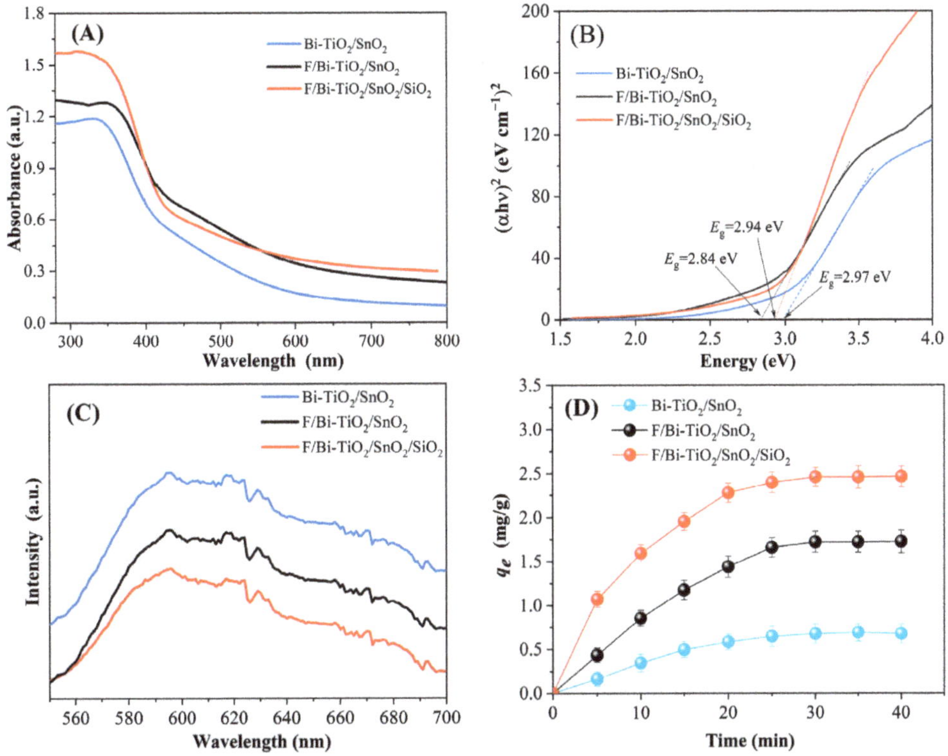

Figure 8. (**A**) UV–vis DRS of Bi-TiO_2/SnO_2, F/Bi-TiO_2/SnO_2, and F/Bi-TiO_2/SnO_2/SiO_2; (**B**) Tauc plot; (**C**) PL spectra; and (**D**) the amount of dark adsorption vs. adsorption time.

2.3.5. Adsorption Performance

To assess the influence of F doping and compositing with SiO_2 on the adsorption performance of the catalyst, the adsorption capacity of RhB in the dark was measured for Bi-TiO_2/SnO_2, F/Bi-TiO_2/SnO_2, and F/Bi-TiO_2/SnO_2/SiO_2, as shown in Figure 8D. The saturated adsorption capacities of the three catalysts for RhB were 0.67, 1.72, and 2.46 mg/g, respectively. It was evident that both F modification and SiO_2 modification increase the adsorption capacity of the catalyst for RhB molecules. The reason for the increased adsorption capacity with F modification may be that, although F did not increase the specific surface area of the catalyst, it increased the surface acidic sites that tend to adsorb basic Rhodamine B molecules. Moreover, the surface acidic sites are in favor of converting adsorbed H_2O to OH, and subsequently, more surface hydroxyl radicals with strong mobility are generated [64]. While SiO_2 modification did not increase the catalyst's absorption of visible light, it contributed to the increase in the specific surface area. This resulted in the dispersion of more active sites on the catalyst surface for adsorbing reactants, thereby increasing the adsorption capacity of reactants and promoting the progression of the reaction.

2.4. Impact of F and SiO_2 Addition on RhB Photocatalytic Degradation

2.4.1. Impact of F Doping on RhB Degradation

The optimal doping amount of F was determined by testing the photocatalytic activity of F/Bi-TiO_2/SnO_2 with different F doping amounts (Figure 9A). As observed, with the increase in F addition amounts, the photocatalytic activity of F/Bi-TiO_2/SnO_2 initially increased and then decreased. The catalyst exhibited the highest activity when the molar ratio of F to Ti was 0.2, achieving an RhB degradation ratio of 96% after 60 min of reaction. The RhB degradation processes were fitted by the pseudo-first-order model (Figure 9B). The results showed that the reaction rate constants (k) for Bi-TiO_2/SnO_2 and F/Bi-TiO_2/SnO_2 (F/Ti = 0.2) were determined to be 0.042 and 0.098 min^{-1} (Figure S2), respectively, indicating that F doping increased the reaction rate constant by 2.3 times compared to the Bi-TiO_2/SnO_2. A small amount of doping (F/Ti = 0.1, 0.2) was more favorable for enhancing catalyst activity. One possible reason is that a small amount of fluoride ions enters the TiO_2 lattice and occupies the O positions, causing lattice defects such as oxygen vacancy. These defects provide locations for electrons and facilitate the transfer of photo-generated carriers. Additionally, the doping ions may introduce impurity states within the bandgap, which hybridize with the energy bands of TiO_2. This ensures the efficient transfer of photo-generated carriers to the surface active sites, thereby increasing the absorption of sunlight. However, an excess of F ions decreased the catalyst's activity, potentially due to the generation of a negative electric potential that inhibited electron migration to the catalyst surface and increased the electron–hole pair recombination. Excessive doping concentrations may also lead to more severe lattice expansion, increased particle size, and a decrease in photocatalytic efficiency.

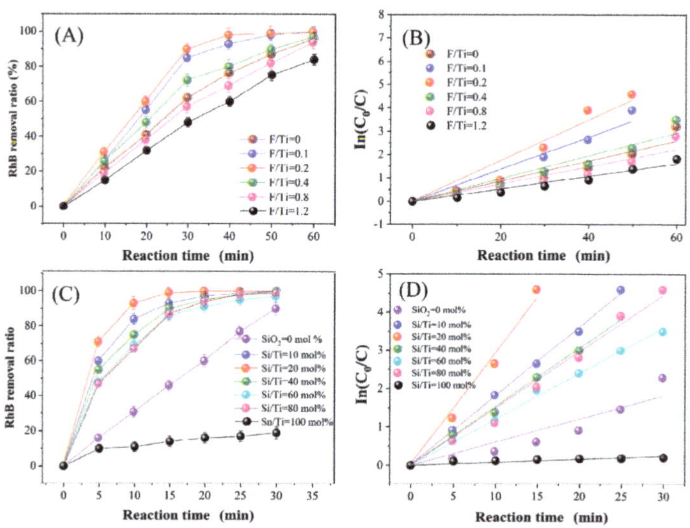

Figure 9. (**A**) Photocatalytic removal curves and (**B**) pseudo-first-order kinetics curves for the degradation of RhB by F/Bi-TiO_2/SnO_2 with different F amounts, and (**C**) photocatalytic removal curves and (**D**) pseudo-first-order kinetics curves of F/Bi-TiO_2/SnO_2/SiO_2 with different SiO_2/SnO_2 molar ratios.

2.4.2. Impact of SiO_2/SnO_2 Molar Ratio on RhB Degradation

The impact of the SiO_2/SnO_2 molar ratio on the photocatalytic activity was investigated. Figure 9C depicts the degradation efficiency of RhB over time for F/Bi-TiO_2/SnO_2/SiO_2 catalysts with different SiO_2/SnO_2 molar ratios. It can be seen that, with an increase in the SiO_2 addition level, the catalytic activity initially increased and then decreased. The catalyst exhibited the highest degradation ratio of RhB when the SiO_2/SnO_2

molar ratio was 20 mol%, reaching over 90% in 10 min with a reaction rate constant of 0.291 min^{-1} based on pseudo-first-order fitting (Figures 9D and S2). This indicated that an appropriate amount of SiO_2 composite was conducive to enhancing photocatalytic activity. Considering the characterization results, this improvement was likely due to the SiO_2/SnO_2 composite having a certain porosity, which improved the porous structure and increased the specific surface area, thereby facilitating the adsorption and degradation of RhB on the catalyst surface. However, excessive SiO_2 may cover the active sites of the catalyst, reducing light absorption and the contact points between active components and the reactants, limiting photocatalytic performance.

By the optimization of the addition amounts of Bi, SnO_2, F, and SiO_2, the RhB degradation percentage was close to 100% after 20 min of irradiation under simulated sunlight. Table 1 lists the catalytic degradation effects of some similar catalysts on Rhodamine B degradation in some studies.

Table 1. Statistics of photocatalytic degradation of RhB with different photocatalysts.

Composite	Light	Concentration	*D* (%)	Time (min)
SnO_2/TiO_2/PVDF [65]	UV	10 mg/L	91.84	270
SnO_2/TiO_2/RGO [66]	UV	10 mg/L	97.60	40
SnO_2–TiO_2 [67]	Visible	10^{-5} mol/L	76.00	180
S doped SnO_2 @TiO_2 [68]	Sunlight	10 mg/L	97.00	200
SnO_2/TiO_2 [69]	UV–vis	10 mg/L	99.00	30
SnO_2/TiO_2 [70]	UV–vis	20 mg/L	94.00	180
SnO_2/TiO_2 [39]	UV–vis	10 mg/L	98.00	120
B-TiO_2/SnO_2 [54]	VisibleUV–vis	10 mg/L	96.62100	9070
O-g-C_3N_4/SnO_2 [71]	Visible	10 mg/L	98.10	30
Bi/Bi_2O_3/TNAs [24]	Sunlight	2 mg/L	35	240
Bi/BPNs/P–BiOCl [72]	Xenon lamp	20 mg/L	100	30
KF/Bi/BC [55]	Xenon lamp	30 mg/L	100	10
C/F–Ag–TiO_2 [73]	Xenon lamp	10 mg/L	84.2	240
N,F-$TiO_{2-\delta}$ [74]	LED lights	10 mg/L	77.2	60
F/Bi-$TiO_2/SnO_2/SiO_2$ (this study)	Xenon lamp	10 mg/L	100	20

2.4.3. Cyclic Use of the Photocatalyst

The stability of the optimal F/Bi-$TiO_2/SnO_2/SiO_2$ was assessed through cyclic RhB degradation tests, with each reaction lasting 30 min. After each reaction, the catalyst was rinsed several times with deionized water, dried, and then reused for subsequent reactions. The results shown in Figure 10A indicated that there was little attenuation in catalytic performance after five repeated reactions. However, the degradation ratio of RhB could still reach 100% within 30 min, indicating the stability of the F/Bi-$TiO_2/SnO_2/SiO_2$ catalyst.

2.4.4. Effect of pH Value on Catalytic Activity

In this experiment, the natural pH of the reaction solution was around 7. The pH of the reaction system was adjusted using 0.01 mol/L NaOH and 0.01 mol/L HNO_3 solutions. The activity test results shown in Figure 10B revealed that the pH value significantly influenced the activity of F/Bi-$TiO_2/SnO_2/SiO_2$ in RhB degradation. The degradation performance of RhB within 30 min followed the order: neutral (pH = 7) > slightly acidic (pH = 5) > slightly alkaline (pH = 9) > highly alkaline (pH = 11) > highly acidic (pH = 3). The reaction rate initially increased and then decreased with an increase in the pH value. The fastest reaction rate was observed at pH = 7, indicating the best performance under neutral conditions, and both highly acidic and highly alkaline environments were unfavorable for the photocatalytic reaction. The reason lies in the impact of pH changes on the charge distribution on the catalyst surface and on the reactant molecular structure [52], which affect the adsorption capacity of the catalyst for RhB molecules. In highly alkaline conditions, hydrogen ions in the surface hydroxyl groups are replaced by cations, resulting in a decrease in the number of surface hydroxyl groups. The hydroxyl groups are in favor of oxidation of organic pollutants. When it comes to highly acidic conditions, the positive charge is carried by the catalyst. Since RhB is a cationic dye [75], the increased positive charge diminishes the adsorption capacity of the catalyst for RhB, thereby hindering the photocatalytic degradation of RhB. Typically, the pH of dye wastewater ranges from 6 to 10. The optimal pH for the $F/Bi\text{-}TiO_2/SnO_2/SiO_2$ photocatalytic reaction in RhB degradation was found to be 7, with a degradation efficiency reaching 93.89% within 10 min. Moreover, pH 7 is more adaptable to the pH variations typically found in dyeing wastewater.

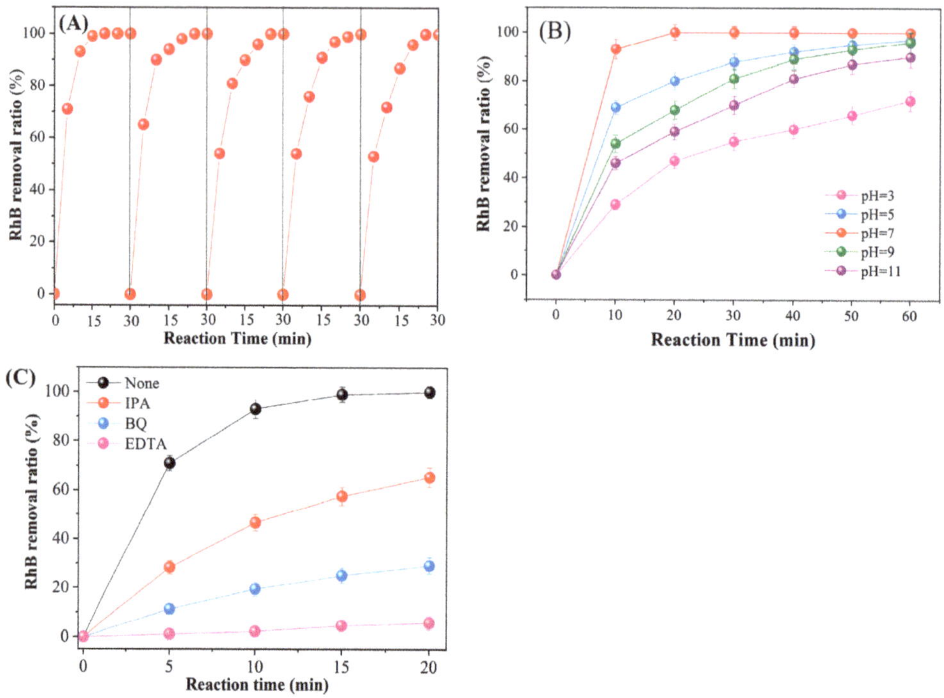

Figure 10. (**A**) Cyclic RhB degradation tests with $F/Bi\text{-}TiO_2/SnO_2/SiO_2$. (**B**) Effect of pH on the degradation of RhB with $F/Bi\text{-}TiO_2/SnO_2/SiO_2$. (**C**) Effects of sacrificial agents on photo-degradation of RhB with $F/Bi\text{-}TiO_2/SnO_2/SiO_2$.

2.4.5. Investigation of Active Species

The optimal amount of $F/Bi\text{-}TiO_2/SnO_2/SiO_2$ was used, and several sacrificial agents were added to the RhB solution to distinguish the main active species during the RhB degradation process. IPA, EDTA, and BQ were selected to be radical quenchers for OH, h^+, and O_2^-, respectively. Figure 10C shows that the RhB degradation ratio after 20 min of reaction decreased from 100% to 65%, 29%, and 6% when IPA, BQ, and EDTA were added, respectively. The results revealed that all of the corresponding •OH, $•O_2^-$, and h^+ participate in the photo-degradation reaction and play an important role. The most crucial active species was h^+ according to the lowest RhB degradation ratio with EDTA addition. It implies that the oxidation of water to introduce OH radicals by a photo-generated h^+ assumes a key role in the process of RhB oxidation. The reaction mechanism was speculated as follows:

$$F/Bi\text{-}TiO_2/SnO_2/SiO_2 + hv \rightarrow h^+ + e^- \quad (1)$$

$$h^+ + H_2O \rightarrow H^+ + •OH \quad (2)$$

$$h^+ + OH^- \rightarrow •OH \quad (3)$$

$$O_2 + e^- \rightarrow \bullet O_2^- \quad (4)$$

$$Rh\,B + (\bullet OH + \bullet O_2^-) \rightarrow CO_2 + H_2O \quad (5)$$

3. Discussion

Based on the characterization test, the impact of the Bi and F doping and compositing with SnO_2 and SiO_2 on the catalyst's performance was analyzed. The results indicated that Bi addition induced a loose and porous surface of the catalyst. Doping Bi into the TiO_2 lattice formed electron interactions with TiO_2 in the form of $Bi^{3+\sigma}$, potentially generating more oxygen vacancies that act as traps for photo-generated electrons and impede the recombination of electron–hole pairs. Additionally, Bi doping improved UV light-harvesting properties, while the combination with SnO_2 can improve the absorption of visible and ultraviolet light. The SnO_2 and F modification suppressed the recombination of photo-induced carriers. The doping of fluorine may elevate the oxygen vacancy and hybrid level, which reduced the bandgap and increased visible light absorption. In addition, the adsorption ability toward RhB molecules on catalyst's surface was advanced by F addition, which may be due to the increase in acidity. The composited SiO_2 contributed by providing a larger specific surface area for the adsorption of more RhB molecules. Benefiting from the effect of doped F and Bi atoms in the combination with SnO_2 and SiO_2, the recombination of photo-excited electrons and holes is inhibited, and the adsorption of the reactant molecules is increased, thus improving the activity of the catalyst.

4. Materials and Methods

4.1. Material and Reagents

Tetrachlorostannane pentahydrate ($SnCl_4{\cdot}5H_2O$, Sigma-Aldrich, St. Louis, MO, USA, AR/99.99%), Bismuth nitrate pentahydrate ($Bi(NO_3){\cdot}5H_2O$, Sigma-Aldrich, St. Louis, MO, USA, AR/99.0%), Polyethylene glycol (PEG-1000, Tianjin Kemiou Chemical Reagent Co., Ltd., Tianjin, China, AR), Urea ($CO(NH_2)_2$, Sinopharm, Beijing, China, AR/99.5%), Tetraethl silicate (TEOS, Tianjin Kemiou Chemical Reagent Co., Ltd., Tianjin, China, AR/99.0%), Ethyl alcohol (EtOH, Sinopharm, Beijing, China, AR/99.5%), Hydrofluoric acid (HF, Tianjin Kemiou Chemical Reagent Co., Ltd., Tianjin, China, AR/50%), Titanium butoxide (TBT, aladdin, Bay City, MI, USA, AR/99%), Hydrochloric acid (HCl, Sinopharm, Beijing, China, AR/36%~38%), Iso-propyl alcohol (IPA, Sigma-Aldrich, St. Louis, MO, USA, AR/99.5%), benzoquinone (BQ, Sigma-Aldrich, St. Louis, MO, USA, AR/99.5%), and EDTA (Sigma-Aldrich, St. Louis, MO, USA, AR/99.5%).

All data processing was conducted using Origin Pro 2018 (OriginLab, Northampton, MA, USA). XRD data analysis was performed using Jade version 5.0 (Materials Data Inc., Livermore, CA, USA).

4.2. Preparation of Photocatalysts

4.2.1. Synthesis of Bi-TiO_2/SnO_2

Firstly, SnO_2 was prepared using a precipitation method described in the Supplementary Materials. Then, Bi-TiO_2/SnO_2 was prepared using a sol–gel method: A certain volume of EtOH and TBT, in a volume ratio of 5:13, was measured. TBT was then slowly added drop by drop into EtOH. Then, the mixture was covered with plastic wrap and magnetically stirred for 20 min to obtain Solution A. After that, EtOH, deionized water, and CH_3COOH were measured in a volume ratio of 21:1:4 and mixed uniformly. $Bi(NO_3)_3{\cdot}5H_2O$ was added to the mixture (Bi/Ti = 0.7 mol%), and the solution was sonicated for 30 min to form Solution B. Solution B was transferred into a pear-shaped funnel and slowly added drop by drop into continuously stirred Solution A. The mixture was then covered with plastic wrap and stirred for 2 h until a wall-sticking phenomenon was observed when gently tilting the beaker. The plastic wrap was removed, and a certain amount of the previously prepared SnO_2 was added based on a Sn/Ti molar ratio of 1:4. The mixture was then stirred until the stir bar could no longer rotate; at which point, a sol was formed. The sol–gel was aged at room temperature for 24 h and then dried in an 80 °C oven for 24 h until a dry gel was obtained. The gel was ground and calcined in a muffle furnace at 400 °C for 2 h to yield the Bi-TiO_2/SnO_2 powder.

4.2.2. Synthesis of F-Bi-TiO_2/SnO_2

The steps were identical to those in Section 4.2.1, with the only difference being the addition of hydrofluoric acid (50%) to Solution B in a certain F/Ti molar ratio as the precursor for fluorine.

4.2.3. Synthesis of F-Bi-$TiO_2/SnO_2/SiO_2$

First, a SnO_2/SiO_2 composite gel was prepared: A mixture solution of TEOS, anhydrous ethanol, and HCl (molar ratio of 1:6:0.02) was prepared. The solution was stirred thoroughly to obtain a transparent SiO_2 sol, which was left at room temperature for 60 min. A 1.5×10^{-3} mol/L $SnCl_4$ aqueous solution was prepared, and $SnCl_4$ was slowly added to the SiO_2 sol in a certain molar ratio. Ammonia solution was dropwise added to the mixture under continuous stirring until the pH reached 7.0. After aging for 2 days at room temperature, the gel was filtered and washed with water until free of Cl^- ions (tested with 0.1 mol/L $AgNO_3$ solution). Finally, the gel was dried in an oven at 60 °C to obtain the SnO_2/TiO_2 composite gel.

Subsequently, the sol–gel method, as described in Section 4.2.2, was used. The difference was that the SnO_2 added was replaced with the SnO_2/SiO_2 composite gel. The final product was F-Bi-$TiO_2/SnO_2/SiO_2$.

4.3. Catalyst Evaluation

An experimental device (Figure S1) was employed for photocatalytic activity evaluation. A self-made quartz glass reactor, with a height of 100 mm and an inner diameter of 62 mm, was used, featuring a sampling port on the side. Continuous stirring of the solution in the reactor was achieved using a magnetic stirrer to ensure sufficient contact between the catalyst and the solution. The light source, i.e., a 350 W spherical xenon lamp, was positioned 15 cm vertically above the reactor to simulate solar light. During illumination, a blower was used for cooling the air in the xenon lamp, and circulating cooling water was utilized to ensure that the reaction occurred at room temperature. Rhodamine B (RhB) was chosen as the target for degradation to investigate the activity of the catalyst, with an initial concentration of 10 mg/L. The prepared catalyst was added to the solution under stirring, with a catalyst dosage of 1 g/L. The reactor was sealed with quartz plates and in darkness. The solution was stirred for 30 min to reach the adsorption–desorption equilibrium. The adsorption equilibrium time was determined through dark adsorption experiments. The xenon lamp was then turned on to initiate irradiation. During the reaction, at regular intervals, 4 mL of solution was withdrawn using a syringe. The concentration of Rhodamine B in the samples was detected using a UV-2450 (Shimadzu, Beijing, China) UV-vis spectrophotometer, and the removal efficiency of RhB was calculated using Equation as follows: $D = (C_0 - C_t)/C_0$. Wherein D is the degradation efficiency of RhB (%); C_0 and C_t represent the RhB concentration at 0 and t min, respectively. In addition, the RhB degradation processes were fitted by a pseudo-first-order model using the following equation: $\text{In}(C_0/C_t) = kt$, where k is the reaction rate constant.

4.4. Catalyst Characterizations

X-ray diffraction (XRD) pattern of the catalysts was recorded using a D/Max 2500PC diffractometer (Rigaku, Akishima, Japan) with Cu Ka radiation. Scans were collected in the 2θ range of 10~80° and a step of 4 °/min. The crystallite size was calculated based on the Scherrer equation. X-ray photoelectron spectroscopy (XPS) measurement was performed with a *K-Alpha*$^+$ (Thermo fisher Scientific, Waltham, MA, USA). Al Kα was used as a radiation source. The binding energy of C1s was used for correction. The micromorphologies of the samples were obtained with a S4800 scanning electron microscopy (SEM) (Rigaku, Japan). The specific surface area, pore size, and pore volume of the catalysts were measured by an automatic constant-volume adsorption instrument (Belsorp II, Bayer Japan Co., Ltd., Chiyoda, Japan). Before nitrogen adsorption, the catalyst was pretreated in vacuum at 400 °C for 3 h. The nitrogen adsorption–desorption process was carried out at the liquid nitrogen temperature (77 K). The light absorption capacity of the catalysts was recorded by barium sulfate tablet method with a UV-2540 UV-vis spectrophotometer (Shimadzu, Tokyo, Japan) in the wavelength range of 200~800 nm. The bandgaps were computed via the Tauc and Menth Equation.

$$\alpha hv = A(hv - E_g)^n \quad (6)$$

where α is the absorptivity, h is the Planck's constant, v is the frequency of radiation, E_g is the bandgap, A is the constant of proportionality, and n is the type of optical transition following photon absorption.

Fluorescence analysis was obtained using a FluoroMax-4 photoluminescence spectrometer (HORIBA Scientific, Irvine, CA, USA) with a wavelength scanning range of 350~700 nm, step size of 0.1 s, and slit size of 2 nm.

5. Conclusions

The TiO_2 was modified by co-doping with Bi and F and compositing with SnO_2 and SiO_2. The synthesized catalyst demonstrated efficient photocatalytic performance toward RhB degradation under simulated sunlight. Through the optimization of the doping levels of Bi and F, as well as the ratio of SnO_2 and SiO_2 composite, the catalyst reached a degradation efficiency of 100% for RhB within 20 min under simulated sunlight, with a first-order reaction rate constant of 0.291 min^{-1}. This optimal catalyst showed stability in the cycle tests. The reaction rate initially increased and then decreased with an increase in the pH value. The active species h^+ assume a key role in the process of RhB oxidation. The combined effects of Bi, F, Sn, and Si modifications include the following: (1) enhanced absorbance intensity in the UV and visible light regions; (2) reduced recombination of photogenerated carriers; and (3) increased specific surface area that provides more adsorption sites for reactant molecules. Because of these beneficial modifications, the catalyst exhibited excellent photocatalytic activity for RhB degradation under simulated sunlight.

Supplementary Materials: The following supporting information can be downloaded at: https://www.mdpi.com/article/10.3390/catal14100735/s1. Refs. [76,77] are cited in Supplementary Materials.

Author Contributions: Conceptualization, L.Q. and R.Z.; methodology, R.Z. and F.O.; software, L.Q. and W.X.; validation, H.L.; formal analysis, H.L.; resources, F.O. and R.Z.; data curation, L.Q. and W.X.; writing—original draft preparation, L.Q. and H.L.; writing—review and editing, L.Q. and R.Z.; visualization, L.Q. and H.L.; funding acquisition, L.Q., H.L. and R.Z. All authors have read and agreed to the published version of the manuscript.

Funding: This study was funded by Guizhou Provincial Department of Science and Technology, China (grant number: [2022] 202), Foundation of Guizhou Minzu University (GZMUZK [2024]QD65), the Special Project for Sustainable Development Science Technology in Shenzhen (No. KCXFZ20201221173000001), and the Basic Scientific Research Project of Liaoning Provincial Education Department for Universities (LJ212410147048).

Data Availability Statement: The original contributions presented in the study are included in the article/Supplementary Materials, further inquiries can be directed to the corresponding author.

Conflicts of Interest: The authors declare no conflicts of interest.

References

1. Raizada, P.; Sudhaik, A.; Singh, P.; Shandilya, P.; Gupta, V.K.; Hosseini-Bandegharaei, A.; Agrawal, S. Ag_3PO_4 modified phosphorus and sulphur co-doped graphitic carbon nitride as a direct Z-scheme photocatalyst for 2, 4-dimethyl phenol degradation. *J. Photochem. Photobiol. A* **2019**, *374*, 22–35. [CrossRef]
2. Mendoza-Diaz, M.-I.; Lecestre, A.; Salvagnac, L.; Bounor, B.; Pech, D.; Djafari-Rouhani, M.; Esteve, A.; Rossi, C. High surface area TiO_2 photocatalyst for H_2 production through silicon micromachining. *Appl. Surf. Sci.* **2022**, *588*, 152919. [CrossRef]

3. Kim, H.; Lee, J.; Yantara, N.; Boix, P.P.; Kulkarni, S.A.; Mhaisalkar, S.; Grätzel, M.; Park, N.G. High Efficiency Solid-State Sensitized Solar Cell-Based on Submicrometer Rutile TiO_2 Nanorod and $CH_3NH_3PbI_3$ Perovskite Sensitizer. *Nano Lett.* **2013**, *13*, 2412–2417. [CrossRef] [PubMed]
4. Cui, X.; Li, H.; Yang, Z.; Li, Y.; Zhang, P.; Zheng, Z.; Wang, Y.; Li, J.; Zhang, X. A novel $CaIn_2S_4/TiO_2$NTAs heterojunction photoanode for highly efficient photocathodic protection performance of 316 SS under visible light. *Nanotechnology* **2021**, *32*, 395702. [CrossRef] [PubMed]
5. Zhang, G.; Zhang, Y.C.; Nadagouda, M.; Han, C.; O'Shea, K.; El-Sheikh, S.M.; Ismail, A.A.; Dionysiou, D.D. Visible light-sensitized S, N and C co-doped polymorphic TiO_2 for photocatalytic destruction of microcystin-LR. *Appl. Catal. B Environ.* **2014**, *144*, 614–621. [CrossRef]
6. Isari, A.A.; Hayati, F.; Kakavandi, B.; Rostami, M.; Motevassel, M.; Dehghanifard, E. N, Cu co-doped TiO_2@functionalized SWCNT photocatalyst coupled with ultrasound and visible-light: An effective sono-photocatalysis process for pharmaceutical wastewaters treatment. *Chem. Eng. J.* **2020**, *392*, 123685. [CrossRef]
7. Ismael, M. Enhanced photocatalytic hydrogen production and degradation of organic pollutants from Fe (III) doped TiO_2 nanoparticles. *J. Environ. Chem. Eng.* **2020**, *8*, 103676. [CrossRef]
8. Meng, A.; Zhang, L.; Cheng, B.; Yu, J. Dual Cocatalysts in TiO_2 Photocatalysis. *Adv. Mater.* **2019**, *31*, 1807660. [CrossRef]
9. Moon, H.S.; Hsiao, K.; Wu, M.; Yun, Y.; Hsu, Y.; Yong, K. Spatial separation of cocatalysts on Z-Scheme organic/inorganic heterostructure hollow spheres for enhanced photocatalytic H_2 evolution and in-depth analysis of the charge-transfer mechanism. *Adv. Mater.* **2022**, *35*, 2200172. [CrossRef]
10. Kumar, A.; Khan, M.; He, J.; Lo, I.M. Recent developments and challenges in practical application of visible-light-driven TiO_2-based heterojunctions for PPCP degradation: A critical review. *Water Res.* **2020**, *170*, 115356. [CrossRef]
11. Meng, A.; Cheng, B.; Tan, H.; Fan, J.; Su, C.; Yu, J. TiO_2/polydopamine S-scheme heterojunction photocatalyst with enhanced CO_2-reduction selectivity. *Appl. Catal. B Environ.* **2021**, *289*, 120039. [CrossRef]
12. Liang, Z.; Bai, X.; Hao, P.; Guo, Y.; Xue, Y.; Tian, J.; Cui, H. Full solar spectrum photocatalytic oxygen evolution by carbon-coated TiO_2 hierarchical nanotubes. *Appl. Catal. B Environ.* **2019**, *243*, 711–720. [CrossRef]
13. Singh, J.; Juneja, S.; Soni, R.K.; Bhattacharya, J. Sunlight mediated enhanced photocatalytic activity of TiO_2 nanoparticles functionalized CuO-Cu_2O nanorods for removal of methylene blue and oxytetracycline hydrochloride. *J. Colloid Interf. Sci.* **2021**, *590*, 60–71. [CrossRef] [PubMed]
14. Yu, C.; Zhou, W.; Yang, K.; Rong, G. Hydrothermal synthesis of hemisphere-like F-doped anatase TiO_2 with visible light photocatalytic activity. *J. Mater. Sci.* **2010**, *45*, 5756–5761. [CrossRef]
15. Wang, Y.; Zhang, Y.; Zhu, X.; Liu, Y.; Wu, Z. Fluorine-induced oxygen vacancies on TiO_2 nanosheets for photocatalytic indoor VOCs degradation. *Appl. Catal. B Environ.* **2022**, *316*, 121610. [CrossRef]
16. Tosoni, S.; Lamiel-Garcia, O.; Fernandez, H.; Doña, J.M.; Illas, F. Electronic Structure of F-Doped Bulk Rutile, Anatase, and Brookite Polymorphs of TiO_2. *J. Phys. Chem. C* **2012**, *116*, 12738–12746. [CrossRef]
17. Pang, D.; Wang, Y.; Ma, X.; Ouyang, F. Fluorine promoted and silica supported TiO_2 for photocatalytic decomposition of acrylonitrile under simulant solar light irradiation. *Chem. Eng. J.* **2014**, *258*, 43–50. [CrossRef]
18. Wen, J.; Li, X.; Liu, W.; Fang, Y.; Xie, J.; Xu, Y. Photocatalysis fundamentals and surface modification of TiO_2 nanomaterials. *Chin. J. Catal.* **2015**, *36*, 2049–2070. [CrossRef]
19. Chakinala, N.; Gogate, P.R.; Chakinala, A.G. Highly efficient bi-metallic bismuth-silver doped TiO_2 photocatalyst for dye degradation. *Korean J. Chem. Eng.* **2021**, *38*, 2468–2478. [CrossRef]
20. Liang, Z.; Cao, Y.; Li, Y.; Xie, J.; Guo, N.; Jia, D. Solid-state chemical synthesis of rod-like fluorine-doped β-Bi_2O_3 and their enhanced photocatalytic property under visible light. *Appl. Surf. Sci.* **2016**, *390*, 78–85. [CrossRef]
21. Cao, W.; Wang, W.; Yang, Z.; Wang, W.; Chen, W.; Wu, K. Enhancing photocathodic protection performance by controlled synthesis of Bi/BiOBr/TiO_2 NTAs Z-scheme heterojunction films. *J. Alloys Compd.* **2023**, *960*, 170675. [CrossRef]
22. Reddy, P.A.K.; Srinivas, B.; Kala, P.; Kumari, V.D.; Subrahmanyam, M. Preparation and characterization of Bi-doped TiO_2 and its solar photocatalytic activity for the degradation of isoproturon herbicide. *Mater. Res. Bull.* **2011**, *46*, 1766–1771. [CrossRef]
23. Huang, J.; Cheuk, W.; Wu, Y.; Lee, F.S.; Ho, W. Fabrication of Bi-Doped TiO_2 spheres with ultrasonic spray pyrolysis and investigation of their visible-light photocatalytic properties. *J. Nanotechnol.* **2012**, *2012*, 214783. [CrossRef]
24. Althamthami, M.; Guettaf Temam, E.; Temam, H.B.; Saâd, R. Improved photocatalytic activity under the sunlight of high transparent hydrophilic Bi-doped TiO_2 thin-films. *J. Photochem. Photobiol. A* **2023**, *443*, 114818. [CrossRef]
25. Murcia-López, S.; Hidalgo, M.C.; Navío, J.A. Synthesis, characterization and photocatalytic activity of Bi-doped TiO_2 photocatalysts under simulated solar irradiation. *Appl. Catal. A Gen.* **2011**, *404*, 59–67. [CrossRef]
26. Li, D.; Zhang, Y.; Zhang, Y.; Zhou, X.; Guo, S. Fabrication of bidirectionally doped β-Bi_2O_3/TiO_2-NTs with enhanced photocatalysis under visible light irradiation. *J. Hazard. Mater.* **2013**, *258–259*, 42–49. [CrossRef]
27. Liu, Z.; Wang, Q.; Tan, X.; Zheng, S.; Zhang, H.; Wang, Y.; Gao, S. Solvothermal preparation of Bi/Bi_2O_3 nanoparticles on TiO_2 NTs for the enhanced photoelectrocatalytic degradation of pollutants. *J. Alloys Compd.* **2020**, *815*, 152478. [CrossRef]
28. Mohamed, I.M.A.; Dao, V.; Yasin, A.S.; Mousa, H.M. Nitrogen-doped&SnO_2-incoportaed TiO_2 nanofibers as novel and effective photoanode for enhanced efficiency dye-sensitized solar cells. *Chem. Eng. J.* **2016**, *304*, 48–60.

29. Chawla, A.; Sudhaik, A.; Raizada, P.; Khan, A.A.P.; Singh, A.; Van Le, Q.; Nguyen, V.H.; Ahamad, T.; Alshehri, S.M.; Asiri, A.M.; et al. An overview of SnO_2 based Z scheme heterojuctions: Fabrication, mechanism and advanced photocatalytic applications. *J. Ind. Eng. Chem.* **2022**, *116*, 515–542. [CrossRef]
30. Roy, H.; Rahman, T.U.; Khan, M.A.J.R.; Al-Mamun, M.R.; Islam, S.Z.; Khaleque, M.A.; Hossain, M.I.; Khan, M.Z.H.; Islam, M.S.; Marwani, H.M.; et al. Toxic dye removal, remediation, and mechanism with doped SnO_2-based nanocomposite photocatalysts: A critical review. *J. Water Process Eng.* **2023**, *54*, 104069. [CrossRef]
31. Pesci, F.M.; Wang, G.; Klug, D.R.; Li, Y.; Cowan, A.J. Efficient suppression of electron-hole recombination in oxygen-deficient hydrogen-treated TiO_2 nanowires for photoelectrochemical water splitting. *J. Phys. Chem. C Nanomater. Interfaces* **2013**, *117*, 25837–25844. [CrossRef] [PubMed]
32. Kusior, A.; Zych, L.; Zakrzewska, K.; Radecka, M. Photocatalytic activity of TiO_2/SnO_2 nanostructures with controlled dimensionality/complexity. *Appl. Surf. Sci.* **2019**, *471*, 973–985. [CrossRef]
33. Alagarasi, A.; Rajalakshmi, P.U.; Shanthi, K.; Selvam, P. Solar-light driven photocatalytic activity of mesoporous nanocrystalline TiO_2, SnO_2, and TiO_2-SnO_2 composites. *Mater. Today Sustain.* **2019**, *5*, 100016. [CrossRef]
34. Magdalane, C.M.; Kanimozhi, K.; Arularasu, M.V.; Ramalingam, G.; Kaviyarasu, K. Self-cleaning mechanism of synthesized SnO_2/TiO_2 nanostructure for photocatalytic activity application for waste water treatment. *Surf. Interfaces* **2019**, *17*, 100346. [CrossRef]
35. De Mendonça, V.R.; Avansi, W.; Arenal, R.; Ribeiro, C. A building blocks strategy for preparing photocatalytically active anatase TiO_2/rutile SnO_2 heterostructures by hydrothermal annealing. *J. Colloid Interface Sci.* **2017**, *505*, 454–459. [CrossRef]
36. Angel, R.D.; Durán-Álvarez, J.; Zanella, R. TiO_2-low band gap semiconductor heterostructures for water treatment using sunlight-Driven photocatalysis. *Titan. Dioxide Mater. Sustain. Environ.* **2018**, *76501*, 305–329.
37. Albornoz, L.L.; Da Silva, S.W.; Bortolozzi, J.P.; Banús, E.D.; Brussino, P.; Ulla, M.A.; Bernardes, A.M. Degradation and mineralization of erythromycin by heterogeneous photocatalysis using SnO_2-doped TiO_2 structured catalysts: Activity and stability. *Chemosphere* **2021**, *268*, 128858. [CrossRef]
38. Rajput, R.B.; Jamble, S.N.; Kale, R.B. A review on TiO_2/SnO_2 heterostructures as a photocatalyst for the degradation of dyes and organic pollutants. *J. Environ. Manag.* **2022**, *307*, 114533. [CrossRef]
39. Hassan, S.M.; Ahmed, A.I.; Mannaa, M.A. Structural, photocatalytic, biological and catalytic properties of SnO_2/TiO_2 nanoparticles. *Ceram. Int.* **2018**, *44*, 6201–6211. [CrossRef]
40. Hassan, S.M.; Ahmed, A.I.; Mannaa, M.A. Preparation and characterization of SnO_2 doped TiO_2 nanoparticles: Effect of phase changes on the photocatalytic and catalytic activity. *J. Sci. Adv. Mater. Dev.* **2019**, *4*, 400–412. [CrossRef]
41. Yang, J.; Zhang, J.; Zou, B.; Zhang, H.; Wang, J.; Schubert, U.; Rui, Y. Black SnO_2-TiO_2 Nanocomposites with High Dispersion for Photocatalytic and Photovoltalic Applications. *ACS Appl. Nano Mater.* **2020**, *3*, 4265–4273. [CrossRef]
42. Lin, C.; Gao, Y.; Zhang, J.; Xue, D.; Fang, H.; Tian, J.; Zhou, C.; Zhang, C.; Li, Y.; Li, H. GO/TiO_2 composites as a highly active photocatalyst for the degradation of methyl orange. *J. Mater. Res.* **2020**, *35*, 1307–1315. [CrossRef]
43. Akti, F.; Balci, S. Synthesis of APTES and alcohol modified Sn/SBA-15 in presence of competitive ion: Test in degradation of Remazol Yellow. *Mater. Res. Bull.* **2022**, *145*, 111496. [CrossRef]
44. Soni, S.S.; Henderson, M.J.; Bardeau, J.F.; Gibaud, A. Visible-light photocatalysis in titania-based mesoporous thin films. *Adv. Mater.* **2008**, *20*, 1493–1498. [CrossRef]
45. Liao, Y.; Huang, C.; Liao, C.; Wu, J.; Wu, K. Synthesis of mesoporous titania thin films (MTTFs) with two different structures as photocatalysts for generating hydrogen from water splitting. *Appl. Energ.* **2012**, *100*, 75–80. [CrossRef]
46. Li, L.; Yan, J.; Wang, T.; Zhao, Z.-J.; Zhang, J.; Gong, J.; Guan, N. Sub-10 nm rutile titanium dioxide nanoparticles for efficient visible-light-driven photocatalytic hydrogen production. *Nat. Commun.* **2015**, *6*, 5881. [CrossRef]
47. Wang, Q.; Rhimi, B.; Wang, H.; Wang, C. Efficient photocatalytic degradation of gaseous toluene over F-doped TiO_2/exfoliated bentonite. *Appl. Surf. Sci.* **2020**, *530*, 147286. [CrossRef]
48. Preeti; Mishra, S.; Chakinala, N.; Chakinala, A.G. Bimetallic Bi/Zn decorated hydrothermally synthesized TiO_2 for efficient photocatalytic degradation of nitrobenzene. *Catal. Commun.* **2022**, *172*, 106538. [CrossRef]
49. Yan, B.; Chen, G.; Ma, B.; Guo, Y.; Zha, Y.; Li, J.; Wang, S.; Liu, J.; Zhao, B.; Xie, H. Construction of surface plasmonic Bi nanoparticles and α-Bi_2O_3 co-modified TiO_2 nanotube arrays for enhanced photocatalytic degradation of ciprofloxacin: Performance, DFT calculation and mechanism. *Sep. Purif. Technol.* **2024**, *330*, 125180. [CrossRef]
50. Bagwasi, S.; Tian, B.; Zhang, J.; Nasir, M. Synthesis, characterization and application of bismuth and boron Co-doped TiO_2: A visible light active photocatalyst. *Chem. Eng. J.* **2013**, *217*, 108–118. [CrossRef]
51. Huy, T.H.; Phata, B.D.; Kang, F.; Wang, Y.F.; Liu, S.H.; Thi, C.M.; You, S.J.; Chang, G.M.; Pham, V.V. SnO_2/TiO_2 nanotube heterojunction: The first investigation of NO degradation by visible light-driven photocatalysis. *Chemosphere* **2019**, *215*, 323–332. [CrossRef] [PubMed]
52. Shi, H.; Zhu, H.; Jin, T.; Chen, L.; Zhang, J.; Qiao, K.; Chen, Z. Construction of Bi/Polyoxometalate doped TiO_2 composite with efficient visible-light photocatalytic performance: Mechanism insight, degradation pathway and toxicity evaluation. *Appl. Surf. Sci.* **2023**, *615*, 156310. [CrossRef]
53. Li, G.; Li, J.; Li, G.; Jiang, G. N and Ti^{3+} co-doped 3D anatase TiO_2 superstructures composed of ultrathin nanosheets with enhanced visible light photocatalytic activity. *J. Mater. Chem. A* **2015**, *3*, 22073. [CrossRef]

54. Li, Y.; Feng, Y.; Bai, H.; Liu, J.; Hu, D.; Fan, J.; Shen, H. Enhanced visible-light photocatalytic performance of black TiO_2/SnO_2 nanoparticles. *J. Alloy. Compd.* **2023**, *960*, 170672. [CrossRef]
55. Song, Y.; Long, A.; Ge, X.; Bao, Z.; Meng, M.; Hu, S.; Gu, Y. Construction of floatable flower-like plasmonic Bi/BiOCl-loaded hollow kapok fiber photocatalyst for efficient degradation of RhB and antibiotics. *Chemosphere* **2023**, *343*, 140240. [CrossRef]
56. Li, J.; Jin, Z.; Zhang, Y.; Liu, D.; Ma, A.; Sun, Y.; Li, X.; Cai, Q.; Gui, J. Ag-induced anatase-rutile TiO_2-x heterojunction facilitating the photogenerated carrier separation in visible-light irradiation. *J. Alloy. Compd.* **2022**, *909*, 164815. [CrossRef]
57. Li, H.; Qiu, L.; Bharti, B.; Dai, F.; Zhu, M.; Ouyang, F.; Lin, L. Efficient photocatalytic degradation of acrylonitrile by Sulfur-Bismuth co-doped F-TiO_2/SiO_2 nanopowder. *Chemosphere* **2020**, *249*, 126135. [CrossRef]
58. Choukaife, A.E.; Aljerf, L. A descriptive study-in vitro: New validated method for checking Hap and Fap Behaviours. *Int. Med. J.* **2017**, *24*, 394–397.
59. Donohue, M.D.; Aranovich, G.L. A new classification of isotherms for Gibbs adsorption of gases on solids. *Fluid Phase Equilibr.* **1999**, *158–160*, 557–563. [CrossRef]
60. Liu, G.; Yang, H.G.; Wang, X.; Cheng, L.; Lu, H.; Wang, L.; Lu, G.; Cheng, H.M. Enhanced Photoactivity of Oxygen-Deficient Anatase TiO_2 Sheets with Dominant {001} Facets. *J. Phys. Chem. C* **2009**, *113*, 21784–21788. [CrossRef]
61. Lin, Z.; Orlov, A.; Lambert, R.M.; Payne, M.C. New insights into the origin of visible light photocatalytic activity of nitrogen-doped and oxygen-deficient anatase TiO_2. *J. Phys. Chem. B* **2005**, *109*, 20948–20952. [CrossRef] [PubMed]
62. Law, M.; Greene, L.E.; Radenovic, A.; Kuykendall, T.; Liphardt, J.; Yang, P. ZnO-Al_2O_3 and ZnO-TiO_2 core-shell nanowire dye-sensitized solar cells. *J. Phys. Chem. B* **2006**, *110*, 22652–22663. [CrossRef] [PubMed]
63. Parayil, S.K.; Kibombo, H.S.; Wu, C.; Peng, R.; Baltrusaitis, J.; Koodali, R.T. Enhanced photocatalytic water splitting activity of carbon-modified TiO_2 composite materials synthesized by a green synthetic approach. *Int. J. Hydrogen Energ.* **2012**, *37*, 8257–8267. [CrossRef]
64. Wang, X.; Yu, J.C.; Liu, P.; Wang, X.; Su, W.; Fu, X. Probing of photocatalytic surface sites on $SO_4{}^{2-}/TiO_2$ solid acids by in situ FT-IR spectroscopy and pyridine adsorption. *J. Photochem. Photobiol. A* **2006**, *179*, 339–347. [CrossRef]
65. Hong, W.; Li, C.; Tang, T.; Xu, H.; Yu, Y.; Liu, G.; Wang, F.; Lie, C.; Zhu, H. The photocatalytic activity of the SnO_2/TiO_2/PVDF composite membrane in rhodamine B degradation. *New J. Chem.* **2021**, *45*, 2631–2642. [CrossRef]
66. Jiang, H.; Wang, R.; Li, Y.; Hong, X.; Liang, B. Glucose-tailored SnO_2/TiO_2/RGO ternary composite for degradation of organic pollutants. *J. Phys. Chem. Solids* **2022**, *161*, 110442. [CrossRef]
67. Stefan, M.; Leostean, C.; Pana, O.; Popa, A.; Toloman, D.; Macavei, S.; Perhaita, I.; Barbu-Tudoran, L.; Silipas, D. Interface tailoring of SnO_2-TiO_2 photocatalysts modified with anionic/cationic surfactants. *J. Mater. Sci.* **2019**, *55*, 3279–3298. [CrossRef]
68. Wang, Q.; Fang, Y.; Meng, H.; Wu, W.; Chu, G.; Zou, H.; Cheng, D.; Chen, J. Enhanced simulated sunlight induced photocatalytic activity by pomegranate-like S doped SnO_2@TiO_2 spheres. *Colloids Surf. A* **2015**, *482*, 529–535. [CrossRef]
69. Shi, H.; Zhou, M.; Song, D.; Pan, X.; Fu, J.; Zhou, J.; Ma, S.; Wang, T. Highly porous SnO_2/TiO_2 electrospun nanofibers with high photocatalytic activities. *Ceram. Int.* **2014**, *40*, 10383–10393. [CrossRef]
70. Gao, Q.; Wu, X.; Fan, Y.; Zhou, X. Low temperature fabrication of nanoflower arrays of rutile TiO_2 on mica particles with enhanced photocatalytic activity. *J. Alloys Compd.* **2013**, *579*, 322–329. [CrossRef]
71. Chen, Y.; Jiang, Y.; Chen, B.; Tang, H.; Li, L.; Ding, Y.; Duan, H.; Wu, D. Insights into the enhanced photocatalytic activity of O-g-C_3N_4 coupled with SnO_2 composites under visible light irradiation. *J. Alloys Compd.* **2022**, *903*, 163739. [CrossRef]
72. Ma, H.; Wang, Y.; Zhang, Z.; Liu, J.; Yu, Y.; Zuo, S.; Li, B. A superior ternary Z-scheme photocatalyst of Bi/Black Phosphorus nanosheets/P-doped BiOCl containing interfacial P–P bond and metallic mediator for H_2O_2 production and RhB degradation. *Chemosphere* **2023**, *330*, 138717. [CrossRef] [PubMed]
73. Chen, Z.; Yu, S.; Liu, J.; Zhang, Y.; Wang, Y.; Yu, J.; Yuan, M.; Zhang, P.; Liu, W.; Zhang, J. C, F co-doping Ag/TiO_2 with visible light photocatalytic performance toward degrading Rhodamine B. *Environ. Res.* **2023**, *232*, 116311. [CrossRef] [PubMed]
74. Chen, Y.; Li, A.; Fu, X.; Peng, Z. Novel F-doped carbon nanotube@(N,F)-co-doped $TiO_{2-\delta}$ nanocomposite: Highly active visible-light-driven photocatalysts for water decontamination. *Appl. Surf. Sci.* **2023**, *609*, 155460. [CrossRef]
75. Khandekar, D.C.; Bhattacharyya, A.R.; Bandyopadhyaya, R. Role of impregnated nano-photocatalyst ($Sn_xTi_{(1-x)}O_2$) inside mesoporous silica (SBA-15) for degradation of organic pollutant (Rhodamine B) under UV light. *J. Environ. Chem. Eng.* **2019**, *7*, 103433. [CrossRef]
76. Wu, Y.M.; Xing, M.Y.; Tian, B.Z.; Zhang, J.L.; Chen, F. Preparation of nitrogen and fluorine co-doped mesoporous TiO_2 microsphere and photodegradation of acid orange 7 under visible light. *Chem. Eng. J.* **2010**, *162*, 710–717. [CrossRef]
77. Yang, G.D.; Jiang, Z.; Shi, H.H.; Jones, M.O.; Xiao, T.C.; Edwards, P.P.; Yan, Z.F. Study on the photocatalysis of F-S co-doped TiO_2 prepared using solvothermal method. *Appl. Catal. B Environ.* **2010**, *96*, 458–465. [CrossRef]

catalysts

Article

Boosting the Efficiency of Titanium Oxides and Accelerating Industrial Pollutant Removal Through Triple-Action Effects to Purify Water by Light

Osama Saber *, Chawki Awada *, Aya Osama, Nagih M. Shaalan, Adil Alshoaibi, Shehab A. Mansour, Mostafa Osama and Ashraf H. Farha

Department of Physics, College of Science, King Faisal University, P.O. Box 400, Al-Ahsa 31982, Saudi Arabia; ayamohamed19984@yahoo.com (A.O.); nmohammed@kfu.edu.sa (N.M.S.); aalshoaibi@kfu.edu.sa (A.A.); samhassan@kfu.edu.sa (S.A.M.); mostafa.osama8664@gmail.com (M.O.); afarha@kfu.edu.sa (A.H.F.)
* Correspondence: osmohamed@kfu.edu.sa (O.S.); cawada@kfu.edu.sa (C.A.); Tel.: +966-13-589-9440 (O.S.)

Abstract: Water is the source of life on Earth. Therefore, water pollution is one of the greatest problems in the world. On this basis, the current study focuses on accelerating industrial pollutant removal from water using light by designing effective photocatalysts. This target was achieved through a triple-action effect. This effect depends on the integration of the doping process with nanotube formation in addition to the surface plasmon resonance of gold for titanium oxides. In this way, titanium oxide nanoparticles were prepared and converted to nanotubes during the doping process. These nanoparticles and nanotubes were supported by gold nanoparticles to use this triple-action effect for increasing charge carriers and active sites of the photocatalysts and preventing recombination reactions. High-resolution transmission electron microscopy (HRTEM) and selected area electron diffraction (SAED), Raman spectra, energy-dispersive X-ray spectrometer (EDX), and X-ray diffraction were used to clarify the triple-action effect on the structure of the photocatalysts. The optical properties and activity of the prepared photocatalysts were studied in terms of the photocatalytic degradation of the green dyes (acid green 1). The experimental results indicated that the triple-action effect has a strong positive role in increasing industrial pollutant removal with or without light. Here, the percentage of photocatalytic decomposition reached 100% after 17 min of light radiation. In addition, 27% of the pollutants were removed without light radiation. In conclusion, the current study indicated that the triple-action effect could solve the drawbacks of titanium oxide by creating new photo-active sites and novel tracks for charge carriers in addition to preventing recombination reactions.

Keywords: photocatalysts based on nanotubes; sodium-doping process; integration between doping process and nanotube shaping; gold–titanium oxide nanoparticles; fast removal of industrial pollutants; photocatalytic decomposition

1. Introduction

Modern life and overpopulation have created one of the greatest problems in the world. It is the pollution of water. Sources of this problem are due to the introduction of contaminants into water that are produced by different kinds of manufacturing activities. The most critical problems are due to non-biodegradable pollutants, which do not decompose naturally. Although different techniques were used and developed for treating wastewater by removing pollutants, some of them have complicated technologies in addition to their high costs. Therefore, water pollution is an important challenge for researchers and scientists, as they need to look for benign, effective, and inexpensive solutions [1]. Dyes are dangerous pollutants because even low concentrations of dyes (below 1 mg/L) are toxic and difficult to be treated by biological methods. In addition, they consist of complex structures of hydrocarbons [2]. Large amounts of these pollutants are produced by food coloring, textile, paper, and cosmetic industries in addition to printing and dyeing units.

The techniques, which are based on light as a source for removing pollutants, are favorable for water remediation [3,4]. Therefore, the photocatalysis technique has attracted the attention of researchers to produce effective and inexpensive methods for water purification [3,4]. Titanium oxides are very famous as photo-active materials for light-induced redox processes because of their effective band gap energies and electronic structures in addition to photostability [5,6]. Many studies focused on modifying and developing titanium oxides to enhance their photo-activities through the doping effect, coupling with other metals, using support materials, or shaping the photocatalyst into nanofiber structures [7,8].

In 2020, the effect of copper on the photo-activity of titanium oxide was studied for the degradation of methylene blue (MB) dye by Khdary et al. [9]. They improved the charge separation and maintained the non-charge recombination processes by chemically combining copper oxides with TiO_2 nanoparticles exhibiting high photo-activity by reducing its band gap energy. In 2021, the photo-activity for the degradation of MB dye was boosted by biochar (BC) nanoparticles over TiO_2 nanotube arrays [10]. This positive effect depended on the amount of BC deposited, indicating that the photocatalytic decomposition of methylene blue dyes increased approximately three times with respect to the bare TiO_2. In the same year, Ma et al. [11] indicated the positive effect of silver nanoparticles for increasing photonic degradation of TiO_2 for the MB dye through enhancing electron–hole separation with UV light. In 2022, the teamwork of Bergamonti et al. [12] used chitosan/Ag/TiO_2 3D-printed scaffolds as an effective strategy to reduce wastewater pollution by pharmaceutical compounds. This research exhibited photocatalytic degradation efficiency higher than 90% within one hour. Ren et al. [13] used carbon nanotubes with titanium oxides to absorb microwaves and conduct electrons to prevent the recombination of photo-generated charge carriers and enhance the separation efficiency of photo-generated electrons and holes. Milidrag et al. [14] achieved 85% magenta dye decolorization by combining titanium oxides with iron oxides. In 2023, Székely and his colleagues [15] used hetero-structures based on titanium oxides to arrive at photocatalytic degradation efficiencies of 80–95% for oxalic acid, phenol, methyl orange, and aspirin. When employing Au/TiO_2/WO_3, the degradation efficiency for methyl orange, oxalic acid, and phenol reached 90, 97, and 99%, respectively. In the same year, Indira et al. [16] increased the photocatalytic performance of titanium oxide by forming nanocomposites with cerium and loading them over chitosan to achieve 95% RhB dye degradation against pure TiO_2 nanoparticles. In 2024, in combination with the cement slurry, it was shown that nano-TiO_2 could be utilized to degrade nitrogen oxides (NO_x) in vehicle exhaust [17]. By supporting these compounds over zeolite, titanium oxides [18] were effective in degrading 65% of sulfamethoxazole during 120 min of UV irradiation. Abbasi et al. [19] indicated that the combination of zinc and titanium oxides achieved complete degradation for methylene blue dye after 150 min of UV radiation. For enhancing the photo-degradation of acetone up to 95%, titanium oxides were supported over cellulose acetate nanofibers and used as photocatalysts for remediating the pollution of volatile organic compounds [20]. To increase the removal efficiency of the dyes from water up to 95%, titanium oxides were combined with biochar and used as efficient photocatalysts [21]. Sifat et al. [22] studied the photo-activities of the different kinds of metal oxides such as titanium dioxide, zinc oxides, zirconium dioxide, iron (III) oxides, copper (II) oxides, copper (I) oxides, and niobium pentoxide for removing crystal violet and triphenyl-methane dyes from water. They concluded that ~98% of the dye removal was achieved after 120 min of UV irradiation by titanium oxides and zinc oxides.

Although many studies have used the mono-action effect (such as the doping effect) or dual-action effect (such as doping with morphology effects) to enhance the photo-activity of titanium oxide, the triple-action effect (doping, shaping the photocatalyst into nanotubes, and gold loading) was rarely used by researchers. The current study focuses on uniting these three effects during the designing process of titanium oxide photocatalysts to create new photo-active sites and different kinds of charge carriers in addition to preventing recombination reactions. By introducing these positive factors, fast removal of industrial pollutants can be achieved by light in a few minutes. In this line, nanoparticles of titanium oxide were prepared and converted to nanotubes during the doping process. Then, by loading gold, the triple-action effect was induced inside the titanium oxide photocatalyst. To study the triple-action effect on the structure of the photocatalysts, high-resolution transmission electron microscopy (HRTEM), selected area electron diffraction (SAED), Raman spectra, energy-dispersive X-ray spectrometer (EDX), and X-ray diffraction were used in detail. The optical properties and activity of the prepared photocatalysts were studied in terms of the photocatalytic degradation of the green dyes.

2. Results

2.1. Powder X-Ray Diffraction

X-ray diffraction diagrams of the prepared nanoparticles of titanium oxide (TONP-1), the gold-supported titanium oxide nanoparticles (Au-TONP-1), and the gold-supported titanium oxide nanotubes (Au-TONT-1) are shown in Figure 1. The diagram of TONP-1 shows reflection peaks of titanium oxide. These reflection peaks attained for TONP-1 are in agreement with the anatase phase through matching and fitting with the entire standard diffraction diagram of JCPDS 21-1272 as shown in Figure 1a.

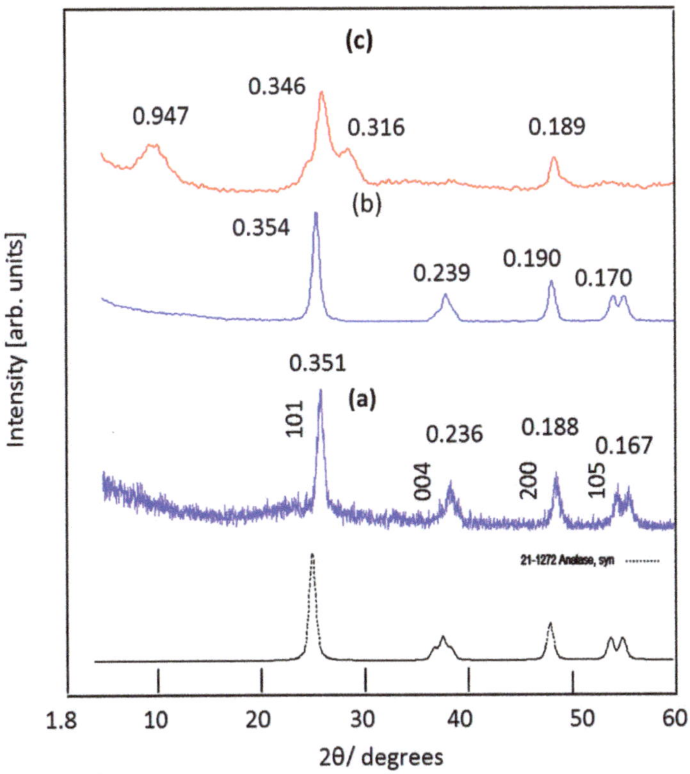

Figure 1. XRD patterns of (**a**) TONP-1, (**b**) Au-TONP-1, and (**c**) Au-TONT-1.

Figure 1a exhibits five wide peaks. This wideness suggested that the prepared particles of titanium oxides are crystallized in the nanoscale. These five peaks were observed at large angles. The reflection planes of (101), (004), (200), (105), and (211) have angles of 25.39°, 37.92°, 48.11°, 54.04°, and 55.08°, respectively.

By supporting the gold nanoparticles, the peaks of the prepared titanium oxide did not change because the gold nanoparticles did not penetrate inside the crystal lattice of titanium oxide. Figure 1b shows the XRD pattern of the sample of Au-TONP-1. The reflections of different lattice planes, (101), (004), (200), (105), and (211), were observed at 2θ = 25.30°, 37.82°, 47.91°, 53.83°, and 54.92°. Also, there are no diffraction peaks detecting metallic gold because of the very low amount of Au used in the synthesis process in addition to the homogenous distribution of gold nanoparticles.

By converting the nanoparticles to nanotubes and supporting gold nanoparticles, a clear phenomenon was observed for Au-TONT-1 as shown in Figure 1c. This phenomenon is an appearance of a new broad peak characteristic of the titanium oxide nanotubes at very low 2θ. Figure 1c shows a broad peak at 2θ = 9.33° agreeing with d-spacing at 0.947 nm. This indicates that the anatase phase has the structure of nanotubes with sodium doping [23,24]. The reflection of anatase crystal near this value is plane (001) with d-spacing = 0.951 nm [24,25]; however, there is a slight deviation between these values. This deviation is due to the inclusion of sodium dopants inside the crystal of titanium oxide. This large inter-planar distance is a typical characteristic in titanium oxide nanotubes and is closely related to the formation of nanotube structures as described in the next section of TEM images. In addition, another new peak was observed at 2θ = 28.4, agreeing with d-spacing = 0.316 nm. Agreeing with Morgan et al. [25], this peak is due to titanium oxide nanotubes. A similar shift was observed for the main peaks of the anatase phase to higher angles as seen in Figure 1c. The reflections of planes (101) and (200) shifted from 25.39° and 48.11° to 25.76° and 48.22°, respectively. This shift is due to the presence of sodium dopants inside the crystal lattice of titanium oxide and its effect on the lattice parameters of titanium oxide.

2.2. Transmission Electron Microscopy

Figure 2a–d show the transmission electron microscopy (TEM) images of TONP-1, TONT-1, Au-TONP-1, and Au-TONT-1. The morphology of titanium oxide in TONP-1 confirmed the presence of a spherical shape with an average size of 50 nm; however, it is a nanotube shape in the case of TONT-1 with an average diameter of 20 nm. After supporting these compounds with gold, the particles became more agglomerated in the case of Au-TONT-1. The gold nanoparticles are not observed in the TEM images due to their very small proportion. To quantify the inter-planar distance of TONP-1 and TONT-1 crystal structures, high-resolution transmission electron microscopy (HRTEM) and selected area electron diffraction (SAED) were performed on them; see Figure 3a–f. From HRTEM, the inter-planar distance was estimated for TONP-1 and TONT-1 with values of 3.3 Å and 2.8 Å, respectively. The two distances correspond to the two crystal planes of the anatase phase, (101) and (004). SAED rings confirm the presence of (101), (004), and (200) crystal planes in TONP-1 and TONT-1.

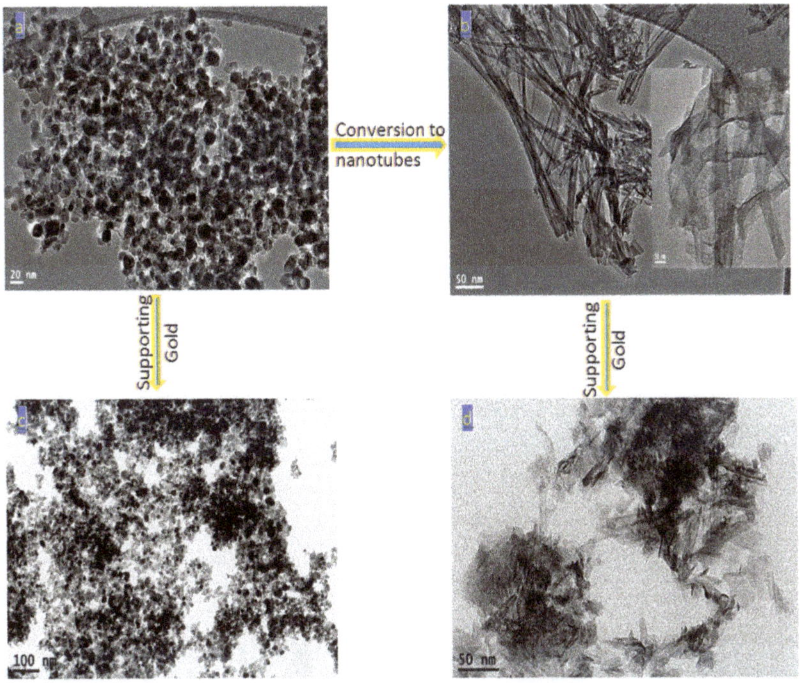

Figure 2. TEM images of (**a**) TONP-1, (**b**) TONT-1, (**c**) Au-TONP-1, and (**d**) Au-TONT-1.

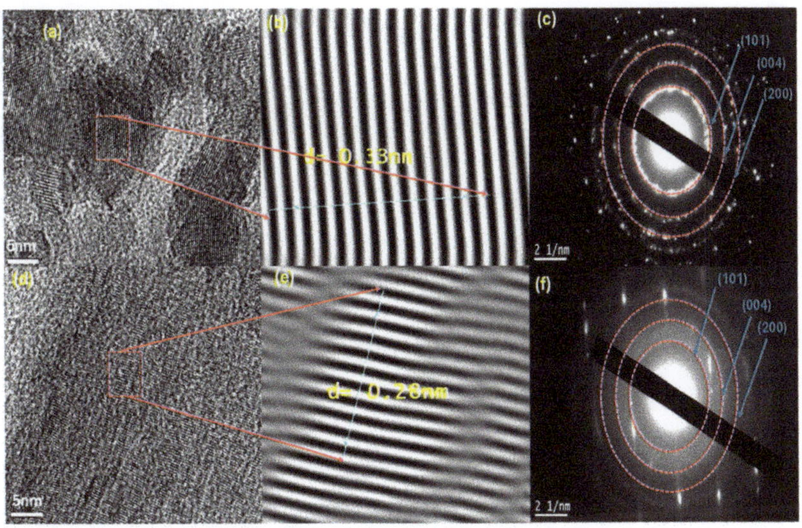

Figure 3. HRTEM and SAED images of (**a–c**)Au-TONP-1 and (**d–f**) Au-TONT-1.

2.3. Raman Spectra

Figure 4 shows Raman spectra of both the nanoparticles and the nanotubes of titanium oxides TONP-1 and TONT-1. As observed in XRD patterns for TONP-1, the Raman spectrum aligns purely with the anatase phase for the observation of phonon modes located at 143 cm^{-1}, 195 cm^{-1}, 396 cm^{-1}, 515 cm^{-1}, and 635 cm^{-1}. The space group for tetragonal anatase is $I4_{1/amd}$ and the point symmetry is D_{2d}. The two TiO_2 units presenting in the primitive cell give six Raman-active fundamentals in the phonon spectrum: three E_g modes, two B_{1g} modes, and one A_{1g} mode [26]. However, for TONT-1, the Raman spectrum predominantly confirms the presence of a rutile phase for the observation of the E_g mode located at 440 cm^{-1} [27]. There is also a mode assigned to the anatase phase located at 190 cm^{-1}. There are other phonon modes located at 278 cm^{-1}, 670 cm^{-1}, 809 cm^{-1}, and 907 cm^{-1} that belong to anatase and rutile phases. The band located at 278 cm^{-1} could be assigned to the multiphonon scattering process [28]. In the literature, the band at 670 cm^{-1} could be assigned to Ti-O-Ti vibrations [29], and the bands located at 809 cm^{-1} and 907 cm^{-1} could be attributed to a covalent Ti-O-H bond and a Ti-O-Na vibration [29,30].

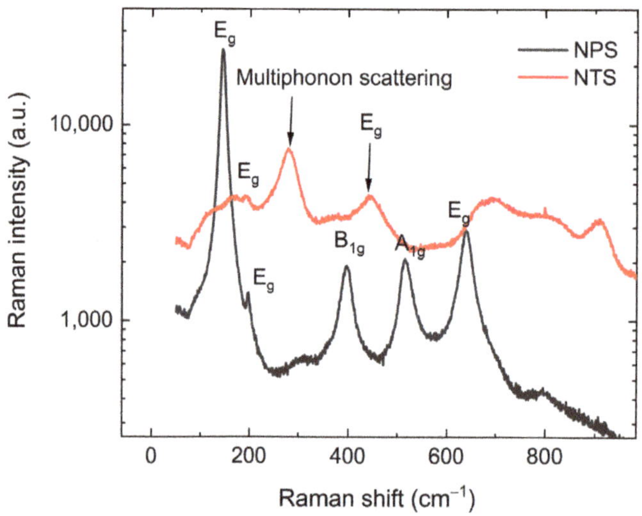

Figure 4. Raman spectra of TONP-1 (——) and TONT-1(——).

2.4. Energy-Dispersive X-Ray Spectrometer

Energy-dispersive X-ray spectrometer (EDX) analysis provides local information on the concentrations of different elements in the outermost layers of the sample of Au-TONP-1. Titanium, oxygen, and gold are clearly identified in Au-TONP-1 as shown in Figure 5.

The scattered molecules of Ti, O, and Au are shown in the point-mapping image in Figure 5, which shows them (the colored points are molecules) being distributed throughout the specific area. Additionally, the points found in the Ti chart are located in similar positions to the points found in the O chart indicating the formation of titanium oxides. Figure 5c indicates that the gold is homogeneously distributed at the nanometer scale over the surface of titanium oxides.

The chemical composition of the Au-TONP-1 sample was determined by EDX analysis as shown in Figure 5d. The EDX spectrum showed sharp peaks for titanium and oxygen with high weight percentages. In addition, Figure 5d shows that the low weight percentage of gold agrees with the percentage of gold supported over titanium oxide during the preparation method. EDX results concluded that the Au-TONP-1 sample is composed of titanium oxide supported with 1% gold.

Figure 6 shows the elements composition of the Au-TONT-1 sample and the mapping images of its different elements. The scattered molecules of Ti and O are shown in the point-mapping image in Figure 6a,b as colored points distributed throughout the specific area. Additionally, the points found in the Ti chart have the shape of nanotubes as shown in Figure 6a by yellow lines. Also, Figure 6b shows similar nanotubes for oxygen molecules agreeing with the titanium molecules located in Figure 6a. In the same line, Figure 6c shows the scattered molecules of gold, indicating similar nanotubes as marked by yellow lines. This finding confirms that Au-TONT-1 has nanotubes of titanium oxide supported by gold. In addition, Figure 6c shows that the gold is homogenously dispersed over titanium oxides. Furthermore, the elements' composition of Au-TONT-1 was determined by EDX analysis as shown in Figure 6d. Strong peaks of titanium and oxygen were identified by the EDX spectrum. In addition, Figure 6d confirmed the doping process for titanium oxide through the observation of a new clear peak of sodium in the EDX diagram. This means that the Au-TONT-1 sample has nanotubes of the Na-doped TiO_2 supported with nanoparticles of gold.

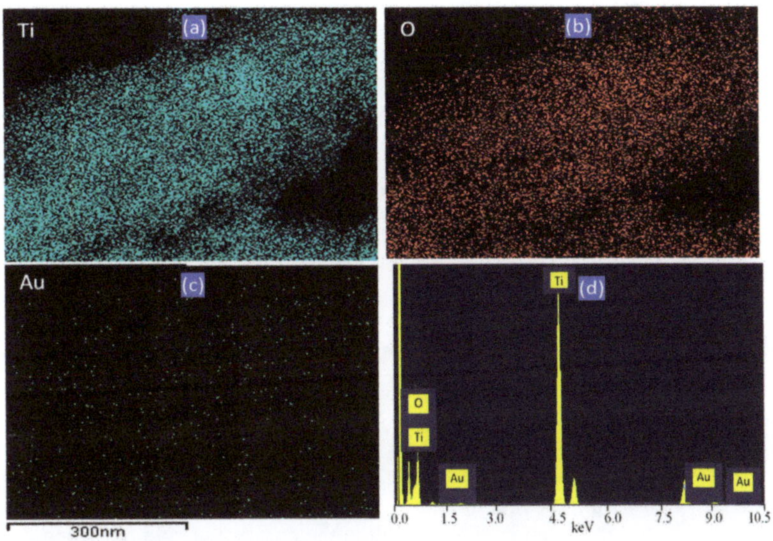

Figure 5. EDX analysis of Au-TONP-1 and mapping images of different elements: (**a**) titanium, (**b**) oxygen, (**c**) gold, and (**d**) elemental composition.

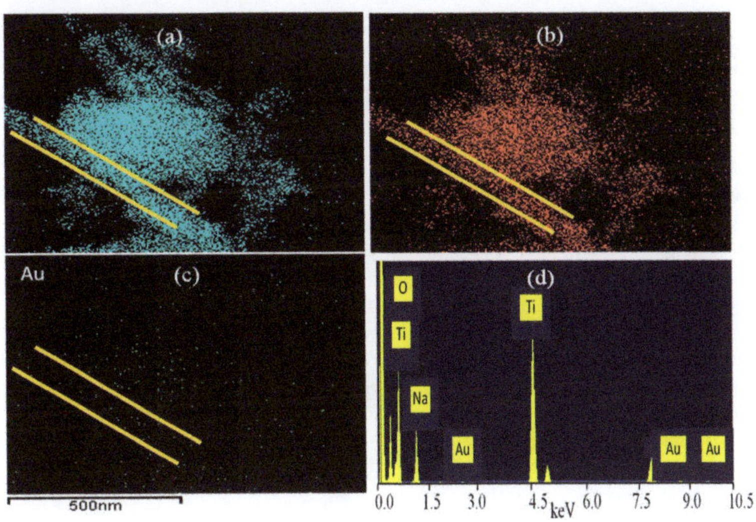

Figure 6. EDX analysis of Au-TONT-1 and mapping images of different elements: (**a**) titanium, (**b**) oxygen, (**c**) gold, and (**d**) elemental composition.

2.5. Optical Properties

The optical properties and applications of titanium oxides are changed according to their different phases and morphologies. Therefore, this study has used UV–Vis absorbance spectroscopy to discuss and analyze the optical parameters of the prepared nanoparticles and nanotubes of titanium oxides. Figure 7a indicates that there is a clear absorbance in the UV region for the TONP-1 sample. Two maxima were observed at 222 nm and 330 nm in the spectrum of the UV–Vis absorbance of TONP-1.

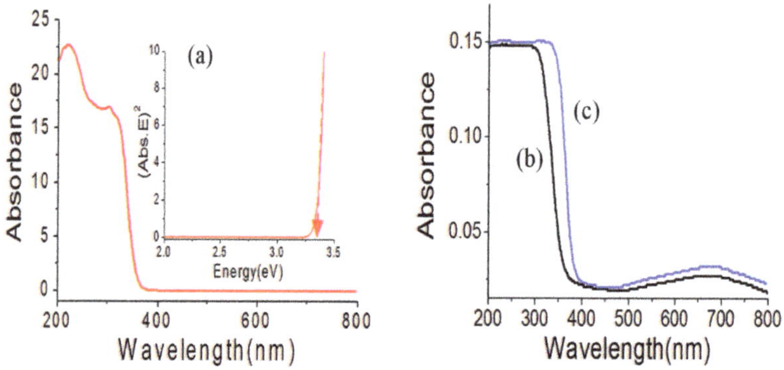

Figure 7. (**a**) The absorbance spectrum of TONP-1 (inset, its band gap energy), (**b**) the absorbance spectrum of Au-TONP-1, and (**c**) the absorbance spectrum of Au-TONT-1.

This UV absorbance was developed by the addition of supporting gold. Through the addition of supporting gold in the TONP-1 sample, two maxima were observed in Figure 7b; one is in the UV region at 300 nm and the other is observed in the visible region as a weak peak at 660 nm. The first maximum of Au-TONP-1 is due to the absorbance of titanium oxide and the second one is a result of the presence of gold. By using the triple-action effect, conversion to nanotubes with the Na-doping process in addition to supporting Au addition, the UV absorbance shifted toward a high wavelength in addition to a maximum at 680 nm as shown in Figure 7c. The effect of the triple-action effect was clear when calculating the band gap energy. The energy of the band gap is the energy between the valance band and the conduction band. It is considered one of the most important parameters of photocatalysts. Depending on the absorbance factor (α) of TONP-1, the band gap energy (E) could be assessed by using the following equation [31]:

$$(\alpha h\nu)^2 = \text{constant}\ (h\nu - E_g) \quad (1)$$

By plotting the values of $(Abs.E)^2 = (\alpha h\nu)^2$ and energy (hν) along the *Y*-axis and *X*-axis, the band gap energy was determined by extending a straight line toward the *X*-axis.

Figure 7a (inset) shows the band gap energy of TONP-1. This band gap energy was 3.4 eV. By adding supporting gold to titanium oxide, Figure 8a indicates that the band gap energy of Au-TONP-1 slightly changed to 3.50 eV. Through the triple-action effect, the conversion to nanotubes with the Na-doping process in addition to Au-supporting caused a reduction in the band gap energy. Figure 8b shows the band gap energy of Au-TONT-1. A clear shift was observed for the band gap energy from 3.50 eV to 3.25 eV. The effect of the triple-action effect was clear during the study of the photo-activity of Au-TONT-1 for photocatalytic degradation processes.

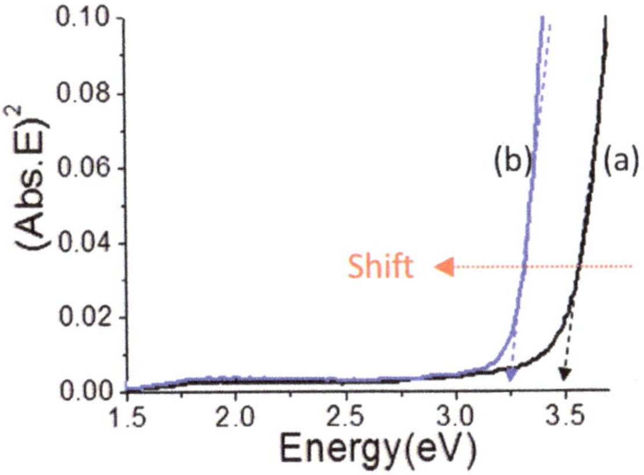

Figure 8. (**a**) The band gap energy of Au-TONP-1 and (**b**) the band gap energy of Au-TONT-1.

2.6. Water Purification

Dyes are dangerous pollutants in water because most of them are not biodegradable. In addition, the color removal from these pollutants is considered a critical problem for water because the low contents of dyes, which are below 1 mg/L, are clearly visible in water [32,33]. Optical materials can be used in light to remove industrial pollutants by converting the hydrocarbons to carbon dioxide and water [34,35]. By using the prepared titanium oxides and their products, the removal processes of acid green 1 dye were studied in the presence and absence of light.

By withdrawing a certain amount of liquid after a few minutes of light radiation, the concentration of dye was determined by measuring its absorbance with a UV–Vis spectrophotometer. The main structures of the pollutants were gradually decomposed leading to a regular reduction in the characteristic peak of the dye at 714 nm as shown in Figures 9 and 10. In addition, a similar reduction was observed in the other peaks at 320 nm, 283 nm, and 232 nm, indicating a complete degradation of the internal parts of the dyes.

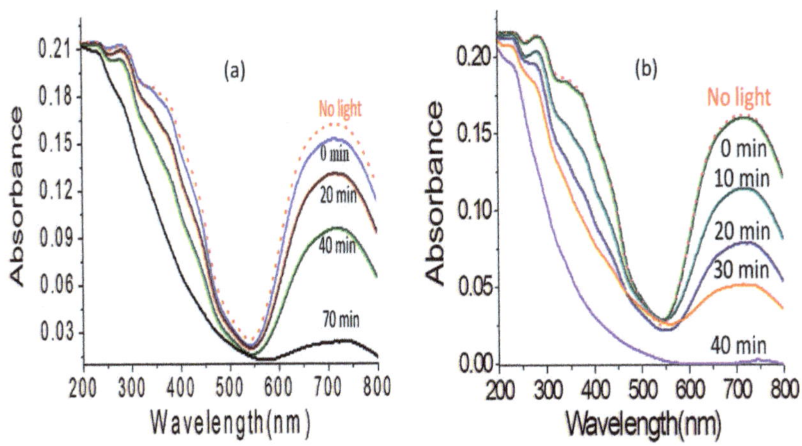

Figure 9. Absorbance spectra of the green dyes with different irradiation times in UV light in the presence of (**a**) TONP-1 and (**b**) Au-TONP-1.

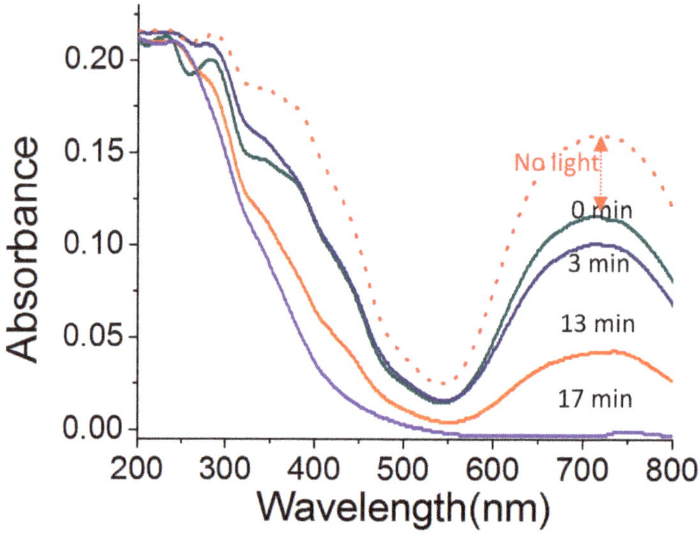

Figure 10. Absorbance spectra of the green dyes with different irradiation times in UV light in the presence of Au-TONT-1.

By exposing the green dye without photocatalysts to the light, its stability was measured using a blank test. The photo-activity of the prepared photocatalysts was determined by studying and analyzing the photocatalytic degradation of the green dye against the light radiation duration. To ascertain the adsorption capability of the used photocatalysts, the concentration of the dye was determined after the green solution of the dye was mixed with the photocatalyst in the dark. The steady loss of the industrial pollutant's green color was indicative of the green dye's photocatalytic breakdown in the presence of UV radiation and one of the prepared photocatalysts. Figures 9 and 10 present illustrations of the acquired results.

Figure 9a shows the effect of the adsorption capacity of TONP-1 on the concentration of AG1 in the absence of light. A small change was observed in the concentration of the dye, indicating 6.1% removal by the adsorption process. In the presence of light, the photocatalytic decomposition of AG1 increased as the radiation time increased. After 70 min, the degradation of the green dye was 84%, as shown in Figure 9a.

By using the gold effect (mono-action effect) for TONP-1, the addition of supporting gold increased the optical activity of Au-TONP-1. Here, the percentage of decomposition reached 100% after 40 min of light radiation, as shown in Figure 9b. The complete disappearance of the green dyes indicates that the mono-action effect is effective for titanium oxide.

By using the triple-action effect, titanium oxide became more effective in the presence and absence of light. Here, Figure 10 shows a stronger effect of the adsorption capacity of Au-TONT-1 on the concentration of AG1 in the absence of light. A large change was observed in the concentration of the dye, indicating 27% removal by the adsorption process.

In the presence of light, the percentage of decomposition reached 100% after 17 min of light radiation, as shown in Figure 10. The complete disappearance of the green dyes indicates that the triple-action effect is very effective for titanium oxide.

3. Discussion

The full degradation and decolorization of the green dyes were achieved through developing titanium oxides by mono- and triple-actions. The mono-action effect was carried out through the addition of supporting gold to titanium oxide to act as a photocatalyst. This mono-action was effective for the photocatalytic degradation of the dyes. During 40 min of light radiation, the removal of the dyes was completed with a full disappearance of the green color from the water. To increase the effectiveness of the photocatalyst, the morphology of titanium oxide was converted to nanotubes. At the same time, titanium oxide was doped with sodium cations. The titanium oxide, after the conversion to nanotubes and the doping process, was supported with gold to increase its efficiency through a triple-action effect. This triple-action effect was very effective because the green color of the dyes disappeared after 17 min of light radiation; additionally, the absorbance peak of the dyes at 714 nm disappeared, as shown in Figure 9b. Also, the absence of the bands at 320 nm, 283 nm, and 232 nm indicated that the components of organic pollutants were fully oxidized and were mineralized into carbon dioxide and water because these peaks are linked to the organic phenyl groups of the dyes. Photocatalysis processes depend on the production of strong oxidizing agents from the photo-active materials using light. These oxidizing agents play a key role in decomposing and removing organic pollutants through advanced oxidation processes. These advanced oxidation processes involve the in situ generation of highly potent chemical oxidants such as the hydroxyl radical ($^{\bullet}OH$) and the superoxide radicals ($^{\bullet}O_2^-$).

The primary process by which photons of light excite the electrons in photo-active materials allows them to separate and move into the conduction band. In the valance band, holes begin to appear in this manner. The oxygen molecules and the excited electrons react to form the superoxide radicals $^{\bullet}O_2^-$. These free radicals are potent oxidizers. Hydroxide radicals are produced as a result of the reaction between the holes and water molecules, which also produces additional oxidizing agents. The $^{\bullet}OH$ radicals are highly oxidizing agents.

By accelerating the mechanism of the photo-active materials through the triple-action effect, high levels of photocatalytic activity were achieved with Au-TONT-1. The triple-action effect enabled the mechanism of the photo-activity of the Au-supported TiO_2 nanotubes to be accelerated. The first action was produced from the nanotubes. The nanotubes of titanium oxide have many advantages and actions. Such nanotubes generate new photo-active centers. Also, they trapped and confined the pollutants inside their tubes to attack different active sites at the same time. This finding was confirmed by increasing the adsorption capacity of Au-TONT-1 to attain 27% removal of the dyes. In addition, one-dimensional nanotubes prevent recombination reactions by increasing the separation processes between the excited electrons and the holes because of the short distance for charge carrier diffusion and light-scattering properties. In addition, the nanotubes look like the wires used for transferring electrons from the conduction band to other sites.

The second action depends on the doping process because the insertion of sodium cations inside the structure of titanium oxide caused a lowering of its band gap energy. The separation step was accelerated to increase the number of holes in the valance band and the number of electrons in the conduction band as a result of the drop in energy from 3.50 eV to 3.25 eV.

The third action relates to the addition of supporting gold to titanium oxide. The presence of gold nanoparticles leads to many positive actions for the photo-activity of titanium oxide because gold has a wide range of light absorption and can act as electron traps. Furthermore, it possesses a strong surface plasmon resonance (SPR) peak that can function as a light-trapping receptor. This causes the SPR peak to be photoexcited, which in turn creates a locally increased electric field near the gold nanoparticles. Moreover, SPR increases light scattering and extends light absorption to longer wavelengths. The enhanced photocatalytic degradation of the dyes and Au-TONT-1's high photocatalytic activity are caused by these triple-action effects. The following mechanism demonstrates how this triple-action effect became a driving force for increasing the production of oxidizing agents and accelerating pollutant removal as shown in Figure 11.

The excited electrons are swiftly transported through the nanotubes and the supporting gold to other sites in the photocatalytic degradation mechanism of the dyes in the presence of Au-TONT-1, preventing recombination events. Furthermore, Figure 11 provides numerous pathways for oxidizing agent production to hasten the dyes' photocatalytic breakdown. The hydroxyl radicals ($^{\bullet}OH$) and superoxide radicals ($^{\bullet}O_2^-$) that were adsorbed on the photocatalyst surface reacted to form adducts in the presence of the dye molecules. This was followed by the molecular structure fragmenting into several intermediate species until total mineralization, which produced CO_2 and H_2O.

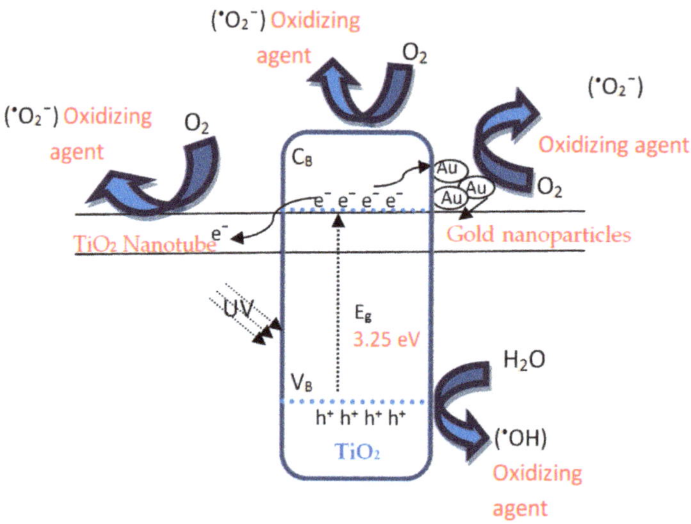

Figure 11. The mechanism of producing oxidizing agents and the photocatalytic degradation of dyes in the presence of Au-TONT-1.

In comparison with the previous results of the researchers, which were recently published, the prepared titanium oxide in the current study based on the triple-action effect is the optimum. Table 1 indicated that the highest performance for photocatalytic degradation of pollutants was observed for the prepared photocatalysts in the current study.

In 2023, pure TiO_2, Ce-TiO_2, and chitosan-loaded Ce-TiO_2 caused 70%, 90%, and 95% of Rhodamine B (RhB) dye degradation after 100 min of light radiation, respectively [16]. Paeng et al. [36] indicated that the Ce-doped titanium oxides and Er-Yb dual-doped titanium oxide caused 98% and 95% photocatalytic degradation of RhB dye after light irradiation for 60 and 240 min, respectively. Li et al. [37] reported that 95% of RhB was achieved within 20 min of light using $MoS_2/MoO_3/TiO_2$ composites. Chen et al. [38] used TiO_2-SiO_2 composites to degrade 99.9% of methyl orange (MO) after 3 h of UV light. By using a 3D N-doped TiO_2 film photocatalyst, 50% of MB degradation was achieved after 3 h of light radiation [39]. For the TiO_2-NH_2-Cu_xO nano-photocatalyst [9], 95% of the MB dye was degraded after 2 h of contact time with light. In the case of TiO_2 nanotube arrays decorated with biochar nanoparticles [40], 90% of the MB dye was degraded after 3 h of UV light. Ag/TiO_2 supported by 3D chitosan caused a complete removal of pharmaceutical compounds in wastewater during 2 h under UV irradiation [10]. Vo et al. [41] reported that the RhB dye was completely degraded by the Ce-TiO_2/chitosan composite catalyst. In the current study, the green dyes were fully oxidized and mineralized after 17 min of light radiation, indicating high performance for titanium oxides which are modified by the triple-action effect.

Table 1. A comparison between the current study and the published data of different photocatalysts based on titanium oxides and their performance.

Type	Photocatalysts	Target Pollutant	Degradation Rate	Radiation Time (min)	Ref.
TiO_2	Pure TiO_2	Rhodamine B	70%,	100	[16]
	Ce-TiO_2	Rhodamine B	90%	100	
	Ce-TiO_2/chitosan	Rhodamine B	95%	100	
TiO_2	Ce-TiO_2	Rhodamine B	98%	60	[36]
	Er-Yb–TiO_2	Rhodamine B	95%	240	
TiO_2	$MoS_2/MoO_3/TiO_2$	Rhodamine B	95%	20	[37]
TiO_2	TiO_2-SiO_2	Methyl orange	99.9%	180	[38]
TiO_2	N-doped TiO_2	Methyl orange	50%	180	[39]
TiO_2	TiO_2-NH_2-Cu_xO	Methyl orange	95%	120	[9]
TiO_2	Biochar/TiO_2 nanotubes	Methyl orange	90%	180	[40]
TiO_2	Ag/TiO_2/chitosan	Pharmaceutical compounds	100%	120	[10]
TiO_2	Ce-TiO_2/chitosan	Methyl orange	100%	120	[41]
TiO_2	Au/Na-TiO_2 nanotubes	Acid green 1	100%	17	The current study
	Au/TiO_2 nanoparticles	Acid green 1	100%	40	

To examine the reusability of the optimum photocatalyst, three cycles of photocatalytic processes were carried out for the Au-TONT-1 sample. The same results were observed for the three cycles. A complete disappearance of the green dyes occurred after 17 min of light radiation. This means that there is a high stability for the activity of Au-TONT-1 for the photocatalytic degradation processes.

The reaction rate of photocatalytic decomposition can be measured by kinetic models. The Langmuir–Hinshelwood mechanism was reported as a model for heterogeneous photocatalysis. The photocatalytic decomposition reaction of the aqueous solution of green dyes is a bimolecular reaction. During the reaction, the concentration of water is constant and the concentration of the dye strongly changes with time. This reaction appears like a first-order reaction (monomolecular reaction). Therefore, the kinetic reactions of the degradation process of the AG1 dye were studied using the following formula: $\ln C_o/C_t = kt$. The concentration of the dyes (C_o) was measured at a time of zero. The concentration of the dyes (C_t) was then determined at different times. The value (k) is the rate constant of the reaction. The kinetic plots of the photocatalytic decomposition of the dye using the prepared photocatalysts TONP-1, Au-TONP-1, and Au-TONT-1 are displayed in Figure 12.

Figure 12a shows the linear relation of the decomposition of the dyes using TONP-1. This linear relation means that this process is a pseudo-first-order kinetic reaction. Additionally, Figure 12a shows that the rate constant of the reaction is 0.0091 min^{-1}. The linear relation of the decomposition reaction of the dyes using Au-TONP-1 is displayed in Figure 12b. The rate constant of the decomposition reaction was found to be 0.037 min^{-1}. For Au-TONT-1, Figure 12c shows that the rate constant of the decomposition reaction increased to 0.0779 min^{-1}. The kinetic parameters of both TONP-1 and Au-TONP-1 revealed the positive effect of gold supporting on increasing the reaction rate of photocatalytic degradation of the green dyes from 0.0091 min^{-1} to 0.037 min^{-1}. The comparison between TONP-1 and Au-TONT-1 indicated that the triple-action effect boosted the photo-activity of titanium oxide to accelerate the reaction rate of the photocatalytic decomposition of the green dyes eight times faster than the pure titanium oxide nanoparticles.

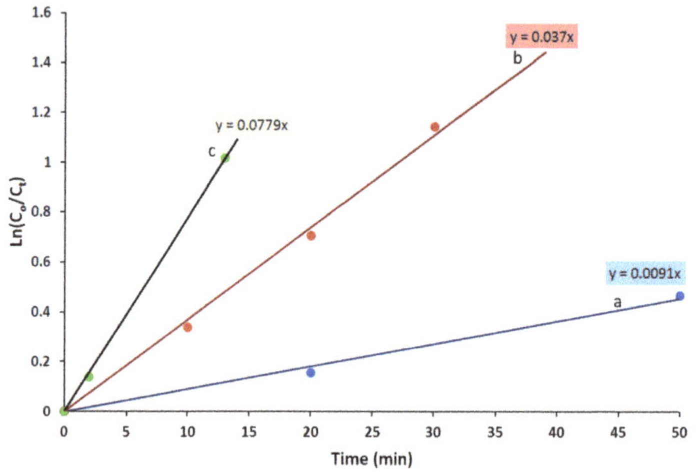

Figure 12. The rate of the photocatalytic decomposition reaction of the dye in the presence of light and (**a**) TONP-1, (**b**) Au-TONP-1, and (**c**) Au-TONT-1.

4. Materials and Methods

4.1. Design of Titanium Oxide Nanoparticles

Titanium oxide nanoparticles were synthesized by the sol–gel technique depending on the alcohol-based processes. The alcoholic mixtures were produced by reacting 20 mL of titanium isoperoxide with 125 mL of ethyl alcohol. The strong electronegativity between titanium and oxygen leads to the highly reactive and polar molecules of titanium isoperoxide. The gel form was produced by adding 125 mL of deionized water to the alcoholic mixtures of titanium isoperoxide through simultaneous hydrolysis and condensation processes.

Filtration and washing processes were used by the deionized water and ethyl alcohol to separate the white product. The products were dried in a vacuum at room temperature for 24 h. It was classified as TONP-1. The product was thermally treated at 400 °C for 3 h.

4.2. Doping and Transforming Nanoparticles into Nanotubes

The titanium oxide nanoparticles, TONP-1, were doped and transformed into nanotubes through one step. Using a hydrothermal technique, 1.50 g of the prepared nanoparticles of titanium oxide was reacted with sodium hydroxide at a high temperature and pressure. The concentrated aqueous solution of sodium oxide (10 M) was added and stirred with the nanoparticles of titanium oxide at room temperature for 2 h. Then, the mixture was thermally treated inside a Teflon-lined autoclave at 130 °C for 24 h. After washing and filtration processes, the product was dried at 60 °C. The final mass of the product was 1.60 g. The product was calcined at 400 °C for 3 h. It was coded as TONT-1.

4.3. Gold Loading

Gold nanoparticles were purchased from Nanoimmunotech Inc., Vigo, Spain. The concentration of gold was 0.5 mM dispersed in 10 mL of water. The diameter of the gold nanoparticles was around 50 nm. The gold was supported over titanium oxides according to the dry impregnation method. Typically, one gram of titanium oxide was mixed with the deionized water. Then, the titanium oxide was weighted after water impregnation. In this way, the amount of water, which can be impregnated by titanium oxide, was calculated through the difference between the weights of titanium oxide before and after water impregnation. After that, one gram of TiO_2 nanoparticles was impregnated with 10 mL of the aqueous solution of gold in order to produce 1 wt.% of gold in the final material. The impregnation process was achieved under stirring for 30 min at room temperature. Finally, the product was dried in a vacuum for 36 h. The product was thermally treated under a flow of nitrogen at 450 °C. It was coded as Au-TONP-1. Using a similar method, the second sample was prepared for the nanotube TONT-1 to produce Au-TONT-1.

4.4. Characterization of the Prepared Samples

To determine the crystalline structures, X-ray powder diffraction (XRD) was used by Rigaku RINT 2200 (Tokyo, Japan) with CuK_ (filtered) as the radiation source at a wavelength = 0.154 nm between angles of 2θ = 4–60°. Transmission electronic microscopy (TEM) and energy-dispersive X-ray spectroscopy (EDX) were performed by using the equipment (JEOL, JEM-2100F, Tokyo, Japan) with a voltage of 200 kV. A solution of 10 μL of nanoparticles was deposited on a carbon-covered copper grid after sonication for 20 min. A confocal Raman microscope (HORIBA; LabRAM HR800, Longjumeau, France SAS) to measure Raman spectra at room temperature with a blue laser (He-Cd) with a wavelength of 442 nm and an output power of 20 mW was used. The Raman spectra were recorded in a backscattering configuration at ambient temperature. A grating of 1800 L/mm, an acquisition time of 90 s, and 2 mW excitation power were used to record Raman spectra. The Shimadzu 3600 UV diffuse reflection technique was used to determine the optical parameters of the prepared nanomaterials. Via spectrophotometer software, the reflectance was converted to absorption using (Shimadzu, Columbia, MD, USA). To measure solids, an ISR-603 spectrophotometer (Integrated Ball Attachment, Shimadzu, Columbia, MD, USA) was used. In order to measure liquid samples, absorption coefficients were measured by conventional UV/VIS.

4.5. Photocatalytic Processes

The photocatalytic degradation of green dyes was used for purifying water from industrial pollutants by light. At the same time, to determine the efficiency and the photoactivity of the prepared titanium oxide nanoparticles, the nanoparticle-supported gold, and the nanotube-supported gold, the photocatalytic reactions were carried out inside the quartz immersion well reactor (RQ400) with a mercury lamp (400 W). This lamp covers a wide range of wavelengths in the UV and visible regions. In the current study, an aqueous solution of green dye (acid green 1) was prepared at an appropriate concentration of 0.0004 M. Two basic experiments were performed; one was without light and the other was carried out without a photocatalyst. In the standard experiment, the dye concentration can be determined by following the characteristic band in the dye spectrum according to the Beer–Lambert law. The concentration change was monitored in the green band color of the 714 nm feature of acid green 1. Small doses of the solution were withdrawn after UV irradiation at different time intervals. The concentration of the remaining dye in the solution was measured using a UV spectrophotometer.

5. Conclusions

The current study focuses on achieving a fast removal of industrial pollutants. This objective was accomplished by engineering titanium oxide through a triple-action effect to become an effective photocatalyst. The nanoparticles of titanium oxide were prepared and converted to nanotubes during the doping process. In addition, the doped nanotubes were supported by gold nanoparticles to induce the triple-action effect inside titanium oxide. Using the triple-action effect, new optical sites were produced through sodium doping and gold nanoparticles. Also, the nanotubes and gold nanoparticles created different tracks for charge carriers to transfer the excited electrons to other locations. In addition, recombination reactions were prevented by separation processes between the hole–electron pairs through the triple-action effect. High-resolution transmission electron microscopy (HRTEM) and selected area electron diffraction (SAED), Raman spectra, energy-dispersive X-ray spectrometer (EDX), and X-ray diffraction confirmed the strong role of the triple-action effect on the structure of the photocatalyst. The results of the photocatalytic degradation of the green dyes showed that the triple-action effect has a strong positive role in increasing industrial pollutant removal with or without light. Here, the percentage of photocatalytic decomposition reached 100% after 17 min of light radiation. In addition, 27% of the pollutants were removed without light radiation. The present study suggests an effective photocatalyst for water purification and environmental remediation.

Author Contributions: Conceptualization, O.S.; methodology, O.S., M.O. and A.O.; software, A.O., A.A. and A.H.F.; validation, O.S., S.A.M. and M.O.; formal analysis, N.M.S., C.A. and O.S.; investigation, A.O., O.S., M.O. and N.M.S.; resources, M.O., O.S., C.A. and A.O.; data curation, A.O., A.A., S.A.M., A.H.F. and M.O.; writing—original draft preparation, M.O., A.O., C.A. and O.S.; writing—review and editing, O.S., A.A. and N.M.S.; visualization, O.S. and M.O.; supervision, O.S. and A.O.; project administration, O.S.; funding acquisition, O.S. All authors have read and agreed to the published version of the manuscript.

Funding: This work was funded by the Deanship of Scientific Research at King Faisal University under grant no. KFU242015, and the APC was funded by the same grant, KFU242015.

Data Availability Statement: Data will be provided by the authors upon request.

Acknowledgments: The authors acknowledge the Deanship of Scientific Research, Vice Presidency for Graduate Studies and Scientific Research, King Faisal University, for financial support under grant no. KFU242015.

Conflicts of Interest: The authors declare no conflicts of interest.

References

1. Veziroglu, S.; Obermann, A.L.; Ullrich, M.; Hussain, M.; Kamp, M.; Kienle, L.; Leißner, T.; Rubahn, H.G.; Polonskyi, O.; Strunskus, T.; et al. Photodeposition of Au Nanoclusters for Enhanced Photocatalytic Dye Degradation over TiO_2 Thin Film. *ACS Appl. Mater. Interfaces* **2020**, *12*, 14983–14992. [CrossRef] [PubMed]
2. Abshari, F.; Veziroglu, S.; Adejube, B.; Vahl, A.; Gerken, M. Photocatalytic Edge Growth of Conductive Gold Lines on Microstructured TiO_2-ITO Substrates. *Langmuir* **2024**, *40*, 19051–19059. [CrossRef] [PubMed]
3. Ayati, A.; Ahmadpour, A.; Bamoharram, F.F.; Tanhaei, B.; Mänttäri, M.; Sillanpää, M. A review on catalytic applications of Au/TiO_2 nanoparticles in the removal of water pollutant. *Chemosphere* **2014**, *107*, 163–174. [CrossRef] [PubMed]
4. Zhang, Y.; Ju, S.; Casals, G.; Tang, J.; Lin, Y.; Li, X.; Liang, L.; Jia, Z.; Zeng, M.; Casals, E. Facile aqueous synthesis and comparative evaluation of TiO_2-semiconductor and TiO_2-metal nanohybrid photocatalysts in antibiotics degradation under visible light. *RSC Adv.* **2023**, *13*, 33187–33203. [CrossRef]
5. Shahzad, R.; Muneer, M.; Khalid, R.; Amin, H.M.A. ZnO-Bi_2O_3 Heterostructured Composite for the Photocatalytic Degradation of Orange 16 Reactive Dye: Synergistic Effect of UV Irradiation and Hydrogen Peroxide. *Catalysts* **2023**, *10*, 1328. [CrossRef]
6. Suherman, A.L.; Zampardi, G.; Kuss, S.; Tanner, E.E.L.; Amin, H.M.A.; Young, N.P.; Compton, R.G. Understanding gold nanoparticle dissolution in cyanide-containing solution via impact-chemistry. *Phys. Chem. Chem. Phys.* **2018**, *20*, 28300–28307. [CrossRef]
7. Richardson, D.; Plewa, J.; Wagner, D.; Schoeny, R.; Demarini, M. Occurrence, genotoxicity, and carcinogenicity of regulated and emerging disinfection by-products in drinking water: A review and roadmap for research. *Mutat. Res.* **2007**, *636*, 178–242. [CrossRef]
8. Slokar, Y.M.; Majcen Le Marechal, A. Methods of decoloration of textile wastewaters. *Dye. Pigment.* **1998**, *37*, 335–356. [CrossRef]
9. Khdary, N.H.; Alkhuraiji, W.S.; Sakthivel, T.S.; Khdary, D.N.; Abdel Salam, M.; Alshihri, S.; Al-Mayman, S.I.; Seal, S. Synthesis of Superior Visible-Light-Driven Nanophotocatalyst Using High Surface Area TiO_2, Nanoparticles Decorated with CuxO Particles. *Catalysts* **2020**, *10*, 872. [CrossRef]
10. Pinna, M.; Binda, G.; Altomare, M.; Marelli, M.; Dossi, C.; Monticelli, D.; Spanu, D.; Recchia, S. Biochar Nanoparticles over TiO_2 Nanotube Arrays: A Green Co-Catalyst to Boost the Photocatalytic Degradation of Organic Pollutants. *Catalysts* **2021**, *11*, 1048. [CrossRef]
11. Ma, Y.; Tao, L.; Bai, S.; Hu, A. Green Synthesis of Ag Nanoparticles for Plasmon-Assisted Photocatalytic Degradation of Methylene Blue. *Catalysts* **2021**, *11*, 1499. [CrossRef]
12. Bergamonti, L.; Graiff, C.; Bergonzi, C.; Potenza, M.; Reverberi, C.; Ossiprandi, M.C.; Lottici, P.P.; Bettini, R.; Elviri, L. Photodegradation of Pharmaceutical Pollutants: New Photocatalytic Systems Based on 3D Printed Scaffold-Supported Ag/TiO_2 Nanocomposite. *Catalysts* **2022**, *12*, 580. [CrossRef]
13. Ren, Y.; Chen, Y.; Li, Q.; Li, H.; Bian, Z. Microwave-Assisted Photocatalytic Degradation of Organic Pollutants via CNTs/TiO_2. *Catalysts* **2022**, *12*, 940. [CrossRef]
14. Milidrag, G.P.; Nikić, J.; Gvoić, V.; Mandić, A.K.; Agbaba, J.; Tomin, M.B.; Kerkez, D. Photocatalytic Degradation of Magenta Effluent Using Magnetite Doped TiO_2 in Solar Parabolic Trough Concentrator. *Catalysts* **2022**, *12*, 986. [CrossRef]

15. Székely, I.; Kovács, Z.; Rusu, M.; Gyulavári, T.; Todea, M.; Focsan, M.; Baia, M.; Pap, Z. Tungsten Oxide Morphology-Dependent Au/TiO_2/WO_3 Heterostructures with Applications in Heterogenous Photocatalysis and Surface-Enhanced Raman Spectroscopy. *Catalysts* **2023**, *13*, 1015. [CrossRef]
16. Indira, A.C.; Muthaian, J.R.; Pandi, M.; Mohammad, F.; Al-Lohedan, H.A.; Soleiman, A.A. Photocatalytic Efficacy and Degradation Kinetics of Chitosan-Loaded Ce-TiO_2 Nanocomposite towards for Rhodamine B Dye. *Catalysts* **2023**, *13*, 1506. [CrossRef]
17. Kuang, Y.; Ding, F.; Peng, Z.; Fan, F.; Zhang, Z.; Ji, X. Photocatalytic Degradation of Vehicle Exhaust by Nano-TiO_2 Cement Slurry: Experimental Factors and Field Application. *Catalysts* **2024**, *14*, 21. [CrossRef]
18. Mergenbayeva, S.; Abitayev, Z.; Batyrbayeva, M.; Vakros, J.; Mantzavinos, D.; Atabaev, T.S.; Poulopoulos, S.G. TiO_2/Zeolite Composites for SMX Degradation under UV Irradiation. *Catalysts* **2024**, *14*, 147. [CrossRef]
19. Abbasi, H.A.; Al Moneef, M.M.; Khan, J.; Hafeez, M.; Hameed, M.U.; Khan, M.A.; Shahida, S.; Abbasi, H.A.; Chang, S.K. Unveiling the Exceptional Performance of ZnO/Zn_2TiO_4 Nanocomposites. *Catalysts* **2024**, *14*, 156. [CrossRef]
20. Ko, Y.T.; Ruiz, J.J.B.; Leron, R.B.; Chang, C.T. Flower-like Titanium Dioxide/Cellulose Acetate Nanofibers for Catalytic Decomposition of Organic Pollutants Including Particulate Matter Removal. *Catalysts* **2024**, *14*, 233. [CrossRef]
21. Liu, J.; Zheng, L.; Gao, Y.; Ji, L.; Yang, Z.; Wang, H.; Shang, M.; Du, J.; Yang, X. TiO_2/p-BC Composite Photocatalyst for Efficient Removal of Tetracycline from Aqueous Solutions under Simulated Sunlight. *Catalysts* **2024**, *14*, 357. [CrossRef]
22. Sifat, M.; Shin, E.; Schevon, A.; Ramos, H.; Pophali, A.; Jung, H.J.; Halada, G.; Meng, Y.; Olynik, N.; Sprouster, D.J.; et al. Photocatalytic Degradation of Crystal Violet (CV) Dye over Metal Oxide (MO_x). *Catalysts* **2024**, *14*, 377. [CrossRef]
23. Sato, H.; Endo, S.; Sugiyama, M.; Kikegawa, T.; Shimomura, O.; Kusaba, K. Baddeleyite-Type High-Pressure Phase of TiO_2. *Science* **1991**, *251*, 786–788. [CrossRef]
24. Sekino, T. Synthesis and Applications of Titanium Oxide Nanotubes. In *Inorganic and Metallic Nanotubular Materials: Recent Technologies and Applications*; Kijima, T., Ed.; Springer: Berlin/Heidelberg, Germany, 2010; Volume 117, pp. 17–32.
25. Morgan, D.L.; Triani, G.; Blackford, M.G.; Raftery, N.A.; Frost, R.L.; Waclawik, E.R. Alkaline hydrothermal kinetics in titanate nanostructure formation. *J. Mater. Sci.* **2011**, *46*, 548–557. [CrossRef]
26. Venkatasubbu, G.D.; Ramakrishnan, V.; Sasirekha, V.; Ramasamy, S.; Kumar, J. Influence of particle size on the phonon confinement of TiO_2 nanoparticles. *J. Exp. Nanosci.* **2014**, *9*, 661–668. [CrossRef]
27. Mathew, S.; Ganguly, P.; Rhatigan, S.; Kumaravel, V.; Byrne, C.; Hinder, S.; Bartlett, J.; Nolan, M.; Pillai, S. Cu-Doped TiO_2: Visible Light Assisted Photocatalytic Antimicrobial Activity. *Appl. Sci.* **2018**, *8*, 2067. [CrossRef]
28. Ma, H.L.; Yang, J.Y.; Dai, Y.; Zhang, Y.B.; Lu, B.; Ma, G.H. Raman study of phase transformation of TiO_2 rutile single crystal irradiated by infrared femtosecond laser. *Appl. Surf. Sci.* **2007**, *253*, 7497–7500. [CrossRef]
29. Bavykin, D.V.; Friedrich, J.M.; Lapkin, A.A.; Walsh, F.C. Stability of Aqueous Suspensions of Titanate Nanotubes. *Chem. Mater.* **2006**, *18*, 1124. [CrossRef]
30. Qian, L.; Du, Z.L.; Yang, S.Y.; Jin, Z.S. Raman study of titania nanotube by soft chemical process. *J. Mol. Struct.* **2005**, *749*, 103. [CrossRef]
31. Saber, O.; Alomair, H.; Abu-Abdeen, M.; Aljaafari, A. Fast Degradation of Green Pollutants Through Nanonets and Nanofibers of the Al-Doped Zinc Oxide. *Acta Metall. Sin.* **2018**, *31*, 533–546. [CrossRef]
32. Saber, O.; Osama, A.; Shaalan, N.M.; Osama, M. Nanolayered Structures and Nanohybrids Based on a Ternary System Co/Ti/Zn for Production of Photo-Active Nanocomposites and Purification of Water Using Light. *Nanomaterials* **2024**, *14*, 93. [CrossRef]
33. Saber, O.; Osama, A.; Shaalan, N.M.; Osama, M. Engineering the Integration of Titanium and Nickel into Zinc Oxide Nanocomposites through Nanolayered Structures and Nanohybrids to Design Effective Photocatalysts for Purifying Water from Industrial Pollutants. *Catalysts* **2024**, *14*, 340. [CrossRef]
34. Saber, O.; Alshoaibi, A.; Al-Yaari, M.; Osama, M. Conversion of Non-Optical Material to Photo-Active Nanocomposites through Non-Conventional Techniques for Water Purification by Solar Energy. *Molecules* **2020**, *25*, 4484. [CrossRef] [PubMed]
35. Saber, O.; Osama, M.; Alshoaibi, A.; Shaalan, N.M.; Osama, D. Designing inorganic–magnetic–organic nanohybrids for producing effective photocatalysts for the purification of water. *RSC Adv.* **2022**, *12*, 18282. [CrossRef]
36. Paeng, D.S.; Huy, B.T.; Phuong, N.T.K.; Dao, V.D.; Lee, Y.I. Photocatalytic activity of Yb, Er, Ce-codoped TiO_2 for degradation of Rhodamine B and 4-chlorophenol. *J. Chem. Technol. Biotechnol.* **2020**, *95*, 2664–2673. [CrossRef]
37. Li, Z.Y.; Cao, F.; Wang, L.; Chen, Z.W.; Ji, X.H. A novel ternary MoS_2/MoO_3/TiO_2 composite for fast photocatalytic degradation of rhodamine B under visible-light irradiation. *New J. Chem.* **2020**, *44*, 537–542. [CrossRef]
38. Chen, Y.F.; Tang, X.N.; Zhang, B.; Luo, Y.; Li, Y. TiO_2@SiO_2 Composites: Preparation and Photocatalytic Antimicrobial Performance. *J. Inorg. Mater.* **2019**, *34*, 1325–1333. [CrossRef]
39. Cho, S.; Ahn, C.; Park, J.; Jeon, S. 3D nanostructured N-doped TiO_2 photocatalysts with enhanced visible absorption. *Nanoscale* **2018**, *10*, 9747–9751. [CrossRef]
40. Saber, O.; Kotb, H.M.; Osama, M.; Khater, H.A. An Effective Photocatalytic Degradation of Industrial Pollutants through Converting Titanium Oxide to Magnetic Nanotubes and Hollow Nanorods by Kirkendall Effect. *Nanomaterials* **2022**, *12*, 440. [CrossRef]
41. Vo, T.L.N.; Dao, T.T.; Duong, A.T.; Bui, V.H.; Nguyen, V.H.; Nguyen, D.L.; Nguyen, D.C.; Nguyen, T.H.; Nguyen, H.T. Enhanced photocatalytic degradation of organic dyes using Ce–doped TiO_2 thin films. *J. Sol-Gel Sci. Technol.* **2023**, *108*, 423–434. [CrossRef]

catalysts

MDPI

Review

Recent Advances in Photocatalytic Degradation of Imidacloprid in Aqueous Solutions Using Solid Catalysts

Song Gao [1], Shanshan Li [2], Shaofan Sun [1] and Maolong Chen [2,*]

[1] The 718th Research Institute of CSSC, Handan 056027, China
[2] School of Food Science and Bioengineering, Changsha University of Science & Technology, Changsha 410114, China
* Correspondence: mlchen@xmu.edu.cn; Tel.: +86-731-85258322; Fax: +86-731-85258733

Abstract: Imidacloprid (IMI), a widely used neonicotinoid pesticide, has led to significant water contamination due to excessive use. As a result, there is an urgent need for effective and straightforward methods to remove IMI residues from water. Photocatalytic technology, an integral part of advanced oxidation processes, is particularly promising due to its renewability, high catalytic efficiency, fast degradation ratio, and cost-effectiveness. This review systematically examines recent progress in the photocatalytic degradation of imidacloprid in aqueous solutions using various solid catalysts. It provides a comparative analysis of key factors affecting catalytic performance, such as catalyst synthesis methods, reaction times, catalyst loading, and IMI concentrations. Among the solid catalysts studied, nano-ZnO achieved a higher degradation rate of IMI in a shorter period and with a reduced catalyst dosage, reaching approximately 95% degradation efficiency within one hour. Additionally, this review explores the types of heterojunctions formed by the catalysts and elucidates the mechanisms involved in the photocatalytic degradation of IMI. In conclusion, this review offers a comprehensive evaluation of solid catalysts for the photocatalytic removal of IMI from water, serving as an important reference for developing innovative catalysts aimed at eliminating organic pollutants from aquatic environments.

Keywords: photocatalytic degradation; imidacloprid; solid catalysts; mechanism

1. Introduction

Imidacloprid (IMI) is a systemic insecticide from the nitroguanidine class, categorized as a neonicotinoid, which represents a new generation of nicotine-like insecticides. This compound is noted for its broad-spectrum activity, high efficacy, low toxicity, and minimal residual presence, leading to its extensive application in pest control, seed treatment, and the management of termites and fleas [1,2]. IMI works by disrupting normal signal transmission in the central nervous system of pests, causing paralysis and, eventually, death. Its effectiveness, combined with a relatively low risk of resistance development in target pests, has made IMI widely adopted in agricultural practices [3,4]. A United Nations report reveals that merely 1% of all pesticides used are effective against target pests, while the remaining 99% infiltrate the environment, causing detrimental effects on ecosystems [5]. The globally accepted safe concentration range for IMI emissions in the environment spans from 0.001 to 320 micrograms per liter. IMIs typically exhibit high persistence in water, leading to contamination of surface water and groundwater, as well as entry into the food chain. Global surveys of surface water have revealed that IMI is detected in 89% to 100% of cases, with concentrations reaching up to 4.50 μg/L, underscoring its pervasive presence [6,7]. Furthermore, studies show that IMI can persist in water for over 100 days [7]. As a result, when IMI enters water bodies through spray drift or rainfall runoff after application, it can pose significant threats to fish and other aquatic species, potentially disrupting entire aquaculture populations and ecosystems [6,7]. On the other hand, excessive use of IMI has been linked to the decline of bee populations, impairing female bees' foraging abilities, increasing competition among males, and affecting overall numbers. The ecological consequences of neonicotinoids have prompted significant concerns, particularly regarding water contamination and the broader impacts on biodiversity [8,9]

The aquatic environment is crucial for human activities and daily living. Therefore, it is imperative to explore sustainable and efficient technologies for the removal of IMI residues from water. Various methods have been developed for the degradation of organic compounds, including wet oxidation, biological oxidation, electrochemical redox processes, and advanced oxidation processes (AOPs) [10]. AOPs are widely acknowledged as the most effective approach for degrading organic pollutants in water due to their high efficiency and manageable reaction conditions [11,12]. The primary types of AOPs include the Fenton reaction and its variants, electrochemical oxidation [13], ozone oxidation [14], and photocatalytic degradation [15]. Among this, photocatalytic technology is essential in the treatment of wastewater pollutants due to its significant advantages, which include remarkable reproducibility, high catalytic efficiency, rapid degradation rates, and cost-effectiveness [16,17].

Photocatalytic technology involves harnessing light energy to enhance the redox capabilities of a photocatalyst, making it effective under various illumination conditions. When exposed to light, the photocatalyst absorbs energy, prompting electrons to transition to elevated energy states and generating electron–hole pairs. These pairs subsequently interact with oxygen and water, leading to the formation of reactive species. Ultimately, these active species, in conjunction with the electron–hole pairs, facilitate the conversion of pollutants into water and carbon dioxide [11]. Consequently, the characteristics of the photocatalyst are crucial for the effective catalytic degradation of IMI in aqueous environments. As shown in Figure 1, this review investigates the effects of various solid photocatalysts on the degradation of IMI in aqueous environments. It provides comprehensive insights into the synthesis methods, reaction times, and degradation efficiencies of different solid catalysts. Additionally, it explores and summarizes the predominant reaction mechanisms that drive the photocatalytic decomposition of IMI.

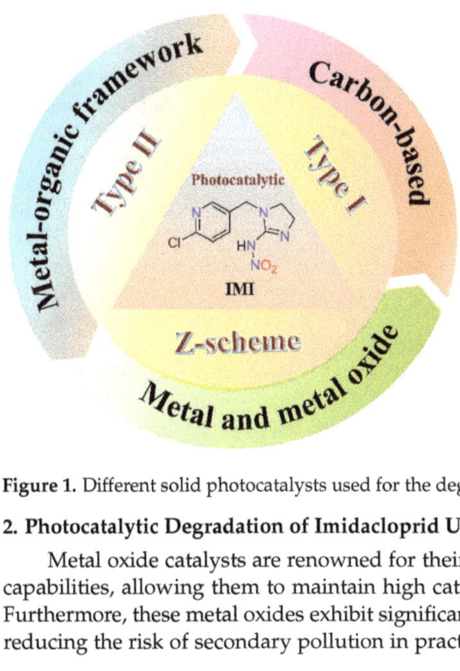

Figure 1. Different solid photocatalysts used for the degradation of IMI.

2. Photocatalytic Degradation of Imidacloprid Using Metal Oxide Photocatalysts

Metal oxide catalysts are renowned for their remarkable stability and regenerative capabilities, allowing them to maintain high catalytic activity even after multiple uses. Furthermore, these metal oxides exhibit significant environmental compatibility, thereby reducing the risk of secondary pollution in practical applications [18]. Moreover, metal oxides possess advantageous light absorption characteristics that broaden the light response spectrum of the catalysts [19]. The photocatalytic degradation of IMI using metals and their oxides catalysts are outlined and summarized in Table 1.

Table 1. Comparison of metal and metal oxide materials in photocatalytic degradation efficiency of IMI.

Photocatalyst	Light Source	Catalysis Time (min)	Catalyst Loading (g/L)	IMI Concentration (mg/L)	Efficiency (%)	Refs.
TiO_2	UV light	360	0.6	20	90.0%	[20]
Black TiO_2	visible light	360	1.0	20	90.0%	[21]
Cu-TiO_2	fluorescent bulb	60	0.5	25	45.0%	[22]
HPW/TiO_2-In_2O_3	Xenon lamp (225 W)	300	3.6	8	83.0%	[23]
ZnO	Xenon lamp	120	0.2	5	92.0%	[24]
nano-ZnO	visible light	30	2.0	50	96.6%	[25]
CuO	UV light	50	0.5	30	99.0%	[26]
Ag-ZnO	UV light	80	0.6	25	65.0%	[27]
Mg-ZnO/Nylon,6/PMMA	UV light	240	2.5	10	78.0%	[28]
N-MgO@Fe_3O_4	Xenon lamps	60	0.15	10	94.7%	[29]
Ag_2S/Fe_3O_4@Ag_3PO_4	Xenon lamp (300 W)	90	0.5	2	73.0%	[30]
Ag/$CuNb_2O_6$/$CuFe_2O_4$	halogen lamp	240	0.5	10	96.0%	[31]
Co_3O_4/PMS	solar irradiation	120	0.4	2.5	99.0%	[32]
CeO_2	light tubes (18 W)	360	0.15	20	30.0%	[33]
PWO/PI	Xenon lamp (225 W)	180	2.5	20	73.0%	[34]
Au-SnO_2-CdS	LED bulb	180	0.03	1.5	95.0%	[35]
In_2S_3/AgI-300	Xenon lamp (300 W)	60	0.5	10	76.2%	[36]
WO_3/SiO_2	UV light	60	0.5	5	59.0%	[37]
$Bi_{12.7}Co_{0.3}O_{19.35}$	visible light	240	1.0	10	96.0%	[38]
TiO_2-Fe-HNT	UV light	300	0.5	8	41.0%	[39]

2.1. Titanium Dioxide (TiO_2)-Based Solid Photocatalysts

TiO_2 is a highly effective photocatalyst known for its non-toxic nature, chemical stability, and significant reactivity. These properties make it suitable for various applications, including water purification, removal of residual pesticides, and degradation of air pollutants [40]. For example, Luminita et al. [20] developed TiO_2 photocatalysts through an eco-friendly and sustainable sol–gel approach, demonstrating effective degradation of IMI under both UV and visible light exposure. In the process of IMI photodegradation, electrons are essential. The introduction of a hole (h^+) scavenger resulted in an increase in the degradation ratio of IMI under UV light from 69% to 90% within a 6 h period. This improvement enhances the interaction between electrons and pollutants, thereby decreasing the recombination ratio of electron–hole pairs. Furthermore, the photocatalytic efficiency of TiO_2 is significantly influenced by the presence of defects and disorder within the material, including oxygen vacancies and Ti^{3+} defects. Consequently, TiO_2 nanoparticles that are rich in defects, such as black TiO_2, serve as optimal candidates for the fabrication of heterojunctions. Therefore, Luminita and co. [21] manipulated defect formation by adjusting the mass ratio of TiO_2 to $NaBH_4$ in an argon environment at 350 °C, which resulted in TiO_2 powders that ranged in color from gray to black. In comparison to the original white titanium dioxide, these modified powders exhibited a remarkable enhancement in photocatalytic activity, particularly in the degradation of IMI. When subjected to visible light irradiation, black TiO_2 demonstrated a photocatalytic degradation efficiency exceeding 90% for IMI. This improvement can be attributed to its increased specific surface area, reduced bandgap, and the prevalence of Ti^{3+} and oxygen vacancies on its surface, all of which promote effective charge carrier separation.

The ability of TiO_2 to effectively absorb visible light from the solar spectrum is limited by its wide band gap, which ranges from 3.0 to 3.2 eV, resulting in catalytic activity that is mainly restricted to ultraviolet radiation. Consequently, there is an urgent need to develop more efficient photocatalysts that maintain catalytic functionality under visible light to address the shortcomings of TiO_2 and enhance overall photocatalytic performance. Recent studies have utilized various metal oxide nanoclusters as cocatalysts to serve as hole trapping centers. Among these, copper (Cu) is an inexpensive alternative that can be used for the surface modification of TiO_2. It induces visible-light photocatalytic activity in a straightforward process without introducing impurities or vacancy levels into the crystal. The photocatalytic activity of Cu-modified TiO_2 is also influenced by the phase composition of TiO_2. Generally, rutile exhibits higher photocatalytic activity compared to anatase and brookite. Additionally, the highly dispersed Cu (I) and Cu (II) sites on the TiO_2 surface contribute to the enhancement of photocatalytic activity. More importantly, Cu (II) nanoclusters grafted onto the TiO_2 surface act as visible-light-sensitive cocatalysts and enhance electron–hole separation [41]. For instance, a nano-TiO_2 supported photocatalyst was prepared by Tihana et al. [22] through electrochemical anodic oxidation and subsequently modified with varying Cu concentrations (0.2–1 M) via wet impregnation. The degradation of IMI was evaluated using an unmodified reference catalyst (0 M $CuTiO_2$) and five Cu-modified catalysts (0.2 M, 0.4 M, 0.6 M, 0.8 M, and 1 M) under UV–visible irradiation in a continuous flow reactor, with 0.8 M $CuTiO_2$ exhibiting the highest photocatalytic activity. In the 0.8 M $CuTiO_2$ sample, the proportion of Ti–O–Cu bonds, which indicate Cu (II) cations directly bonded to the TiO_2 surface, versus Cu (II) present as CuO nanoclusters on the TiO_2 surface, is optimal for enhancing photocatalytic activity. Conversely, when the copper concentration is either too low (0.2 M $CuTiO_2$, 0.4 M $CuTiO_2$, and 0.6 M $CuTiO_2$) or too high (1 M $CuTiO_2$), the quantity of beneficial Ti–O–Cu bonds is not adequate to boost photocatalytic activity. Additionally, Cu–O–Cu bonds signify the formation of CuO nanoparticles, which can trap the holes necessary for the oxidation of IMI Therefore, a higher concentration of CuO on the TiO_2 surface (as in 1 M $CuTiO_2$) inhibits photocatalytic activity.

Although $H_3PW_{12}O_{40}$(HPW) demonstrates strong photocatalytic activity in homogeneous systems, its practical applications are significantly enhanced by immobilizing it on supports such as TiO_2. This immobilization leverages the substantial synergistic effect between HPW and TiO_2, thereby improving the catalyst's photocatalytic performance. Moreover, indium oxide (In_2O_3) is another n-type semiconductor with an indirect band gap of 2.8 eV, making it responsive to visible light. Building on these findings, Liu et al. [21] developed an innovative type II heterojunction visible light catalyst, HPW/TiO_2-In_2O_3, characterized by a ternary composite structure using a sol–gel method. They investigated its catalytic efficiency for degrading IMI under visible light. The HPW/TiO_2-In_2O_3 catalyst demonstrated superior photocatalytic degradation performance for IMI, achieving efficiencies that are 5.6, 9.3, and 12.5 times greater than those of HPW/TiO_2, TiO_2-In_2O_3, and TiO_2, respectively. The enhanced photocatalytic activity of HPW/TiO_2-In_2O_3 under visible light is attributed to the efficient separation of photogenerated charge carriers facilitated by the type II heterojunction, along with the suppression of carrier recombination due to the ternary composite structure.

2.2. Zinc Oxide (ZnO)-Based Solid Photocatalysts

Zinc oxide (ZnO), recognized for its chemical stability and availability, is another extensively researched photocatalyst. It has a band gap similar to that of titanium dioxide (TiO_2) and demonstrates outstanding photocatalytic activity under ultraviolet light [42]. For instance, a study by Maria's team [24] investigated the performance and efficiency of ZnO powder prepared by the sol–gel method for photocatalytic applications. Under simulated solar radiation, the study assessed the removal efficacy of IMI and found that, at an initial IMI concentration of 5 mg/L, the photocatalytic degradation efficiency reached 92% after 4 h of light treatment. Additionally, Dhiraj et al. [25] utilized three synthesis methods, including both template-assisted and non-template techniques, with xanthan gum (XG) serving as a sacrificial template. By modulating the growth conditions, they successfully fabricated ZnO nanostructures that exhibited high specific surface area and low bandgap energy, achieving a degradation efficiency of 96.09% for IMI under UV irradiation. Although ZnO possesses numerous advantages, its wide band gap limits its efficiency in the visible light spectrum, as it mainly absorbs ultraviolet light. Additionally, the rapid recombination of photogenerated electrons and holes adversely affects its photocatalytic performance. Therefore, targeted modifications and the development of composite materials have become essential strategies to overcome these limitations [43]. For example, on the basis of ZnO materials, Amna and colleagues [26] synthesized nano-copper oxide (CuO) and nano-zinc oxide (ZnO) from okra (Abelmoschus esculentus) using the sol–gel method. These nanoparticles demonstrated excellent photocatalytic activity, achieving a removal ratio of 99% for CuO and 81% for ZnO within 60 min of exposure to IMI. In another study, Mahwish et al. [27] synthesized Ag-ZnO composite materials via a hydrothermal technique to explore their efficacy in the complete mineralization of IMI. The findings revealed that the Ag-ZnO composite demonstrated superior photocatalytic activity in degrading IMI, achieving a degradation efficiency of approximately 65% within 80 min, outperforming pure ZnO. Similarly, Khalid et al. [28] prepared Mg-ZnO/Nylon 6,6/PMMA ternary nanocomposites via a solution casting method, investigating their performance in the photocatalytic degradation of IMI. The results showed a 78% increase in photodegradation rate compared to pure Nylon 6,6 and ZnO/Nylon 6,6/PMMA composites. The Mg-ZnO/Nylon 6,6/PMMA composite exhibited high degradation activity, primarily due to the presence of Mg ions, which suppress particle growth and agglomeration, resulting in smaller particle sizes and larger surface areas. This increased surface area provides more active sites for adsorption and catalytic reactions, thereby enhancing photocatalytic efficiency.

2.3. Magnetic Solid Photocatalysts

Magnetic catalysts present multiple advantages, such as straightforward separation and recovery, minimal waste generation, high catalytic performance, and wide-ranging applicability. These features allow them to overcome the limitations associated with traditional stationary catalysts, particularly regarding recycling and reusability. For example, Stefanos et al. [29] developed an innovative nitrogen-doped magnetite-based MgO nanocatalyst (N-MgO@Fe_3O_4) for the photocatalytic activation of persulfate (PMS) under visible light, achieving remarkable degradation of IMI. The degradation ratio for a concentration of 10 mg/L IMI reached approximately 95% within 60 min. In another research, Li et al. [30] utilized a solvothermal approach to deposit Ag_3PO_4 onto Fe_3O_4 nanoparticles, subsequently synthesizing Ag_2S-doped Fe_3O_4@Ag_3PO_4 nanocomposite films through an anion exchange reaction with Na_2S (Figure 2A). This process yielded a novel photocatalyst, Ag_2S/Fe_3O_4@Ag_3PO_4, which demonstrated remarkable catalytic activity under visible light irradiation, facilitating the rapid and efficient degradation of IMI. Notably, the leaching of silver was significantly minimized compared to both Fe_3O_4@Ag_3PO_4 and pure Ag_3PO_4. After four cycles, Ag_2S/Fe_3O_4@Ag_3PO_4 still achieved approximately 83.9% degradation of IMI. The superior photocatalytic activity and stability of the Ag_3PO_4 films can be primarily attributed to the presence of Ag_2S crystals, which effectively enhanced the separation of photogenerated charge carriers. Due to the rapid recombination of photogenerated electrons and holes, the photocatalytic activity of $CuNb_2O_6$ semiconductor material is relatively low. Consequently, introducing noble metal nanoparticles to construct metal-semiconductor composites is considered an effective solution. For instance, Zhang et al. [31] synthesized an $Ag/CuNb_2O_6/CuFe_2O_4$ ternary heterojunction composite photocatalyst (Figure 2B), which exhibits high efficiency and ease of recovery. The $Ag/CuNb_2O_6/CuFe_2O_4$ ternary heterojunction photocatalyst demonstrates superior photocatalytic activity, primarily attributed to its enhanced absorption in the visible light region and the rapid separation of photogenerated charge carriers. Compared to $CuNb_2O_6$, $Ag/CuNb_2O_6$, and $CuNb_2O_6/CuFe_2O_4$, the $Ag/CuNb_2O_6/CuFe_2O_4$ composite shows significantly improved photocatalytic degradation activity for IMI under visible light irradiation. Furthermore, the $Ag/CuNb_2O_6/CuFe_2O_4$ composite can be effortlessly separated and recycled using an external magnetic field. Consequently, its remarkable photoactivity and magnetic recoverability make the $Ag/CuNb_2O_6/CuFe_2O_4$ composite a promising candidate for applications in the environmental sector.

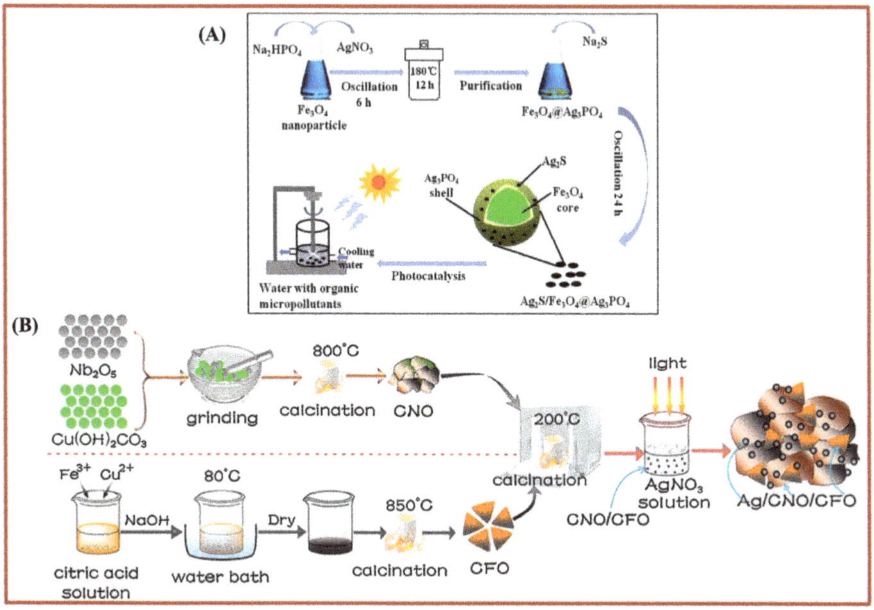

Figure 2. (**A**) Synthesis and photocatalytic reaction of the Ag_2S-doped core–shell nanostructures of $Fe_3O_4@Ag_3PO_4$ ultrathin film in the present study. (**B**) Schematic illustration of the synthetic process of $Ag/CuNb_2O_6/CuFe_2O_4$ (reproduced with permission from Elsevier, copyright 2019 [30] and copyright 2019 [31]).

2.4. Other Metal Oxide Solid Photocatalysts

The activation of PMS is efficiently catalyzed by Co_3O_4, which also exhibits photocatalytic properties driven by visible light. Upon exposure to solar radiation, PMS is synergistically activated by the cobalt species and conduction band electrons within Co_3O_4, resulting in the production of sulfate radicals and reactive oxygen species (ROS). Roberta R.M. et al. [32] prepared Co_3O_4 coatings using precipitation and vacuum filtration methods and applied this photocatalytic material to degrade IMI under continuous flow conditions. Under optimal conditions, the photodegradation ratio of IMI reached 99% after 2 h of operation. The excellent performance of the Co_3O_4/PMS/solar radiation system is attributed to the synergistic interaction between the Co_3O_4 catalyst and the Co^{2+} and Co^{3+} species in the UV component of solar radiation. CeO_2 serves as an effective photocatalyst under visible light, and prior research has demonstrated the feasibility of utilizing CeO_2 nanocrystals for wastewater treatment. In their study, Luminita et al. [33] described the performance of transparent CeO_2 films applied to glass substrates, achieving a degradation ratio of 30% for IMI within 6 h. Furthermore, the degradation ratio exhibited a linear increase with extended light exposure, ultimately reaching complete degradation after 48 h. In another study, Liu et al. [34] developed a novel direct Z-type phosphorous tungsten trioxide/polyimide (PWO/PI) photocatalyst through an in situ solid-state polymerization method to enhance the visible light photocatalytic oxidation capability of polyimide (PI). The study investigated the effects of polymerization temperature and PWO content on the physicochemical properties of the PWO/PI composite and its performance in the photocatalytic degradation of IMI. The results indicated that the PWO/PI composite exhibited visible light photocatalytic degradation efficiency for IMI approximately 3.2 times greater than that of commercial P25, with the corresponding pseudo-first-order reaction rate constant being about 2.9 times that of pure PI. The n-n heterojunction formed between SnO_2 nanoparticles and CdS nanoparticles is an excellent material for visible light degradation. CdS, as a narrow bandgap material, effectively absorbs visible light, while its interface with SnO_2 helps suppress charge recombination caused by junction potential. Therefore, Mohanta and colleagues [35] synthesized a novel Au-SnO_2-CdS ternary nanocomposite catalyst and investigated its photocatalytic degradation activity against IMI under an LED light source. The absorption capabilities of Au and CdS for visible light, along with the heterojunction structure's role in impeding charge recombination, facilitate the photocatalytic degradation of imidacloprid under LED light induction. The results indicated that the degradation efficiency of the Au-SnO_2-CdS nano-catalyst was 1.2, 1.4, and 2.1 times higher than that of the original Au, CdS, and SnO_2 nanomaterials, respectively, achieving approximately 95% degradation efficiency.

According to the data presented in Table 1, the catalysts nano-ZnO and N-MgO @ Fe_3O_4 demonstrated superior photocatalytic performance. They achieved a higher degradation ratio of IMI in a shorter time and with a lower catalyst dosage, reaching approximately 95% degradation efficiency within 1 h. However, their inherent drawbacks cannot be overlooked. For instance, their surface reaction activity is relatively low, necessitating modification or combination with other materials to enhance catalytic performance. Many metal oxides possess wide band gaps, limiting their light absorption within the visible spectrum. Additionally, photogenerated electrons and holes tend to recombine easily, thereby reducing photocatalytic efficiency.

In contrast, carbon materials (such as graphene and carbon nanotubes) and certain metal–organic frameworks (MOFs) exhibit significantly higher specific surface areas and effectively absorb visible light. Additionally, they offer excellent electrical conductivity, which facilitates the rapid transfer of photogenerated electrons and reduces electron–hole recombination. The structure of MOFs further facilitates electron separation and transport, thereby enhancing the photocatalytic performance. In the following, the application of high-efficiency photocatalysts prepared by metal-oxide-based carbon materials and MOFs in the removal of imidacloprid in water will be reviewed.

3. Photocatalytic Degradation of IMI Using Carbon-Based Photocatalysts

Carbon-based photocatalysts, such as graphitic carbon nitride, graphene oxide, and carbon quantum dots, can produce reactive species when exposed to light. These photocatalysts offer several benefits, including a high surface area, activity under visible light, stability, and recyclability. The following is a review of recent articles on the degradation of IMI in water using carbon-based photocatalysts, providing a reference for future research.

3.1. Graphitic Carbon Nitride (g-C_3N_4)-Based Solid Photocatalysts

Graphitic carbon nitride, characterized by its appropriate band structure, metal-free composition, facile synthesis, and outstanding chemical stability, stands out as a highly attractive photocatalyst for pesticide degradation [44–47]. However, g-C_3N_4 also has some drawbacks that need to be addressed, such as the rapid recombination of charge carriers, low surface area, and visible-light-harvesting efficiency, which hinder its application [48–50]. Currently, various modifications of pure g-C_3N_4 have been investigated; including alterations to its electronic structure, surface modifications through element doping, metal or metal oxide doping, the introduction of defects, and composite design, in order to enhance its photocatalytic application efficiency. Different g-C_3N_4-based solid photocatalysts for the IMI degradation are listed in Table 2.

Table 2. Comparison of carbon-based and composite materials in photocatalytic degradation efficiency of IMI.

Photocatalyst	Light Source	Catalysis Time (min)	Catalyst Loading (g/L)	IMI Concentration (mg/L)	Highest Efficiency (%)	Ref.
$SOCN_8$	xenon lamp (300 W)	300	1.0	20	91.4	[51]
CNT/PCN	LED light	540	0.6	10	93.0	[52]
Au @PPy-C/g-C_3N_4	visible light (250 W)	25	3.0	20	96.0	[53]
g-C_3N_4/ZnO	UV light	35	0.6	20	95.6	[54]
g-C_3N_4/TiO_2	W lamp (300 W)	150	1.0	10	93.0	[55]
Ag_2O/g-C_3N_4	halogen lamp (1000 W)	120	1.0	10	80.0	[56]
Ag-BO/GCN	LED light	600	0.5	10	93.0	[57]
g-C_3N_4@BiOCl	sunlight	180	0.3	10	73.4	[58]
g-C_3N_4/$BiVO_4$	UV light	30	0.06	20	94.2	[59]
OCN	xenon lamp (500 W)	120	0.5	3.0	94.5	[60]
HPW/ACN	CEL-LAX500 xenon lamp	350	0.6	10	90.0	[61]
MCN450/HPW	CEL-LAX500 xenon lamp	180	0.7	10	96.0	[62]
g-C_3N_4/KPW-0.2	xenon lamp (300 W)	180	2.0	20	91.7	[63]
13TCN-390	xenon lamp (225 W)	180	2.0	20	~90.0	[64]
CN-PANI-CQDs	xenon lamp (500 W)	70	1.0	10	80.1	[65]

For example, a series of sulfur–oxygen co-doped carbon nitrides ($SOCN_8$) were synthesized by Ma et al. [51] using a one-step thermal polymerization method and, for the first time, were employed for the photodegradation of IMI. Notably, specific doping ratios of sulfur and oxygen led to the formation of a distinctive tubular structure, which significantly improved the mass transfer capabilities and specific surface area of g-C_3N_4. Under visible light irradiation, $SOCN_8$ exhibited a 57.6% increase in degradation efficiency of the target insecticide compared to pure g-C_3N_4, achieving a maximum degradation rate exceeding 90%. In another study, C60/PCN nanocomposites with varying C60 ratios were synthesized by coupling phosphorus-doped g-C_3N_4 (PCN) with fullerene (C60) through an ultrasonic method by Anita et al. [66], and these nanocomposites were subsequently applied to the degradation of IMI. In the C60/PCN catalyst, phosphorus doping extended the solar light response range, prolonged the lifetime of charge carriers, and facilitated charge separation and transfer. Photodegradation experiments demonstrated that the C60/PCN/H_2O_2 and C60/PCN nanocomposites achieved removal efficiencies of 95% and 91% for IMI, respectively, significantly outperforming the degradation efficiency of the individual material catalysts. Similar to this, Anita et al. [52] successfully synthesized PCN through thermal polycondensation, utilizing melamine, ammonium oxalate, and a phosphorus dopant as precursors. The resulting PCN was then combined with carbon nanotubes (CNT), resulting in CNT/PCN nanocomposites that demonstrated enhanced photodegradation activity. The findings indicated that the CNT/PCN photocatalyst demonstrated remarkable efficiency in the photodegradation of IMI, achieving a removal ratio of 97% with CNT/PCN/H_2O_2 and 93% with CNT/PCN nanocomposites. The fabrication of an ultrafast and highly efficient visible light-responsive ternary photocatalyst was investigated by Farid A et al. [53] using an efficient, simple, and straightforward methodology. The photocatalyst contained g-C_3N_4 nanostructures and was combined with doped polypyrrole carbon black (PPy-C) and Au nanoparticles. The Au@PPy-C/g-C_3N_4 ternary photocatalyst demonstrated a 96.0% removal ratio of the target analyte IMI, achieving approximately 2.91 times greater efficiency compared to bare g-C_3N_4. This study presents a promising and effective approach for the degradation of highly toxic pollutants, thereby addressing pressing environmental challenges.

The converter slag composed of $Ca_2Fe_2O_5$ exhibits considerable potential as a photocatalyst; therefore, Keiko et al. [67] synthesized a composite photocatalyst of graphitized carbon nitride and converter slag (g-C_3N_4/CS) using a co-calcination method, and conducted photocatalytic degradation experiments on IMI under visible light irradiation. The optimized g-C_3N_4/CS composite, which incorporated 11.58% converter slag, demonstrated a 2.5-fold increase in the degradation ratio of IMI over a 2 h period compared to pure g-C_3N_4. Due to the tunability of the band gap in ZnO catalysts through the incorporation of metals, non-metals, metal oxides, and other semiconductors, Bansal and co-workers [54] developed g-C_3N_4/ZnO hybrid nano-catalysts with varying compositions. The g-C_3N_4/ZnO semiconductor hybrid nano-catalysts demonstrated remarkable degradation efficiency, achieving a ratio of 95.6% within just 35 min, outperforming pure ZnO and g-C_3N_4 nano-catalysts. Furthermore, total organic carbon (TOC) analysis indicated that the mineralization ratio reached 95.5% after 90 min. Analogously, Puangrat's team [55] developed a g-C_3N_4/TiO_2 composite material using a hydrothermal method and investigated its effectiveness in the photocatalytic degradation of the IMI pesticide. This composite creates a heterojunction between g-C_3N_4 and TiO_2, which leads to a reduction in the bandgap energy and inhibits the recombination of photogenerated carriers. The g-C_3N_4/TiO_2 composite demonstrates remarkable photocatalytic activity, achieving a removal ratio of 93% for IMI within 150 min, which is 2.2 times more effective than pure g-C_3N_4.

$Ag_4V_2O_7$ has garnered significant attention due to its unique photoelectric and chemical properties. Nevertheless, its photocatalytic performance remains relatively limited in its unmodified form. As a result, $Ag_2VO_2PO_4$/g-C_3N_4 composites were synthesized by Pu's team [68] using a hydrothermal method, which led to the formation of a novel $Ag_2VO_2PO_4$/g-C_3N_4 heterojunction. This newly formed heterostructure demonstrated significantly enhanced photocatalytic degradation activity for IMI under visible light irradiation, achieving degradation efficiencies that were 24.3 times higher than those of the pure catalyst and the $Ag_2VO_2PO_4$/g-C_3N_4 composites. Similarly, Pu et al. [56] utilized a chemical precipitation method to synthesize a range of p-n junction Ag_2O/g-C_3N_4 photocatalysts with varying Ag_2O concentrations. The resulting Ag_2O/g-C_3N_4 photocatalysts demonstrated remarkable photocatalytic activity for the degradation of IMI under both visible and near-infrared light, achieving an IMI degradation efficiency of approximately 80% within 2 h. Through both thermal polymerization and a one-step hydrothermal method, g-C_3N_4 nanocomposites were synthesized, resulting in the formation of silver-supported Bi_2O_3/g-C_3N_4 (Ag-BO/g-C_3N_4) nanocomposites [57]. When exposed to visible light, the produced Ag-BO/g-C_3N_4 exhibited a remarkable degradation efficiency of up to 93% for IMI, significantly surpassing the performance of the individual Ag-BO and g-C_3N_4 components. Furthermore, the Ag-BO/g-C_3N_4 nanocomposite showed impressive stability, with the IMI degradation ratio reducing from 98% to merely 88% over ten consecutive catalytic cycles.

Due to the overlapping energy levels in the band structure of BiOCl and g-C_3N_4, they are well-suited for the preparation of visible-light-responsive heterojunction catalysts. Building on this premise, Soumen et al. [58] employed a wet chemical method to mix g-C_3N_4 nanosheets with BiOCl nanosheets, successfully synthesizing a series of g-C_3N_4@BiOCl composites with varying mass ratios, thereby enhancing their photocatalytic performance. The g-C_3N_4@BiOCl catalyst achieved a degradation efficiency of 73.4% for IMI within 1 h. Similarly, $BiVO_4$ has attracted significant attention as a photocatalyst due to its low bandgap energy and high efficiency in degrading organic and inorganic pollutants. Consequently, Ajay et al. [59] effectively synthesized $BiVO_4$/g-C_3N_4 composites by modifying the g-C_3N_4 composition using a hydrothermal technique. This composite exhibited a specific surface area 5.4 times greater than that of $BiVO_4$. Although its bandgap slightly increased, the photoluminescence intensity decreased, which enhanced its photocatalytic performance. Compared to pure $BiVO_4$ and g-C3N4, the $BiVO_4$/g-C_3N_4 composite demonstrated superior catalytic ability, achieving a degradation efficiency of 94.2% for IMI within 30 min.

Oxygen-doped carbon nitride has been demonstrated to be an effective method for enhancing its photocatalytic or non-photochemical activation performance while maintaining its metal-free characteristics. For example, Zhang et al. [60] successfully synthesized a bifunctional oxygen-doped graphite carbon nitride (OCN) catalyst through a one-step thermal polymerization method, using oxalic acid as the oxygen source and urea as the precursor for g-C_3N_4. This catalyst was designed to activate PMS for the degradation of IMI. The modifications in the electronic structure of OCN promoted electron transfer and the formation of redox sites, which in turn improved its light absorption capacity, enhanced the separation efficiency of photogenerated charge carriers, and increased their migration speed. After 2.0 h, the removal ratio of IMI achieved 94.5% with the OCN-10/PMS system. Incorporating HPW or its salts into semiconductor photocatalysts serves as an effective strategy to diminish the recombination of electron–hole pairs [69]. HPW and acidified carbon nitride (ACN) were utilized by Liu et al. [61] as raw materials. By leveraging the hydrogen bonding between the oxygen-containing functional groups on the carbon nitride's surface and the oxygen at the terminal of the HPW, HPW was successfully doped into the carbon nitride, resulting in the preparation of HPW/ACN composite materials. The hydrogen-bonding interaction between HPW and ACN enhances the photogenerated hole–electron separation efficiency of the composite material, thereby improving its photocatalytic activity. Under visible light irradiation, the degradation ratio of HPW/ACN reached 5.8×10^{-3} min^{-1}, which is 16 times that of ACN, and achieved a degradation ratio of over 90% within 6 h. Similarly, Liu et al. [62] synthesized nitrogen carbon (CN) rich in uncondensed amino and carbonyl groups using urea as a precursor through a low-temperature thermal condensation method. Building on this, they performed a reversible nucleophilic addition reaction between the carbonyl of formic acid and the amino groups of CN, resulting in the formation of modified carbon nitride (MCN). Subsequently, MCN was combined with HPW to create a novel photocatalyst, MCN/HPW. Research findings indicate that MCN450/HPW exhibits outstanding performance in the photocatalytic degradation of IMI, achieving a degradation ratio constant 6.4 times greater than that of CN, with a degradation efficiency of 96% within 3 h. In another study, a g-C_3N_4/KPW-0.2 composite photocatalyst was synthesized by Liu et al. [63] using potassium phosphotungstate (KPW) as the raw material through a post-impregnation activation method, as illustrated in Figure 3A. In photocatalytic processes, KPW serves as both an electron acceptor and mediator, thereby enhancing visible light absorption and minimizing the recombination of electrons and holes. The g-C_3N_4/KPW-0.2 catalyst demonstrates outstanding photocatalytic activity, achieving a degradation ratio of 91.72% for IMI under visible light irradiation, which is 2.8 times greater than that of g-C_3N_4 alone. In a separate study, Liu and his team [64] synthesized a series of carbonitride/tungstophosphate acid (TCN) photocatalysts by utilizing phosphotungstic melamine (MPW) as the precursor and employing an in situ solid-phase thermal conversion technique (Figure 3B). The melamine molecules within the MPW hybrids can participate in polymerization reactions, resulting in the formation of carbon nitrides characterized by heptazine structural defects. In comparison to the MPW hybrid, the TCN photocatalyst demonstrates remarkable visible light catalytic activity, achieving photocatalytic degradation efficiencies and rate constants (k) for IMI that are 6.38 and 13.50 times greater than those of CN-390, respectively.

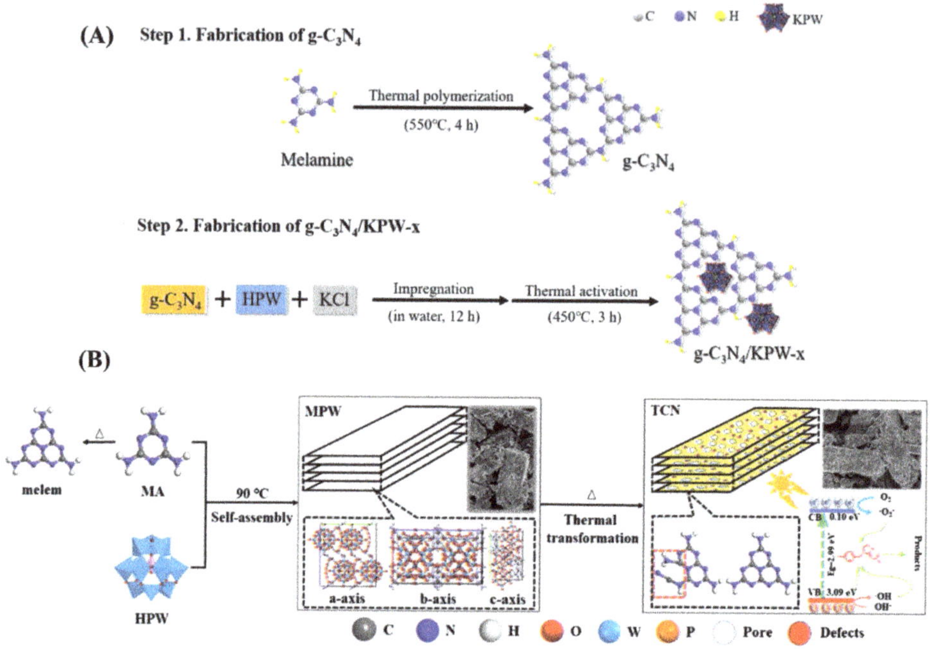

Figure 3. (**A**) Schematic diagram of the synthesis process of g-C_3N_4/KPW-x; (**B**) schematic illustration of constructing TCN photocatalysts and the proposed photocatalytic mechanism in the 13TCN-390 under visible light irradiation (reproduced with permission from Elsevier, copyright 2024 [63] and copyright 2019 [64]).

3.2. Graphene-Oxide-Based (RGO) and Carbon Quantum Dots (CQD)-Based Solid Photocatalysts

Graphene is a two-dimensional material made up of a single layer of carbon atoms arranged in a hexagonal honeycomb lattice through sp^2 hybridization. This remarkable material exhibits a range of exceptional properties, including high mechanical strength, low production costs, excellent transparency, outstanding electrical conductivity, significant specific surface area, high adsorption capacity, and superior thermal conductivity (Figure 4A) [70]. Graphene oxide (GO) is a derivative of graphene, typically synthesized by reacting graphite with strong oxidants in an acidic environment. During this process, carbon atoms are modified with various oxygen-containing functional groups, such as hydroxyl, carboxyl, and epoxy groups [71]. GO possesses excellent properties as a catalyst material due to its large specific surface area and rich functionality, along with stable thermal conductivity and high electron transfer rates, making it suitable for the photocatalytic degradation of organic compounds (Figure 4B) [72]. By removing the oxygen-containing functional groups, GO can be reduced to reduced graphene oxide (RGO). RGO exhibits improved chemical stability and electronic conductivity compared to GO. Additionally, the hydrophobic nature of RGO facilitates its aggregation, which can alter its morphology and specific surface area. This transformation enhances its potential applications in various fields, including catalysis, energy storage, and environmental remediation [73]. The unique properties of both GO and RGO make them valuable materials for advanced photocatalytic processes and other technological applications [74]. Table 3 presents various solid photocatalysts based on GO and RGO that have been utilized for the degradation of IMI.

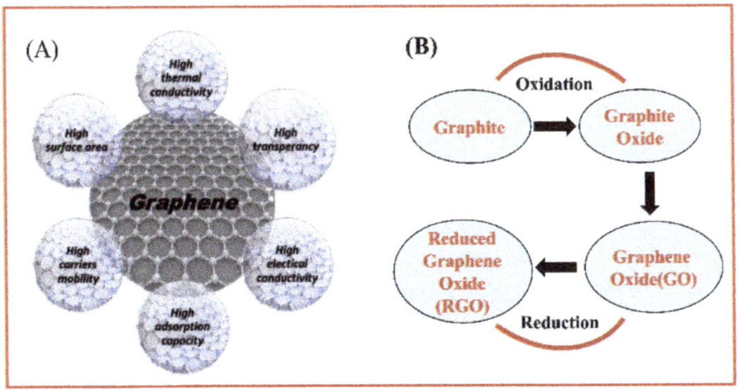

Figure 4. (**A**) The exceptional properties of pure graphene. (**B**) Flowchart of GO and RGO preparation from graphite (reproduced with permission from Elsevier, copyright 2023 [70]).

Table 3. Comparison of photocatalytic degradation efficiency of IMI by RGO- and CQD-based composite materials.

Catalyst	Light Source	Catalysis Time (min)	Catalyst Loading (g/L)	IMI Concentration (mg/L)	Highest Efficiency (%)	Refs.
GO@TiO_2	UV light	30	0.5	100	92.6	[75]
$BiVO_4$/RGO-TNT	UV lamp (40 W)	30	1.4	80	73.0	[76]
GO@TiO_2·ZnO·Ag HNM	visible light	115	0.5	10	50.0	[77]
Ce-TiO_2/RGO	UV light	480	0.5	10	85.0	[78]
TiO_2/RGO/Cu_2O-CuS	Full-spectrum fluorescent bulb solar (160 W)	360	1.0	20	>95.0	[79]
GO@PdO@rGO.SrO	visible light	210	3.5	30	86.0	[80]
GO/Fe_3O_4/TiO_2-NiO	visible light	30	2.5	10	97.3	[81]
CdS/MIPs	Hg light (400 W)	90	1.0	10	84.0	[82]

Based on the unique properties of GO and semiconductor TiO_2, Maged et al. [75] synthesized GO@TiO_2 nanocomposites and applied them for the photocatalytic degradation of IMI. In this composite, GO acts as an acceptor and transporter of the photogenerated electrons produced by the TiO_2 nanoparticles, effectively reducing the recombination of photogenerated electrons and holes within the TiO_2 nanoparticles. Compared to pure TiO_2 nanoparticles, the synthesized GO@TiO_2 nanocomposite exhibited higher efficiency in the photocatalytic degradation of IMI, achieving a degradation ratio of approximately 93% within 30 min. In another study, Teng et al. [76] synthesized $BiVO_4$/RGO-TNT (TiO_2 nanotubes) photocatalysts using a one-step hydrothermal method. The photocatalytic efficiency of the composite material is significantly improved, and the degradation ratio of IMI can reach 73% within 30 min. Similarly, hybrid nanomaterials consisting of GO@TiO_2, ZnO, and Ag composites were synthesized by Nagi M et al. [77], where graphene oxide (GO) was used to incorporate silver nanoparticles (Ag NPs), zinc oxide (ZnO NPs), and titanium dioxide (TiO_2 NPs). The carbon nanotubes and GO served as electron acceptors from the conduction band of the nanoparticles, thereby enhancing the photocatalytic efficiency, resulting in a degradation ratio of approximately 50% within about 2 h.

Recognizing the dual functional properties of cerium-doped titanium dioxide, the Sasmita team [78] employed a hydrothermal method to synthesize cerium-doped titanium dioxide nanoparticles (Ce-TiO_2) that exhibit excellent water dispersion. Using a deposition technique, they integrated Ce-TiO_2 with reduced graphene oxide (RGO) to create a novel ternary photocatalyst, Ce-TiO_2/RGO. Compared to conventional TiO_2 catalysts, this photocatalyst exhibited superior degradation efficiency for IMI under visible light exposure. In another study, Luminita et al. [79] investigated the performance of TiO_2@Cu_2O-CuS heterostructures enhanced by RGO in photocatalytic degradation of organic pollutants using IMI as the object. This heterojunction structure significantly enhances photocatalytic activity through the complementary properties of its constituent materials and RGO, surpassing the effect of each component used alone, exhibiting a significant photocatalytic degradation efficiency of over 95% for IMI. The composite photocatalyst PdO-GO-SrO-RGO was successfully synthesized by Nagi M et al. [80] through the integration of palladium oxide (PdO) nanoparticles with n-type semiconductor strontium oxide (SrO) nanoparticles, which function as electron carriers for RGO and GO nanosheets. When applied to the degradation of IMI using an adsorption–photocatalysis method, the degradation ratio was found to be 0.0086 min^{-1}, achieving a photocatalytic efficiency of 86%. To address the challenges of separation often faced by nanocomposites due to their extremely small dimensions, Ali et al. [81] synthesized TiO_2-NiO magnetic nanosheets on nanostructured graphene oxide using a sol–gel method. These nanosheets were then utilized for the visible-light-induced degradation of the IMI. Under optimal photocatalytic conditions, the catalyst of GO/Fe_3O_4/TiO_2-NiO achieved a degradation ratio of 97% for IMI within 30 min. Furthermore, even after at least four cycles of reuse, the catalyst maintained a degradation ratio of approximately 85% for IMI.

Carbon quantum dots (CQDs) derived from biomass are carbon-based materials characterized by remarkable optical and electrical properties. CQDs possess the ability to efficiently capture sunlight, adjust their photoluminescence, and transfer light-excited electrons effectively, making them ideal for luminescent applications [83,84]. Sourced from sustainable and renewable materials, CQDs exhibit significant potential for various applications in energy and environmental sectors, owing to their excellent water solubility and biocompatibility [85]. The addition of carbon quantum dots (CQDs) to graphitic carbon nitride with polyaniline (PANI) can improve its light absorption abilities and reduce the recombination of holes and electrons. Based on this, novel CQDs decorated on PANI with hollow porous graphitic carbon nitride (CN) were fabricated via an in situ polymerization followed by an ultra-sonication [65]. The nanocomposite of CN-PANI-CQDs exhibited the high visible light absorption with a high specific surface area. Under the visible light, the degradation was higher than 80.1% within 70 min, which can be attributed to its higher charge separation and destruction of recombination rate through the heterojunction of excited electrons among CN, PANI, and CQDs. In another study, Deepak et al. [82] utilized gingerol as a template molecule, ethylene glycol dimethacrylate (EGDMA) as a crosslinking agent, and azobisisobutyronitrile (AIBN) as an initiator to synthesize CdS quantum dots through a simple precipitation method. These quantum dots were then embedded into a molecularly imprinted polymer matrix, resulting in the novel nanocomposite CdS/MIPs. The study found that CdS/MIPs exhibited significant photocatalytic degradation of IMI under visible light irradiation, achieving a maximum degradation ratio of 84%.

Based on the above analysis and the data in Tables 2 and 3, carbon-based catalysts exhibit a marked advantage in catalytic efficiency, with degradation ratio typically exceeding 80% or 90%. This superior performance is largely due to their high specific surface area, which offers a wealth of active sites. Additionally, carbon-based materials possess good stability, giving them a clear advantage in terms of recovery and repeated use. Consequently, utilizing carbon-based photocatalysts for the photocatalytic degradation of IMI represents a promising approach to reducing the environmental impact of this pesticide. Ongoing research and development in this area could optimize photocatalytic systems, thereby fostering more sustainable agricultural practices and promoting environmental protection.

4. Photocatalytic Degradation of IMI Using Metal–Organic-Framework-Based Solid Photocatalysts

Metal–organic frameworks (MOFs) represent a category of porous coordination polymers formed from metal nodes and organic ligands. These materials are characterized by tunable pore sizes, high surface areas, exceptional thermal stability, and strong capabilities for photocatalysis and adsorption. Recent research has demonstrated that MOFs can produce electrons (e^-) and holes (h^+) when exposed to light, indicating their significant potential and wide-ranging applications in the photocatalytic degradation and removal of various organic pollutants [86,87]. Table 4 summarizes the research progress on the photocatalytic degradation of IMI using metal–organic framework-based catalysts.

Table 4. Comparison of photocatalytic degradation efficiency of IMI by metal–organic framework-based solid photocatalysts.

Catalyst	Light Source	Catalysis Time (min)	Catalyst Loading (g/L)	IMI Concentration (mg/L)	Highest Efficiency (%)	Refs.
TiO_2/ZIF-8	UV light	240	1.0	10	55.0	[88]
MIL-101(Fe)	Blue LED light	30	-	40	100.0	[89]
MOF/BWZTO/RGO	UV visible	300	0.6	15	90.0	[90]
Bi_2WO_6/NH_2-MIL-88B(Fe)	xenon lamp	180	0.4	10	84.5	[91]
SAO/NH_2-UiO-66	xenon lamp	120	0.6	20	97.0	[92]

For example, Lucija et al. [88] synthesized a composite material of zeolitic imidazolate framework (ZIF-8) and isopropyl titanate using a mechanochemical method and then utilized a hydrothermal method to prepare TiO_2 composite photocatalysts (TiO_2/ZIF-8) with three different mass fractions. The study demonstrated that TiO_2/ZIF-8 exhibited approximately 55% degradation efficiency in the photocatalytic degradation of IMI. In another study, Meral's research team [89] successfully synthesized iron-based metal–organic framework compounds, specifically MIL-101(Fe) and its amino-functionalized derivative NH_2-MIL-101(Fe), utilizing a conventional solvothermal approach. Additionally, they produced another significant iron-based framework, MIL-121(Fe), along with its amino-functionalized counterpart, NH_2-MIL-121(Fe). These metal–organic frameworks were employed in photocatalytic applications to facilitate the photodegradation of the pesticide IMI. Under optimal photocatalytic conditions, both catalysts achieved a complete IMI removal ratio of 100% within a span of 30 min. Moreover, the study assessed the adsorption capabilities of the catalysts, demonstrating that NH_2-MIL-101(Fe) displayed enhanced adsorption efficiency compared to MIL-101(Fe). In a similar manner, Chen et al. [93] investigated amino-functionalized metal–organic framework materials NH_2-MIL-88B(Fe) and NH_2-MIL-101(Fe). Both iron-based MOFs exhibited significant adsorption capacity and Fenton-like degradation ability for IMI, making them suitable for the removal of IMI in aqueous solutions. Under optimal experimental conditions, NH_2-MIL-88B(Fe) achieved a total removal ratio of 93% for IMI, while NH_2-MIL-101(Fe) reached a removal ratio of 97%. In another study, Taher et al. [90] successfully synthesized the ternary perovskite photocatalyst $Bi_2WZnTiO_9$ for the first time. To enhance its photocatalytic performance, they introduced metal–organic framework particles (PCN-222(Fe)) as a co-catalyst alongside the primary photocatalyst. Furthermore, they integrated carbon materials (CMs) as non-metal co-catalysts, leveraging their low-energy band gap and high surface area. This combination led to the development of a ternary composite nano-catalyst, designated as MOF/BWZTO/RGO. Under optimal experimental conditions, the MOF/BWZTO/RGO exhibited a photocatalytic degradation efficiency of 90% for the pesticide IMI and demonstrated reusability for up to five cycles. MOFs can enhance the light absorption and degradation efficiency of Bi_2WO_6 semiconductors. In a novel approach, Chen's research group [91] exploited the synergistic interaction between Bi_2WO_6 and NH_2-MIL-88B(Fe) to develop a highly effective photocatalyst known as the Bi_2WO_6/NH_2-MIL-88B(Fe) (BNM) heterojunction. The optimized BNM catalyst demonstrated significant degradation rates under visible light due to the production of reactive hydroxyl radicals (·OH), achieving an impressive degradation efficiency of 84% for IMI within just 3 h.

Amino UiO-66 has gained significant attention in the field of photocatalytic degradation due to its superior adsorption characteristics. However, when utilized in isolation, NH_2-UiO-66 demonstrates limited degradation efficiency, primarily attributed to its sluggish electron transfer rate. To overcome this limitation, Li et al. [92] successfully developed a straightforward binary composite, SAO/NH_2-UiO-66, through a solvothermal approach. This composite exhibited remarkable photocatalytic performance in degrading the pesticide IMI. Importantly, the addition of the long afterglow material, SAO, effectively mitigated the challenge of inadequate illumination during photocatalysis, enabling the degradation of nearly 15% of IMI even five hours post removal of the light source. Additionally, mass spectrometry and zebrafish embryo experiments were performed to confirm the structures of potential photocatalytic degradation intermediates and assess their low toxicity (Figure 5).

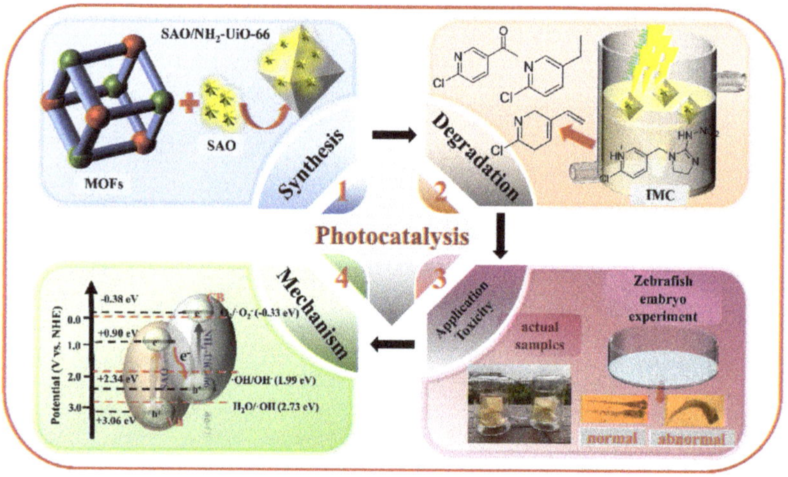

Figure 5. Synthesis of catalyst and photocatalytic degradation of IMI in water (reproduced with permission from Elsevier, copyright 2023 [92]).

In summary, solid photocatalysts excel in light absorption, multifunctionality, and catalytic efficiency. However, in practical applications, these catalysts also face some unavoidable limitations. The first issue is the high cost. Many efficient catalysts, especially noble metal catalysts (such as platinum and palladium), are expensive, which increases the economic burden for large-scale applications. Additionally, the synthesis processes of certain catalysts may consume significant resources and time, further driving up costs. The second issue is stability. During the reaction process, catalysts may degrade or become inactive, particularly under extreme reaction conditions (such as high temperatures, high pressures, or strong acidic or basic environments). The stability of the catalyst directly affects its lifespan and reaction efficiency; frequent deactivation can lead to decreased reaction efficiency and increased frequency and costs of catalyst replacement. Lastly, there is the issue of reusability. Although some catalysts can be recovered and reused after the reaction, their performance may decline over multiple uses. Particularly, in the presence of pollutants or by-products, the catalyst's surface may become covered or poisoned, reducing its catalytic activity. Therefore, despite the important role of catalysts in enhancing the efficiency and selectivity of chemical reactions, challenges related to high costs, stability, and reusability still need to be addressed through further research and technological innovation.

5. Possible Mechanism of Photocatalytic Degradation of IMI

The photocatalytic efficiency of catalysts is primarily constrained by two key factors: (1) the energy of the incident photons (hv) must exceed the bandgap (Eg), suggesting that a smaller Eg enhances the effective utilization of solar energy; (2) the redox potential of the reaction should fall within the range defined by the top of the valence band (VB) and the bottom of the conduction band (CB) [94,95]. In the current literature on the photocatalytic degradation of IMI, the degradation mechanisms are predominantly classified into three categories: Type I, Type II, and Z-scheme. Type I photocatalysts utilize single-component systems, which often exhibit limited redox capabilities and suboptimal light absorption across a broad bandgap. Furthermore, these catalysts are susceptible to the recombination of photogenerated electrons (e^-) and holes (h^+) within the valence band (VB), as illustrated in Figure 6a. In Type II heterojunctions, excited electrons (e^-) from the conduction band (CB) of catalyst S2 migrate to the CB of catalyst S1, while holes (h^+) from the VB of S1 transfer to the VB of S2, as shown in Figure 6b. This mechanism enhances the separation of photogenerated charge carriers. However, the charge transfer in Type II heterojunctions may diminish both oxidation and reduction capabilities, inadvertently overlooking the electrostatic repulsion between like charges. Additionally, these systems often encounter difficulties in achieving efficient light absorption across a wide bandgap, effective charge carrier separation, and robust redox performance simultaneously [96].

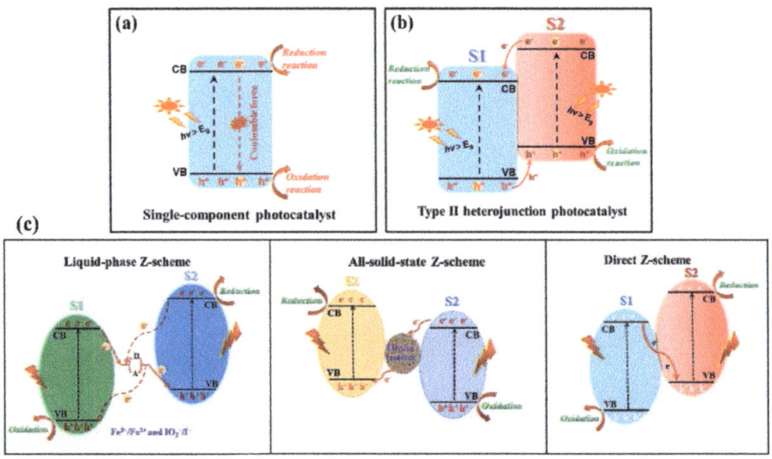

Figure 6. Charge transfer in (**a**) single-component photocatalyst, (**b**) type II heterojunction photocatalytic, and (**c**) Z-scheme heterojunctions system (reproduced with permission from Elsevier, copyright 2023 [95]).

Given the constraints of conventional Type I and Type II heterostructures, there is a pressing need to create a novel heterojunction that can effectively enhance light absorption, optimize charge carrier separation, and maintain robust redox capabilities from both theoretical and practical viewpoints. Unlike Type I and Type II heterojunctions, Z-scheme heterojunctions successfully reconcile the opposing demands of narrow bandgap and strong redox properties, thereby offering an efficient redox system. As a result, Z-scheme heterojunctions hold significant potential for applications in photocatalysis and various energy conversion processes [97,98]. The formation of Z-scheme heterojunctions is inspired by the process of photosynthesis in green plants. In nature, the unique structure of green plants enables them to efficiently harness solar energy to convert carbon dioxide and water into carbohydrates and oxygen (O_2). The generation of oxygen and reduction of cofactors occur in two distinct regions of the chloroplast, with these processes taking place simultaneously. The pathway of charge transfer in this mechanism resembles the letter "Z", effectively facilitating the separation and utilization of charge carriers to enhance the overall efficiency of the photosynthetic process (Figure 6c) [95].

5.1. Type I and Type II Heterojunction

For example, Figure 7A illustrates the photocatalytic degradation of IMI by $Bi_{12.7}Co_{0.3}O_{19.35}$ under varying pH conditions [38]. At a pH of 1.86, an abundance of free H^+ ions in the wastewater attaches to the catalyst's surface. The IMI molecule, being a nicotine derivative, engages in electrostatic interactions with these H^+ ions, leading to its adsorption on the catalyst. In contrast, at a pH of 11.0, the accumulation of free OH^- groups on the catalyst's surface results in the repulsion of IMI molecules, thereby explaining the absence of adsorption during the dark reaction. In acidic conditions, the photocatalytic degradation of IMI primarily involves oxidation, where the IMI molecules are oxidized to smaller organic compounds, which are subsequently mineralized into H_2O, CO_2, and inorganic salts. Conversely, in alkaline conditions, reduction reactions predominate. Figure 7B illustrates the photocatalytic reaction mechanism of the Ag/CNO/CFO ternary photocatalyst [31]. Under visible light irradiation, both CNO and CFO generate photogenerated electrons and holes. Due to the more negative CB energy level of CNO compared to that of CFO, the photogenerated electrons produced in CNO migrate to the CB of CFO and subsequently transfer to the Ag nanoparticles. Meanwhile, the photogenerated holes can transfer from the VB of CNO to the VB of CFO. Consequently, the electron transfer processes from CNO to Ag and/or from CNO to CFO and then to Ag significantly enhance the separation efficiency of photogenerated charge carriers in the Ag/CNO/CFO ternary heterojunction photocatalyst. In Figure 7C, RGO nanoparticles are stimulated by visible light, leading to the generation of e^- and h^+ [90]. The electrons excited by light are transferred to the CB of the BWZTO triple perovskite, while dissolved oxygen is converted into reactive oxygen species (ROS). Simultaneously, the positively charged h^+ that accumulate on the surface of the MOF particles facilitate the oxidation of water molecules, producing hydroxyl radicals. Ultimately, the generated hydroxyl (·OH) and superoxide ($\cdot O^-{}_2$) species contribute to the degradation of pollutants, such as IMI, into non-toxic byproducts. The photocatalytic degradation mechanism of IMI, as depicted in Figure 7D, involves the excitation of the Bi_2WO_6 semiconductor and NH_2-MIL-88B(Fe) under visible light irradiation, leading to the formation of electron–hole pairs through the promotion of electrons from the VB to the CB. Notably, the VB potential of NH_2-MIL-88B(Fe) (2.68 eV) exceeds that of $OH^-/\cdot OH$ (2.38 eV), facilitating the generation of·OH by OH^- under visible light conditions [91].

5.2. Z-Scheme Photocatalytic Mechanism

Compared to traditional Type I and Type II heterojunction photocatalysts, Z-type heterojunction photocatalysts exhibit broader applications due to their wide light response range and excellent redox capabilities (Figure 8). To construct a visible-light-driven Z-type photocatalyst, the selected semiconductor components must be able to be excited by visible light, and their energy band potentials need to be well-matched to ensure effective recombination of photogenerated electrons (e^-) in the conduction band (CB) of one semiconductor with holes (h^+) in the valence band (VB) of the other semiconductor. For example, the photocatalytic efficiency of the Z-scheme g-C_3N_4/TiO_2 composite is markedly superior to that of TiO_2 or g-C_3N_4 when used independently, as illustrated in Figure 8A. When exposed to UV–Vis light, TiO_2 absorbs photon energy, which excites electrons from the valence band (VB) to the conduction band (CB). Concurrently, the holes generated in the valence band of TiO_2 remain there, while the electrons excited into the conduction band can swiftly transfer to the valence band of g-C_3N_4 due to their close proximity. This transfer allows for the electrons in the valence band of g-C_3N_4 to be further excited into its conduction band. This mechanism effectively enhances the separation of photo-induced electron–hole pairs, thereby improving their redox capabilities.

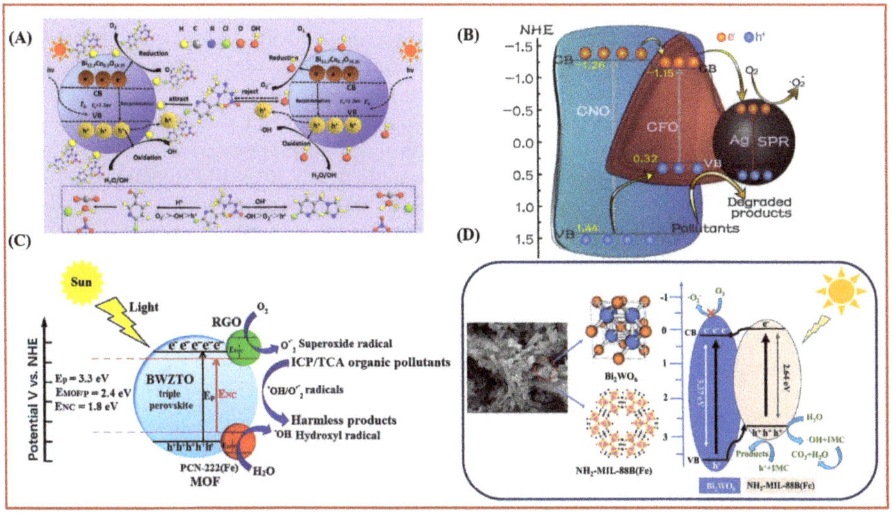

Figure 7. (**A**) Possible mechanisms for $Bi_{12.7}Co_{0.3}O_{19.35}$ photocatalytic degradation of IMI with different pH values under visible light irradiation. (**B**) Schematic illustration of the charge transfer in Ag/CNO/CFO under visible light irradiation. (**C**) Photodegradation process of IMI poison and tetracycline antibiotic organic pollutants by MOFs (PCN-222(Fe)/BWZTO/RGO NC), schematically. (**D**) Possible schematic diagram of the photocatalytic mechanism of 1:2 BNM composite photocatalyst (reproduced with permission from Elsevier, copyright 2022 [38] and copyright 2019 [31] and copyright 2023 [90], and Springer Nature, copyright 2024 [91]).

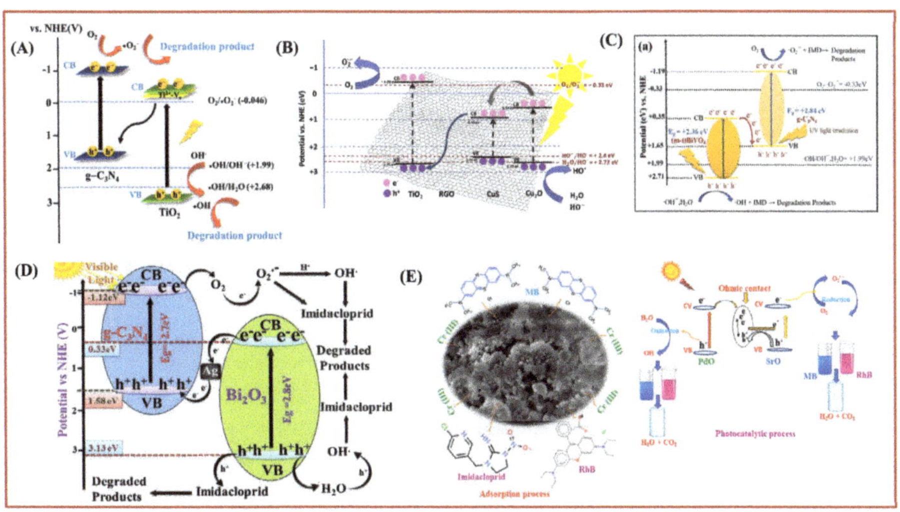

Figure 8. (**A**) Photocatalytic mechanism of 0.5CNS/TiO_2 for degradation of IMI [55]. (**B**) Proposed mechanism of the TiO_2@Cu_2O-CuS heterostructure mediated by RGO for IMD photodegradation under full-spectrum irradiation. (**C**) Photocatalytic mechanism scheme of BVCN10 sample under UV-light irradiation. (**D**) Mechanistic view of IMP degradation using Ag-BO/GCN photocatalyst. (**E**) Photocatalytic and adsorption mechanism of NCP against pollutants compound for clean water production (reproduced with permission from Elsevier, copyright 2024 [79] and copyright 2023 [59] and copyright 2019 [57], and copyright 2024 [80]).

As illustrated in Figure 9A, electrons excited from the valence band (VB) of $Ca_2Fe_2O_5$ to the conduction band bottom (CB) can effectively migrate to the VB of g-CN through a well-constructed Z-scheme heterojunction between g-CN and $Ca_2Fe_2O_5$ [67]. This migration occurs via the CB of $Ca_2Fe_2O_5$, which hinders the recombination of electrons and holes. Consequently, the formation of interface heterojunctions within g-CN/CS composites enhances photocatalytic activity by facilitating the separation of generated electrons and holes. Additionally, the development of AgI nanoparticles (NPs) on In_2S_3 hollow tubes to create a stable Z-scheme heterojunction presents an innovative approach to enhance the intrinsic activity of In_2S_3 hollow tubes while significantly increasing the accessibility of active sites. When photoexcited, electrons from the conduction band of AgI will transfer to the valence band of In_2S_3, where they recombine with positive holes, thereby facilitating the effective separation of conduction band electrons from In_2S_3 and valence band holes from AgI (Figure 9B) [36]. In the Z-type heterojunction shown in Figure 9C, the electrons (e^-) in the CB of Ag_3PO_4 transfer to the VB of Ag_2S, where they recombine with holes (h^+) [30]. The electrons in the conduction band of Ag_2S react with oxygen (O_2) to generate superoxide anions ($\cdot O_2^-$), which subsequently form hydrogen peroxide (H_2O_2). H_2O_2 has a degradation effect on IMI, while the holes in the valence band of Ag_3PO_4 also possess the capability to directly oxidize IMI. In the Z-scheme photocatalytic system illustrated in Figure 9D, the photocatalytic degradation of IMI using the PWO/PI photocatalyst involves not only the direct oxidation of IMI by H+ but also the reaction of conduction band e^- from PI with O_2 to produce $\cdot O_2^-$, which further contributes to the oxidation of IMI [34].

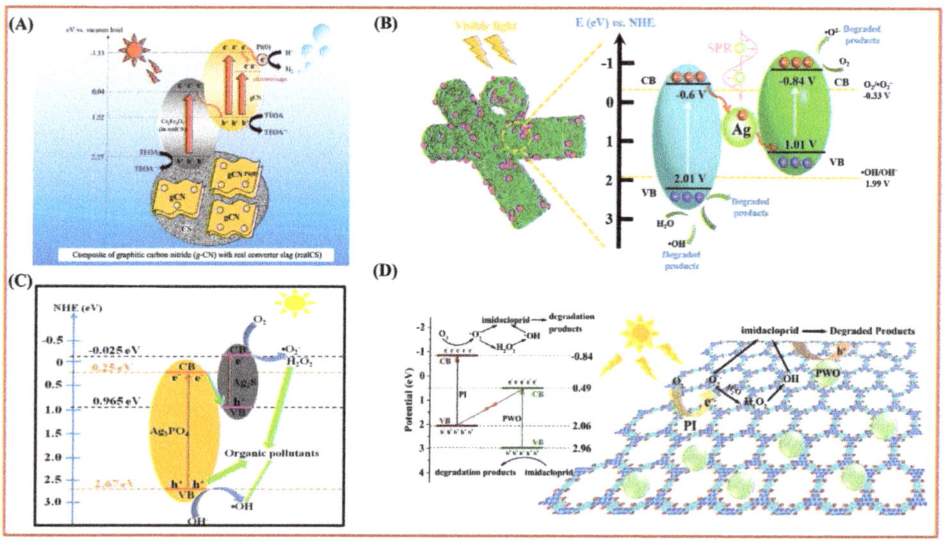

Figure 9. (**A**) Proposed mechanism of photocatalytic hydrogen evolution over g-CN/CS composite. (**B**) Schematic diagram of the interfacial charge migration mechanism over In_2S_3/AgI-300 composite. (**C**) Proposed mechanisms for the $Ag_2S/Fe_3O_4@Ag_3PO_4$ nano-structure. (**D**) Schematic illustration of photocatalytic degradation of IMI over PWO/PI under visible light irradiation (reproduced with permission from Elsevier, copyright 2023 [67] and copyright 2023 [36] and copyright 2019 [30], and copyright 2018 [34]).

Through the analysis above, it is clear that in Type I heterojunctions, the conduction band and valence band of one material are entirely within the energy bands of another material. This structure leads to a high likelihood of photogenerated electrons and holes recombining, which reduces the number of effective photogenerated charge carriers and consequently lowers photocatalytic efficiency. In contrast, Type II heterojunctions feature staggered energy levels between the conduction and valence bands of the two semiconductor materials. In this arrangement, photogenerated electrons and holes migrate to different materials, which partially reduces carrier recombination. However, this can also result in the accumulation of electrons and holes within their respective materials, potentially leading to inefficiencies. Z-type heterojunctions connect the valence band of one semiconductor to the conduction band of another, allowing photogenerated electrons and holes to reside in materials with stronger reducing and oxidizing properties. This configuration maximizes the redox capabilities of the photogenerated electrons and holes while effectively suppressing carrier recombination, significantly enhancing photocatalytic efficiency. Overall, the design of the Z-type heterojunction plays a crucial role in optimizing the performance of photocatalytic materials.

6. Conclusions and Future Perspectives

This review offers a comprehensive examination of the research progress on various solid photocatalysts used for the photocatalytic removal of the pesticide imidacloprid (IMI) from water. It compares different photocatalysts based on their synthesis methods, light sources, and reaction times. Furthermore, this review discusses the mechanisms involved in the photocatalytic degradation of IMI, highlighting the effectiveness of various solid catalysts in this process. By addressing these aspects, this review aims to enhance our understanding of the degradation mechanisms, ultimately guiding future research and development in the field of photocatalytic remediation of water pollutants.

Building on this foundation, future research could prioritize the development of catalysts with enhanced stability to fulfill practical application requirements, improve cost-effectiveness for large-scale deployments, or investigate hybrid or modified catalysts that integrate the discussed materials' properties. These directions will facilitate the advancement of photocatalytic technology in environmental remediation and practical applications. Moreover, future research should investigate the synergistic effects of various solid catalysts to improve degradation efficiency. Focus should be directed towards comprehensive studies of photocatalytic reaction mechanisms to clarify the distinct roles of different catalysts during the process. This will serve as a reference for optimizing catalyst design and enhancing their effectiveness in practical applications. Utilizing advanced characterization techniques can further deepen our understanding of the relationship between catalyst surface properties and reaction performance, providing critical insights for the development of future catalysts.

Author Contributions: S.G.: writing—review and editing; S.L.: writing—review and editing; S.S.: writing—review and editing; M.C.: validation, writing—review and editing. All authors have read and agreed to the published version of the manuscript.

Funding: This work was supported by the Science and Technology Innovation Program of Hunan Province: 2022RC1146; Science Fund for Excellent Young Scholars of Hunan Province: 2022JJ20039; Project of Technical Service: 2024ZKHX383.

Data Availability Statement: No new data were created or analyzed in this study.

Conflicts of Interest: Authors Song Gao and Shao-Fan Sun were employed by the company The 718th Research Institute of CSSC. The remaining authors declare that the research was conducted in the absence of any commercial or financial relationships that could be construed as a potential conflict of interest.

Abbreviations

List of Abbreviations	
ACN	acidified carbon nitride
AIBN	azobisisobutyronitrile
AOPs	advanced oxidation processes
CB	conduction band
CNT	carbon nanotubes
CQD	carbon quantum dots
e^-	electrons
Eg	bandgap
EGDMA	ethylene glycol dimethacrylate
g-CN	graphitic carbon nitride
GO	graphene oxide
h+	holes
HPW	$H_3PW_{12}O_{40}$
IIhv	incident photons
IMI	Imidacloprid
KPW	potassium phosphotungstate
MCN	modified carbon nitride
MIPs	molecularly imprinted polymer
MOFs	metal–organic frameworks
MPW	phosphotungstic melamine
NPs	nanoparticles
OCN	oxygen-doped graphite carbon nitride
PANI	polyaniline
PCN	phosphorus-doped g-CN
PI	polyimide
PMS	peroxymonosulfate
PWO	phosphorous tungsten trioxide
RGO	reduced graphene oxide
ROS	reactive oxygen species
TCN	carbonitride/tungstophosphate acid
TOC	total organic carbon
VB	valence band
ZIF	zeolitic imidazolate framework

References

1. Paleolog, J.; Wilde, J.; Gancarz, M.; Wiącek, D.; Nawrocka, A.; Strachecka, A. Imidacloprid Pesticide Causes Unexpectedly Severe Bioelement Deficiencies and Imbalance in Honey Bees Even at Sublethal Doses. *Animals* **2023**, *12*, 615. [CrossRef] [PubMed]
2. Keshvardoostchokami, M.; Bigverdi, P.; Zamani, A.; Parizanganeh, A.; Piri, F. Silver@ graphene oxide nanocomposite: Synthesize and application in removal of imidacloprid from contaminated waters. *Environ. Sci. Pollut. Res.* **2017**, *25*, 6751–6761. [CrossRef] [PubMed]
3. Singla, A.; Barmota, H.; Kumar Sahoo, S.; Kaur Kang, B. Influence of neonicotinoids on pollinators: A review. *J. Apic. Res.* **2020**, *60*, 19–32. [CrossRef]
4. Kalhor, M.M.; Rafati, A.A.; Rafati, L.; Rafati, A.A. Synthesis, characterization and adsorption studies of amino functionalized silica nano hollow sphere as an efficient adsorbent for removal of imidacloprid pesticide. *J. Mol. Liq.* **2018**, *266*, 453–459. [CrossRef]
5. Bhende, R.S.; Jhariya, U.; Srivastava, S.; Bombaywala, S.; Das, S.; Dafale, N.A. Environmental Distribution, Metabolic Fate, and Degradation Mechanism of Chlorpyrifos: Recent and Future Perspectives. *Biotechnol. Appl. Biochem.* **2022**, *194*, 2301–2335. [CrossRef] [PubMed]
6. Liu, S.; Wang, T.; Lu, J.; Li, Z. Seawater quality criteria derivation and ecological risk assessment for the neonicotinoid insecticide imidacloprid in China. *Mar. Pollut. Bull.* **2023**, *190*, 114871. [CrossRef] [PubMed]
7. Bhende, R.S.; Dafale, N.A. Insights into the ubiquity, persistence and microbial intervention of imidacloprid. *Arch. Microbiol.* **2023**, *205*, 215. [CrossRef] [PubMed]
8. Anderson, N.L.; Harmon-Threatt, A.N. Chronic contact with imidacloprid during development may decrease female solitary bee foraging ability and increase male competitive ability for mates. *Chemosphere* **2021**, *283*, 131177. [CrossRef]

9. Baihetiyaer, B.; Jiang, N.; Li, X.; Song, J.; Wang, J.; Fan, X.; Zuo, Y.; Yin, X. Exploring the toxicity of biodegradable microplastics and imidacloprid to earthworms (*Eisenia fetida*) from morphological and gut microbial perspectives. *Environ. Pollut.* **2023**, *337*, 122547. [CrossRef] [PubMed]
10. Tran, H.D.; Nguyen, D.Q.; Do, P.T.; Tran, U.N.P. Kinetics of photocatalytic degradation of organic compounds: A mini-review and new approach. *RSC Adv.* **2023**, *13*, 16915–16925. [CrossRef] [PubMed]
11. Xin, Y.; Wang, Y.; Jiang, Z.; Deng, B.; Jiang, Z.-J. Advances in the Removal of Organic Pollutants from Water by Photocatalytic Activation of Persulfate: Photocatalyst Modification Strategy and Reaction Mechanism. *ChemSusChem* **2024**, *17*, e202400254. [CrossRef]
12. Mirhosseini, H.; Mostafavi, A.; Shamspur, T.; Sargazi, G. Preparation of novel ternary g-C_3N_4/WO_3/ZnO nanocomposite adsorbent with highly effective imidacloprid removal: Optimization design and a controllable systematic study. *J. Mater. Sci. Mater. Electron.* **2020**, *31*, 17903–17920. [CrossRef]
13. Palariya, D.; Mehtab, S.; Aziz, M.; Zaidi, M.G.H. Exploring Rare Earth Element Doped Nanocomposites as Promising Photocatalysts for Dyes Degradation in Water. *Water Air Soil Pollut.* **2024**, *235*, 286. [CrossRef]
14. Nikbakht Fini, M.; Madsen, H.T.; Muff, J. The effect of water matrix, feed concentration and recovery on the rejection of pesticides using NF/RO membranes in water treatment. *Sep. Purif. Technol.* **2019**, *215*, 521–527. [CrossRef]
15. Xie, Y.; Chen, C.; Ren, X.; Tan, X.; Song, G.; Chen, D.; Alsaedi, A.; Hayat, T. Coupling g-C_3N_4 nanosheets with metal-organic frameworks as 2D/3D composite for the synergetic removal of uranyl ions from aqueous solution. *J. Colloid Interface Sci.* **2019**, *550*, 117–127. [CrossRef] [PubMed]
16. Koe, W.S.; Lee, J.W.; Chong, W.C.; Pang, Y.L.; Sim, L.C. An overview of photocatalytic degradation: Photocatalysts, mechanisms, and development of photocatalytic membrane. *Environ. Sci. Pollut. Res.* **2019**, *27*, 2522–2565. [CrossRef]
17. Faka, V.; Griniezaki, M.; Kiriakidis, G.; Grilla, E.; Mantzavinos, D.; Mao, S.; Shen, S.; Frontistis, Z.; Binas, V. Solar light induced photocatalytic degradation of sulfamethoxazole by $ZnWO_4$/CNNs nanocomposites. *J. Photochem. Photobiol. A* **2022**, *432*, 114108. [CrossRef]
18. Senthil Rathi, B.; Ewe, L.S.; Sanjay, S.; Sujatha, S.; Yew, W.K.; Baskaran, R.; Tiong, S.K. Recent trends and advancement in metal oxide nanoparticles for the degradation of dyes: Synthesis, mechanism, types and its application. *Nanotoxicology* **2024**, *18*, 272–298. [CrossRef] [PubMed]
19. Sahoo, J.K.; Paikra, S.K.; Mishra, M.; Sahoo, H. Amine functionalized magnetic iron oxide nanoparticles: Synthesis, antibacterial activity and rapid removal of Congo red dye. *J. Mol. Liq.* **2019**, *282*, 428–440. [CrossRef]
20. Andronic, L.; Vladescu, A.; Enesca, A. Synthesis, Characterisation, Photocatalytic Activity, and Aquatic Toxicity Evaluation of TiO_2 Nanoparticles. *Nanomaterials* **2021**, *11*, 3197. [CrossRef]
21. Andronic, L.; Lelis, M.; Enesca, A.; Karazhanov, S. Photocatalytic activity of defective black-titanium oxide photocatalysts towards pesticide degradation under UV/VIS irradiation. *Surf. Interfaces* **2022**, *32*, 102123. [CrossRef]
22. Tihana, Č.; Ivana, P.; Ivana, C.; Andreja, G. Nanostructured TiO_2 photocatalyst modified with Cu for improved imidacloprid degradation. *Appl. Surf. Sci.* **2021**, *569*, 151026.
23. Heng, H.; Gan, Q.; Meng, P.; Liu, X. $H_3PW_{12}O_{40}$/TiO_2-In_2O_3: A visible light driven type-II heterojunction photocatalyst for the photocatalytic degradation of imidacloprid. *RSC Adv.* **2016**, *6*, 73301–73307. [CrossRef]
24. Bogatu, C.; Covei, M.; Polo-López, M.I.; Duta, A.; Malato, S. Novel ZnO Photocatalysts for Pollutants' Abatement under Solar Radiation at Pilot Plant Scale. *Catal. Today* **2022**, *413*, 113947. [CrossRef]
25. Kaur, A.; Mehta, V.S.; Kaur, G.; Sud, D. Biopolymer templated strategized greener protocols for fabrication of ZnO nanostructures and their application in photocatalytic technology for phasing out priority pollutants. *Environ. Sci. Pollut. Res.* **2023**, *30*, 25663–25681. [CrossRef]
26. Iqbal, A.; ul Haq, A.; Rios-Aspajo, L.; Iturriaga-Chavez, A. Bio-inspired synthesis of CuO and ZnO nanoparticles by hydrothermal method: Characterization and evaluation as photocatalytic degradation of imidacloprid pesticide. *Glob. Nest J.* **2023**, *25*, 150–158.
27. Kanwal, M.; Tariq, S.R.; Chotana, G.A. Photocatalytic degradation of imidacloprid by Ag-ZnO composite. *Environ. Sci. Pollut. Res.* **2018**, *25*, 27307–27320. [CrossRef] [PubMed]
28. Ullah, I.; Gul, T.; Ali, M.; Khan, I.; Khan, W.; Asghar, H.; Saeed, K. Preparation, Analysis and UV-Accelerated Photocatalytic Degradation of Pesticide Over Mg Doped ZnO/Nylon 6,6/PMMA Ternary Blend. *J. Inorg. Organomet. Polym. Mater.* **2023**, *33*, 3441–3453. [CrossRef]
29. Akbari, S.; Moussavi, G.; Decker, J.; Marin, M.L.; Bosca, F.; Giannakis, S. Superior visible light-mediated catalytic activity of a novel N-doped, Fe_3O_4-incorporating MgO nanosheet in presence of PMS: Imidacloprid degradation and implications on simultaneous bacterial inactivation. *Appl. Catal. B Environ.* **2022**, *317*, 12137. [CrossRef]
30. Shi, E.; Xu, Z.; Wang, W.; Xu, Y.; Zhang, Y.; Yang, X.; Liu, Q.; Zeng, T.; Song, S.; Jiang, Y.; et al. Ag_2S-doped core-shell nanostructures of Fe_3O_4@Ag_3PO_4 ultrathin film: Major role of hole in rapid degradation of pollutants under visible light irradiation. *Chem. Eng. J.* **2019**, *366*, 123–132. [CrossRef]
31. Tang, Y.; Zhang, D.; Li, Y.; Huang, B.; Li, H.; Pu, X.; Geng, Y. Fabrication of magnetically recoverable Ag/$CuNb_2O_6$/$CuFe_2O_4$ ternary heterojunction composite for highly efficient photocatalytic degradation of pollutants. *Sep. Purif. Technol.* **2019**, *220*, 78–88. [CrossRef]

32. Silva, R.R.M.; Valenzuela, L.; Rosal, R.; Ruotolo, L.A.M.; Nogueira, F.G.E.; Bahamonde, A. Peroxymonosulfate activation by Co_3O_4 coatings for imidacloprid degradation in a continuous flow-cell reactor under simulated solar irradiation. *J. Environ. Chem. Eng.* **2023**, *11*, 109265. [CrossRef]
33. Andronic, L.; Mamedov, D.; Cazan, C.; Popa, M.; Chifiriuc, M.C.; Allaniyazov, A.; Palencsar, S.; Karazhanov, S.Z. Cerium oxide thin films: Synthesis, characterization, photocatalytic activity and influence on microbial growth. *Biofouling* **2022**, *38*, 865–875. [CrossRef]
34. Meng, P.; Heng, H.; Sun, Y.; Liu, X. In situ polymerization synthesis of Z-scheme tungsten trioxide/polyimide photocatalyst with enhanced visible-light photocatalytic activity. *Appl. Surf. Sci.* **2018**, *428*, 1130–1140. [CrossRef]
35. Mohanta, D.; Ahmaruzzaman, M. Au-SnO_2-CdS ternary nanoheterojunction composite for enhanced visible light-induced photodegradation of imidacloprid. *Environ. Res.* **2021**, *201*, 111586. [CrossRef]
36. Weng, J.; Chen, J.; Xu, Y.; Hu, X.; Guo, C.; Yang, Y.; Sun, J.; Fu, L.; Wang, Q.; Wei, J.; et al. Engineering highly dispersed AgI nanoparticles on hierarchical In_2S_3 hollow nanotube to construct Z-scheme heterojunction for efficient photodegradation of insecticide imidacloprid. *J. Colloid Interface Sci.* **2023**, *652*, 1367–1380. [CrossRef]
37. Sharma, S.; Basu, S. Highly reusable visible light active hierarchical porous WO_3/SiO_2 monolith in centimeter length scale for enhanced photocatalytic degradation of toxic pollutants. *Sep. Purif. Technol.* **2019**, *231*, 115916. [CrossRef]
38. Fang, L.; Yu-Hang, Z.; Bing, H.; Jiao, Y.; Qi, S.; Fa-Nian, S. Enhanced photocatalytic degradation of imidacloprid and RhB by the precursor derived $Bi_{12.7}Co_{0.3}O_{19.35}$ under different pH value. *J. Phys. Chem. Solids* **2022**, *164*, 110638.
39. Nyankson, E.; Agyei-Tuffour, B.; Adjasoo, J.; Ebenezer, A.; Dodoo-Arhin, D.; Yaya, A.; Mensah, B.; Efavi, J.K. Synthesis and Application of Fe-Doped TiO_2-Halloysite Nanotubes Composite and Their Potential Application in Water Treatment. *Adv. Mater. Sci. Eng.* **2019**, *2019*, 4270310. [CrossRef]
40. Lee, K.; Mazare, A.; Schmuki, P. One-Dimensional Titanium Dioxide Nanomaterials: Nanotubes. *Chem. Rev.* **2014**, *114*, 9385–9454. [CrossRef] [PubMed]
41. Liu, M.; Inde, R.; Nishikawa, M.; Qiu, X.; Atarashi, D.; Sakai, E.; Nosaka, Y.; Hashimoto, K.; Miyauchi, M. Enhanced Photoactivity with Nanocluster-Grafted Titanium Dioxide Photocatalysts. *ACS Nano* **2014**, *8*, 7229–7238. [CrossRef] [PubMed]
42. Eley, C.; Li, T.; Liao, F.; Fairclough, S.M.; Smith, J.M.; Smith, G.; Tsang, S.C.E. Nanojunction-Mediated Photocatalytic Enhancement in Heterostructured CdS/ZnO, CdSe/ZnO, and CdTe/ZnO Nanocrystals. *Angew. Chem. Int. Ed.* **2014**, *53*, 7838–7842. [CrossRef] [PubMed]
43. Liu, J.; Wang, H.; Wu, H.; Yang, Y.; Wang, C.; Wang, Q.; Jia, B.; Zheng, J. Research progress on zinc oxide-based heterojunction photocatalysts. *J. Mater. Chem. A* **2024**, *12*, 20838–20867. [CrossRef]
44. Chen, M.-L.; Li, S.-S.; Wen, L.; Xu, Z.; Li, H.-H.; Ding, L.; Cheng, Y.-H. Exploration of double Z-type ternary composite long-afterglow/graphitic carbon nitride@metal–organic framework for photocatalytic degradation of methylene blue. *J. Colloid Interface Sci.* **2023**, *629*, 409–421. [CrossRef] [PubMed]
45. Xu, H.; Yi, J.; She, X.; Liu, Q.; Song, L.; Chen, S.; Yang, Y.; Song, Y.; Vajtai, R.; Lou, J.; et al. 2D heterostructure comprised of metallic 1T-MoS_2 Monolayer O-g-C_3N_4 towards efficient photocatalytic hydrogen evolution. *Appl. Catal. B* **2018**, *220*, 379–385. [CrossRef]
46. Thaweesak, S.; Lyu, M.; Peerakiatkhajohn, P.; Butburee, T.; Luo, B.; Chen, H.; Wang, L. Two-dimensional g-$C_3N_4/Ca_2Nb_2TaO_{10}$ nanosheet composites for efficient visible light photocatalytic hydrogen evolution. *Appl. Catal. B* **2017**, *202*, 184–190. [CrossRef]
47. Stefa, S.; Skliri, E.; Gagaoudakis, E.; Kiriakidis, G.; Kotzias, D.; Papagiannakopoulos, P.; Konsolakis, M.; Mao, S.; Binas, V. Visible light photocatalytic oxidation of NO using g-C_3N_4 nanosheets: Stability, kinetics, and effect of humidity. *Appl. Phys. A* **2024**, *130*, 451. [CrossRef]
48. Sh, X.; Wu, J.; Zhong, J.; Xu, H.; Yang, Y.; Vajtai, R.; Lou, J.; Liu, Y.; Du, D.; Li, H.; et al. Oxygenated monolayer carbon nitride for excellent photocatalytic hydrogen evolution and external quantum efficiency. *Nano Energy* **2016**, *27*, 138–146. [CrossRef]
49. Fu, J.; Yu, J.; Jiang, C.; Cheng, B. g-C_3N_4-Based Heterostructured Photocatalysts. *Adv. Energy Mater.* **2017**, *8*, 1701503. [CrossRef]
50. Stefa, S.; Zografaki, M.; Dimitropoulos, M.; Paterakis, G.; Galiotis, C.; Sangeetha, P.; Kiriakidis, G.; Konsolakis, M.; Binas, V. High surface area g-C_3N_4 nanosheets as superior solar-light photocatalyst for the degradation of parabens. *Appl. Phys. A* **2023**, *129*, 754. [CrossRef]
51. Liu, X.; Li, C.; Zhang, B.; Yuan, M.; Ma, Y.; Kong, F. A facile strategy for photocatalytic degradation of seven neonicotinoids over sulfur and oxygen co-doped carbon nitride. *Chemosphere* **2020**, *253*, 126672. [CrossRef] [PubMed]
52. Sudhaik, A.; Raizada, P.; Thakur, S.; Saini, R.V.; Saini, A.K.; Singh, P.; Kumar Thakur, V.; Nguyen, V.-H.; Khan, A.A.P.; Asiri, A.M. Synergistic photocatalytic mitigation of imidacloprid pesticide and antibacterial activity using carbon nanotube decorated phosphorus doped graphitic carbon nitride photocatalyst. *J. Taiwan Inst. Chem. Eng.* **2020**, *113*, 142–154. [CrossRef]
53. Faisal, M.; Rashed, M.A.; Ahmed, J.; Alsaiari, M.; Jalalah, M.; Alsareii, S.A.; Harraz, F.A. Au nanoparticles decorated polypyrrole-carbon black/g-C_3N_4 nanocomposite as ultrafast and efficient visible light photocatalys. *Chemosphere* **2021**, *287*, 131984. [CrossRef] [PubMed]
54. Garg, R.; Gupta, R.; Bansal, A. Photocatalytic degradation of imidacloprid using semiconductor hybrid nano-catalyst: Kinetics, surface reactions and degradation pathways. *Int. J. Environ. Sci. Te* **2020**, *18*, 1425–1442. [CrossRef]
55. Kobkeatthawin, T.; Trakulmututa, J.; Amornsakchai, T.; Kajitvichyanukul, P.; Smith, S.M. Identification of Active Species in Photodegradation of Aqueous Imidacloprid over g-C_3N_4/TiO_2 Nanocomposites. *Catalysis* **2022**, *12*, 120. [CrossRef]
56. Liang, S.; Zhang, D.; Pu, X.; Yao, X.; Han, R.; Yin, J.; Ren, X. A novel Ag_2O/g-C_3N_4 p-n heterojunction photocatalysts with enhanced visible and near-infrared light activity. *Sep. Purif. Technol.* **2019**, *210*, 786–797. [CrossRef]

57. Raizada, P.; Sudhaik, A.; Singh, P.; Hosseini-Bandegharaei, A.; Gupta, V.K.; Agarwal, S. Silver-mediated Bi_2O_3 and graphitic carbon nitride nanocomposite as all solid state Z scheme photocatalyst for imidacloprid pesticide abatement from water. *Desalination Water Treat.* **2019**, *171*, 344–355. [CrossRef]
58. Kundu, A.; Sharma, S.; Basu, S. ModulModulated BiOCl nanoplates with porous g-C_3N_4 nanosheets for photocatalytic degradation of color/colorless pollutants in natural sunlight. *J. Phys. Chem. Solids* **2021**, *154*, 100064. [CrossRef]
59. Patial, B.; Bansal, A.; Gupta, R.; Mittal, S.K. Hydrothermal synthesis of (m-t)$BiVO_4$/g-C_3N_4 heterojunction for enhancement in photocatalytic degradation of imidacloprid. *J. Environ. Chem. Eng.* **2023**, *11*, 111138. [CrossRef]
60. Tan, J.; Li, Z.; Li, J.; Meng, Y.; Yao, X.; Wang, Y.; Lu, Y.; Zhang, T. Visible-light-assisted peroxymonosulfate activation by metal-free bifunctional oxygen-doped graphitic carbon nitride for enhanced degradation of imidacloprid: Role of non-photochemical and photocatalytic activation pathway. *J. Hazard. Mater.* **2021**, *423*, 127048. [CrossRef]
61. Sun, Y.; Meng, P.; Liu, X. Self-assembly of tungstophosphoric acid/acidified carbon nitride hybrids with enhanced visible-light-driven photocatalytic activity for the degradation of imidacloprid and acetamiprid. *Appl. Surf. Sci.* **2018**, *456*, 259–269. [CrossRef]
62. Sun, Y.; Liu, X. Efficient visible-light photocatalytic degradation of imidacloprid and acetamiprid using a modified carbon nitride/tungstophosphoric acid composite induced by a nucleophilic addition reaction. *Appl. Surf. Sci.* **2019**, *485*, 423–431. [CrossRef]
63. Xiao, Z.; Huang, X.; Tang, H.; Liu, X. Modified carbon nitride incorporating Keggin-type potassium phosphotungstate enhances photocatalytic degradation of imidacloprid under visible light. *Mater. Sci. Semicond. Process* **2024**, *171*, 108001. [CrossRef]
64. Meng, P.; Huang, J.; Liu, X. In-situ solid phase thermal transformation of self-assembled melamine phosphotungstates produce efficient visible light photocatalysts. *J. Colloid Interface Sci.* **2019**, *551*, 208–218. [CrossRef] [PubMed]
65. Balakumar, V.; Ramalingam, M.; Sekar, K.; Chuaicham, C.; Sasaki, K. Fabrication and characterization of carbon quantum dots decorated hollow porous graphitic carbon nitride through polyaniline for photocatalysis. *Chem. Eng. J.* **2021**, *426*, 131739. [CrossRef]
66. Sudhaik, A.; Raizada, P.; Singh, P.; Hosseini-Bandegharaei, A.; Thakur, V.K.; Nguyen, V.-H. Highly effective degradation of imidacloprid by H_2O_2/fullerene decorated P-doped g-C_3N_4 photocatalyst. *J. Environ. Chem. Eng.* **2020**, *8*, 104483. [CrossRef]
67. Inoue, T.; Chuaicham, C.; Saito, N.; Ohtani, B.; Sasaki, K. Z-scheme heterojunction of graphitic carbon nitride and calcium ferrite in converter slag for the photocatalytic imidacloprid degradation and hydrogen evolution. *J. Photochem. Photobiol. A* **2023**, *440*, 114644. [CrossRef]
68. Zhang, T.; Zhao, D.; Wang, Y.; Chang, Y.; Zhang, D.; Tang, Y.; Pu, X.; Shao, X. Facial synthesis of a novel $Ag_4V_2O_7$/g-C_3N_4 heterostructure with highly efficient photoactivity. *J. Am. Ceram.* **2019**, *102*, 3897–3907. [CrossRef]
69. Huang, X.; Wang, W.; Liu, X. $H_3PW_{12}O_{40}$-doped pyromellitic diimide prepared via thermal transformation as an efficient visible-light photocatalyst. *J. Mater. Sci.* **2020**, *55*, 8502–8512. [CrossRef]
70. Abuzeyad, O.H.; El-Khawaga, A.M.; Tantawy, H.; Elsayed, M.A. An evaluation of the improved catalytic performance of rGO/GO-hybrid-nanomaterials in photocatalytic degradation and antibacterial activity processes for wastewater treatment: A review. *J. Mol. Struct.* **2023**, *1288*, 135787. [CrossRef]
71. Xia, X.; Song, M.; Wang, H.; Zhang, X.; Sui, N.; Zhang, Q.; Colvin, V.L.; Yu, W.W. Latest progress in constructing solid-state Z scheme photocatalysts for water splitting. *Nanoscale* **2019**, *11*, 11071–11082. [CrossRef] [PubMed]
72. Gul, M.M.; Ahmad, K.S. Review elucidating graphene derivatives (GO/rGO) supported metal sulfides based hybrid nanocomposites for efficient photocatalytic dye degradation. *Rev. Inorg. Chem.* **2021**, *42*, 337–354. [CrossRef]
73. Feng, J.; Ye, Y.; Xiao, M.; Wu, G.; Ke, Y. Synthetic routes of the reduced graphene oxide. *Chem. Pap.* **2020**, *74*, 3767–3783. [CrossRef]
74. Wang, Z.-M.; Hirotsu, T.; Wu, H.; Kanoh, H. Advantaging Synergy Photocatalysis with Graphene-Related Carbon as a Counterpart Player of Titania. *Chem. Rec.* **2018**, *19*, 1393–1406. [CrossRef] [PubMed]
75. El-Shafai, N.M.; El-Khouly, M.E.; El-Kemary, M.; Ramadan, M.S.; Derbalah, A.S.; Masoud, M.S. Fabrication and characterization of graphene oxide-titanium dioxide nanocomposite for degradation of some toxic insecticides. *J. Ind. Eng. Chem.* **2019**, *69*, 315–323. [CrossRef]
76. Piao, M.; Sun, Y.; Wang, Y.; Teng, H. Preparation of $BiVO_4$/RGO-TNT Nanomaterials for Efficient and Recyclable Photocatalysis of Imidacloprid Insecticide. *Chem. Sel.* **2022**, *7*, e202200182. [CrossRef]
77. El-Shafai, N.M.; El-Shaer, A.; Eraky, M.R.; Ibrahim, M.M.; Ramadan, M.S.; El-Mehasseb, I.M. Enhancing electron density, electrochemical, and dielectric properties of nanohybrid materials for advanced photocatalytic antifouling and energy storage. *Diam. Relat. Mater.* **2021**, *119*, 108543. [CrossRef]
78. Behera, L.; Barik, B.; Mohapatra, S. Improved photodegradation and antimicrobial activity of hydrothermally synthesized 0.2Ce-TiO_2/RGO under visible light. *Colloid Surf. A* **2021**, *620*, 126553. [CrossRef]
79. Andronic, L.; Abreu-Jaureguí, C.; Silvestre-Albero, J. Construction of TiO_2@Cu_2O-CuS heterostructures integrating RGO for enhanced full-spectrum photocatalytic degradation of organic pollutants. *J. Alloys Compd.* **2024**, *994*, 174628. [CrossRef]
80. El-Gohary, R.M.; El-Shafai, N.M.; El-Mehasseb, I.M.; Mostafa, Y.S.; Alamri, S.A.; Beltagi, A.M. Removal of pollutants through photocatalysis, adsorption, and electrochemical sensing by a unique plasmonic structure of palladium and strontium oxide nanoparticles sandwiched between 2D nanolayers. *J. Environ. Manag.* **2024**, *363*, 121257. [CrossRef]
81. Soltani-nezhad, F.; Saljooqi, A.; Shamspur, T.; Mostafavi, A. Photocatalytic degradation of imidacloprid using GO/Fe_3O_4/TiO_2-NiO under visible radiation: Optimization by response level method. *Polyhedron* **2019**, *165*, 188–196. [CrossRef]

82. Malik, A.Q.; Tabasum, S.; Rani, S.; Lokhande, P.; Singh, P.P.; Mooney, J.; Singh, J.; Alberto, H.-A.C.; Sharma, A.; Aepuru, R.; et al. Fluorescent CdS QDs Modified With Molecular Imprinted Polymer for the Photodegradation of Imidacloprid and Buprofezin Pesticides Under Visible Light. *J. Inorg. Organomet. Polym. Mater.* **2023**, *33*, 3468–3484. [CrossRef]
83. Bressi, V.; Balu, A.M.; Iannazzo, D.; Espro, C. Recent advances in the synthesis of carbon dots from renewable biomass by high-efficient hydrothermal and microwave green approaches. *Curr. Opin. Green Sustain. Chem.* **2022**, *40*, 100742. [CrossRef]
84. Wang, L.; Wang, T.; Hao, R.; Wang, Y. Synthesis and applications of biomass-derived porous carbon materials in energy utilization and environmental remediation. *Chemosphere* **2023**, *339*, 139635. [CrossRef]
85. Roopan, S.M.; Prakash, S.H.; Manjupriya, R.; Afridha, M.S.H.F.; Rajesh, A.; Sneha, R.; Kumar, P.V.; Shobika, M. Biomass-derived carbon quantum dots-supported metal oxide composite for the photocatalytic degradation of toxic pollutants. *Biomass Convers. Biorefinery* **2024**, 054227. [CrossRef]
86. Wang, X.; Zhang, W.; Wei, C.; Li, R.; Guo, J.; Liu, B. Titanium incorporated and g-C_3N_4-coated NH_2-UiO-66 for enhanced photocatalytic hydrogen evolution. *Appl. Phys. A* **2020**, *126*, 594. [CrossRef]
87. Guo, S.; Kong, L.-H.; Wang, P.; Yao, S.; Lu, T.-B.; Zhang, Z.-M. Switching Excited State Distribution of Metal–Organic Framework for Dramatically Boosting Photocatalysis. *Angew. Chem. Int. Ed.* **2022**, *64*, e202206193.
88. Bogdan, L.; Palčić, A.; Duplančić, M.; Leskovac, M.; Tomašić, V. Eco-Friendly Synthesis of TiO_2/ZIF-8 Composites: Characterization and Application for the Removal of Imidacloprid from Wastewater. *Processes* **2023**, *11*, 963. [CrossRef]
89. Gecgel, C.; Simsek, U.B.; Gozmen, B.; Turabik, M. Comparison of MIL-101(Fe) and amine-functionalized MIL-101(Fe) as photocatalysts for the removal of imidacloprid in aqueous solution. *J. Iran. Chem. Soc.* **2019**, *16*, 1735–1748. [CrossRef]
90. Kadkhodayan, H.; Alizadeh, T. Manufacturing visible-light-driven heterojunction photocatalyst based on MOFs/$Bi_2WZnTiO_9$ triple perovskite/carbonous materials for efficient removal of poisons, antibiotics, and inorganic pollutants. *J. Phys. Chem. Solids* **2023**, *183*, 111620. [CrossRef]
91. Chen, M.-L.; Lu, T.-H.; Li, S.-S.; Wen, L.; Xu, Z.; Cheng, Y.-H. Photocatalytic degradation of imidacloprid by optimized Bi_2WO_6/NH_2-MIL-88B(Fe) composite under visible light. *Environ. Sci. Pollut.* **2022**, *29*, 19583–19593. [CrossRef] [PubMed]
92. Li, S.-S.; Wen, L.; He, S.-W.; Xu, Z.; Ding, L.; Cheng, Y.-H.; Chen, M.-L. Enhanced photocatalytic degradation of imidacloprid by a simple Z-type binary heterojunction composite of long afterglow with metal-organic framework. *Catal. Commun.* **2023**, *183*, 106775. [CrossRef]
93. Chen, M.-L.; Lu, T.-H.; Long, L.-L.; Xu, Z.; Ding, L.; Cheng, Y.-H. NH_2-Fe-MILs for effective adsorption and Fenton-like degradation of imidacloprid: Removal performance and mechanism investigation. *Environ. Eng. Res.* **2022**, *27*, 200702. [CrossRef]
94. Luo, J.; Zhang, X.; Lu, J.; Zhang, J. Fine Tuning the Redox Potentials of Carbazolic Porous Organic Frameworks for Visible-Light Photoredox Catalytic Degradation of Lignin β-O-4 Models. *ACS Catal.* **2017**, *7*, 5062–5070. [CrossRef]
95. Deng, F.; Peng, J.; Li, X.; Luo, X.; Ganguly, P.; Pillai, S.C.; Ren, B.; Ding, L.; Dionysiou, D.D. Metal sulfide-based Z-scheme heterojunctions in photocatalytic removal of contaminants, H_2 evolution and CO_2 reduction: Current status and future perspectives. *J. Clean. Prod.* **2023**, *416*, 137957. [CrossRef]
96. Yuan, Y.; Guo, R.T.; Hong, L.F.; Ji, X.Y.; Lin, Z.D.; Li, Z.S.; Pan, W.G. A review of metal oxide-based Z-scheme heterojunction photocatalysts: Actualities and developments. *Mater. Today Energy* **2021**, *21*, 100829. [CrossRef]
97. Li, H.; Tu, W.; Zhou, Y.; Zou, Z. Z-Scheme Photocatalytic Systems for Promoting Photocatalytic Performance: Recent Progress and Future Challenges. *Adv. Sci.* **2016**, *3*, 1500389. [CrossRef] [PubMed]
98. Xu, Q.; Zhang, L.; Yu, J.; Wageh, S.; Al-Ghamdi, A.A.; Jaroniec, M. Direct Z-scheme photocatalysts: Principles, synthesis, and applications. *Mater. Today* **2018**, *21*, 1042–1063. [CrossRef]

catalysts

MDPI

Article

Photocatalytic Degradation of Tetracycline Hydrochloride Using TiO_2/CdS on Nickel Foam Under Visible Light and RSM–BBD Optimization

Kefu Zhu [1], Lizhe Ma [1,2], Jieli Duan [1,*], Zhiyong Fang [1] and Zhou Yang [1,3,*]

1 College of Engineering, South China Agricultural University, Guangzhou 510642, China; prezkf@163.com (K.Z.); lizhe_ma@163.com (L.M.); fangzhiyong0207@163.com (Z.F.)
2 School of Intelligent Engineering, Shaoguan University, Shaoguan 512005, China
3 School of Mechanical Engineering, Guangdong Ocean University, Zhanjiang 524088, China
* Correspondence: duanjieli@scau.edu.cn (J.D.); yangzhou@scau.edu.cn (Z.Y.)

Abstract: This study investigates the photocatalytic degradation of tetracycline hydrochloride (TCH) using a TiO_2/CdS composite nanocatalyst synthesized on flexible nickel foam via a dipping–pull method. By comparing the photocatalytic degradation of TCH by TiO_2/CdS with different precursor ratios, it was found that TiO_2/CdS-1.43% exhibited better photocatalytic degradation performance. The X-ray diffraction (XRD) pattern of the TiO_2/CdS composite retains the characteristic peaks of both TiO_2 and CdS, indicating the successful formation of the composite. According to the analysis of ultraviolet–visible spectroscopy (UV–Vis), the absorption edge of TiO_2/CdS is approximately 530 nm. The transmission electron microscopy (TEM) images show Cd and S evenly, densely distributed in TiO_2/CdS, further validating its successful synthesis. X-ray photoelectron spectroscopy (XPS) analysis reveals that Cd and Ti elements exist in the forms of Cd^{2+} and Ti^{4+}, respectively. TiO_2/CdS loading uniformity on the nickel foam was assessed using super-depth microscopy. The removal efficiency of 10 L of 20 mg/L TCH solution achieved 53.89%, respectively, under response surface methodology—Box–Behnken design (RSM–BBD) optimal conditions (28 g catalyst, 325 rpm, pH = 9.04 within 150 min). Furthermore, five successive cycling experiments demonstrated strong stability, with a catalyst loss of only 4.44%. Finally, free radical scavenging experiments revealed that $\cdot O_2^-$ radicals are the primary active species. This study highlights the potential of TiO_2/CdS composites supported on nickel foam for efficient photocatalytic degradation of antibiotic pollutants in water.

Keywords: photocatalysis; antibiotics degradation; tetracycline hydrochloride; TiO_2/CdS; nickel foam; RSM–BBD optimization

1. Introduction

Antibiotic pollution threatens global water environments, particularly with TCH, widely used in medicine and agriculture, significantly impacting human health and the ecological environment [1,2]. The persistence of TCH in water bodies can foster antibiotic-resistant bacteria, leading to long-term risks [3]. Various technological approaches are currently employed, including membrane filtration, material adsorption, and photocatalytic oxidation. Membrane filtration technology utilizes porous membranes to efficiently remove antibiotics from wastewater [4,5]. Material adsorption techniques use modified activated carbon, biochar, and other materials to adsorb antibiotics from water [6,7]. Photocatalytic oxidation is considered one of the most promising methods for removing antibiotics from wastewater due to its simplicity, lack of secondary pollution, and strong oxidation-reduction capabilities [8,9].

TiO_2 has garnered extensive research attention due to its high efficiency, low cost, non-toxicity, and strong physical stability [10–12]. For instance, TiO_2 demonstrates excellent photocatalytic efficiency, with reaction rates often exceeding 10^{-2} to 10^{-3} mol·L^{-1}·min^{-1} in the degradation of organic pollutants like phenol and methyl orange. Additionally, TiO_2 exhibits a high quantum efficiency of 50–80% under UV light, indicating its ability to convert a significant proportion of incoming photons into useful electron–hole pairs, thereby boosting catalytic activity. TiO_2 maintains its integrity in extreme conditions, with a melting point of 1832 °C and a pH stability range of pH 3–12, enabling its use in a variety of environments, from acidic to alkaline. These properties, combined with TiO_2's non-toxic nature, make it an ideal material for applications in photocatalysis. However, TiO_2 primarily acts as a photocatalyst activated by ultraviolet light, and ultraviolet energy constitutes only a portion of solar energy. Moreover, the recombination of light-induced electron–hole pairs is a major cause of energy loss in photocatalytic reactions, leading to reduced photocatalytic efficiency and energy utilization [13–15]. Combining TiO_2 with other semiconductors to form heterojunctions enhances the light absorption range and the generation of photoinduced electron–hole pairs, making it one of the effective methods to address the low photocatalytic activity of TiO_2 under visible light [16,17].

CdS is a photocatalyst dopant that exhibits strong visible light responsiveness, high photocatalytic activity, and excellent compatibility [18,19]. This is primarily due to CdS favorable bandgap, which is typically around 2.4–2.5 eV, allowing them to absorb visible light (around 400–500 nm) efficiently. Integrating CdS into TiO_2 results in a co-doping system with a compatible band structure, which introduces new energy levels within the TiO_2 bandgap, narrows the bandgap width, and creates oxygen vacancies that boost catalytic activity [20–22]. Compared with traditional photocatalysts, TiO_2/CdS features smaller microcrystal sizes, larger surface areas, a more favorable distribution of defect density, higher surface-active species density, and improved charge carrier separation [23–25].

Nickel foam, a three-dimensional metal foam structure, is characterized by high electrical conductivity that enables rapid electron transfer, along with excellent stability and flexibility, making it well suited for use in various structured photocatalytic reactors [26–28]. To overcome challenges associated with powdered catalysts, such as loss and clogging in water systems, TiO_2/CdS can be effectively loaded onto nickel foam, thereby improving both catalyst recoverability and catalytic performance [29]. Additionally, the porous structure of nickel foam significantly increases the specific surface area of the catalyst, providing more active sites; the porosity of nickel foam facilitates the uniform distribution of reactants and light, promoting contact between the reactants and the catalyst [30–32].

The TiO_2/CdS nanocomposite, formed by integrating TiO_2 and CdS, capitalizes on the remarkable stability of TiO_2 and the efficient visible—light absorption capacity of CdS. When irradiated by visible or UV light, electron–hole pairs are generated in CdS, and the photoexcited electrons can transfer to TiO_2, improving the charge separation efficiency and minimizing recombination. Remarkable achievements have been made in recent research. For instance, an efficient IO-TiO_2-CdS composite with a honeycomb-like 3D inverse opal TiO_2 structure was synthesized, which could degrade over 99% of tetracycline hydrochloride (30 mg/L) within 20 min under visible light irradiation [33]. Another study focused on the beta zeolite-supported Ti^{3+}-TiO_2/CdS heterojunction. Zeolite served as an effective support, leading to a relatively high surface area of approximately 63.2 m^2 g^{-1}. This heterojunction achieved a photocatalytic degradation efficiency of 78% in merely 3 min for tetracycline, with a corresponding rate constant as high as 0.788 min^{-1}, far exceeding that of the individual components [34]. Using carbon fibers (CFs) as the fixing substrate, TiO_2/CdS heterojunction was constructed on the CF surface. The resulting CFs/TiO_2/CdS bundles could be woven into macroscopic cloth, which demonstrated remarkable photocatalytic activities, degrading a variety of pollutants such as 95.44% methylene blue, 64.95% acid orange 7, 91.37% tetracycline hydrochloride, and removing 90.70% hexavalent chromium after 120 min under visible light irradiation [35].

In this study, we applied a TiO_2/CdS film onto nickel foam using the dipping–pull method. TiO_2/CdS composite was characterized through XRD, UV–Vis, TEM, and XPS analyses. The loading state of TiO_2/CdS on the nickel foam was observed using super-depth microscopy. The photocatalytic performance of the TiO_2/CdS catalyst for degrading organic pollutants, with TCH as the target contaminant, was evaluated in a medium-scale reactor. The effect of the TiO_2 to CdS precursor ratio on the degradation of the TCH solution was investigated. To maximize the degradation rate, RSM–BBD was employed to optimize key operational parameters in the water treatment process, including catalyst dosage, stirring speed, and pH, while establishing relevant models. Furthermore, the reusability, recoverability, and major active species of the TiO_2/CdS composite were assessed. Based on this, the photocatalytic reaction mechanism of TiO_2/CdS was analyzed. This research provides insights into the practical application of TiO_2/CdS composites supported on nickel foam for the photocatalytic degradation of antibiotic pollutants in water.

2. Experiment

2.1. Chemicals and Materials

CdS (purity 99%), $Ti(OBu)_4$ (purity 99%), and TCH were purchased from Shanghai Macklin Biochemical Technology Co., Ltd. (Shanghai, China). Ethylene glycol (purity 99%) was purchased from Tianjin Fuyu Fine Chemicals Co., Ltd. (Tianjin, China). Glacial acetic acid (purity 99%) was purchased from Shanghai Runjie Technology Development Co., Ltd. (Shanghai, China). NaOH (purity 96%) was purchased from Tianjin Fuchen Chemical Reagents Co., Ltd. (Tianjin, China). HCl (purity 10%) was purchased from Shenzhen Bida Environmental Protection Technology Co., Ltd. (Shenzhen, China). Nickel foam mesh was sourced from Quanzhou Yunzongcheng New Materials Co., Ltd. (Quanzhou, China). Benzoquinone (BQ), isopropanol alcohol (IPA), and sodium oxalate (AO) were purchased from Aladdin Reagent Co., Ltd. (Shanghai, China). All chemicals are analytical grade and used without further purification. The light source used in the experiment was purchased from Shenzhen Qinming Technology Co. Ltd. (Shenzhen, China). Deionized water used in the experiments was made in our lab.

2.2. Preparation Process

The preparation process is shown in Figure 1. Titanium tetrabutoxide was mixed with anhydrous ethanol in a volume ratio of 1:4 and subjected to ultrasonic oscillation for 5 min. A specific mass of cadmium sulfide powder was weighed using an electronic balance, ensuring that the mass ratio of CdS in the mixed solution was 1.43%. The TiO_2 used in this experiment was maintained at a mass fraction of 7.02% in the mixed solution. Ice acetic acid solution was added to adjust the pH to approximately 4, followed by continuous ultrasonic treatment for 10 min. A magnetic stirrer was used to stir the solution at a constant temperature of 60 °C for 30 min. Then, it was sealed and aged for over 24 h to obtain a stable, homogeneous, and clear orange–yellow solution.

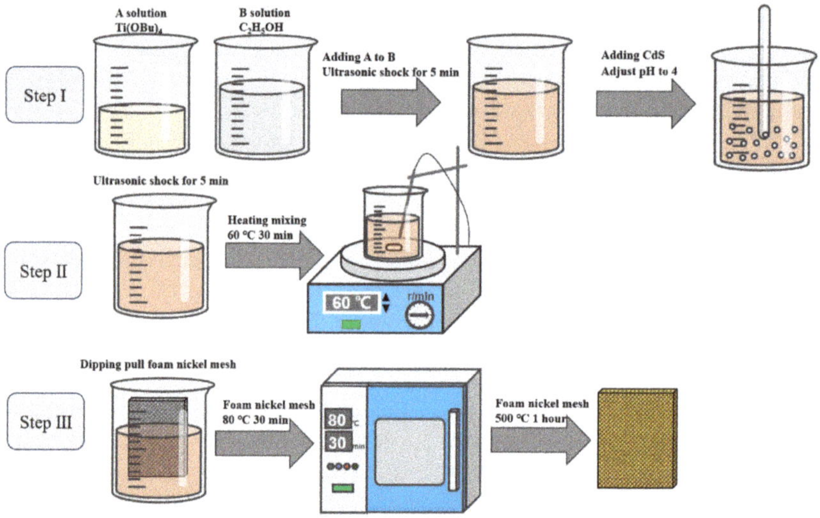

Figure 1. Schematic diagram of the preparation of TiO_2/CdS and its loading onto nickel foam.

The nickel foam was cut into a rectangle measuring 215 mm × 140 mm. Then, the nickel foam was soaked in anhydrous ethanol for 1 h under ultrasonic oscillation to clean it and remove the oxidation layer from the surface. It was then kept in anhydrous ethanol for an additional 24 h. After soaking, the nickel foam was placed in a beaker containing deionized water and subjected to ultrasonic treatment for 30 min for further cleaning. The nickel foam was then heat-treated in a muffle furnace at 400 °C for 10 min. Comparing before and after treatment, the pores on the nickel foam became more permeable, allowing for a larger surface area to load the catalyst. The film coating was applied to the nickel foam using the impregnation method. After coating, the foam was placed in a drying oven at 80 °C to dry, and the coating process was repeated to achieve the desired catalyst loading. Finally, the nickel foam loaded with the catalyst was placed in the muffle furnace and held at 500 °C for 60 min to enhance the bonding strength, successfully loading TiO_2/CdS onto the nickel foam.

2.3. Characterization

The powder X-ray diffraction (XRD) patterns were recorded using an XD-3 (Shimadzu, Kyoto, Japan) with Cu-Kα radiation. The UV–Vis diffuse reflectance spectra were obtained directly using a UV–Vis spectrophotometer (UV-2550, Shimadzu, Kyoto, Japan) with $BaSO_4$ as a reference. X-ray surface morphology and dispersibility of the samples were recorded by high-resolution transmission electron microscopy (Tecnai F20, Portland, OR, America). Photoelectron spectroscopy (XPS) was conducted using an AXIS-ULTRA DLD (Shimadzu, Kyoto, Japan) with an Al Kα X-ray source (hν = 1486.8 eV). The nickel foam loaded withTiO_2/CdS was observed and analyzed using a super depth microscope. (VHX-6000, Keyence Corporation, Osaka, Japan).

2.4. Photocatalytic Experiments

The photocatalytic degradation experiment was conducted in a 10 L cylindrical acrylic glass photoreactor, which was made of polymethyl methacrylate, as shown in Figure 2. A 24 W visible-light LED array, with its emission wavelength peaking at around 450 nm, was selected as the light source. The central layout of the LED array ensures that the visible light uniformly illuminates the surrounding nickel foam, improving the uniformity of light exposure during the reaction. A magnetic stirrer drives the stirring bar to mix the TCH solution, facilitating contact between the TCH and the active sites of TiO_2/CdS on the nickel foam. The nickel foam loaded with the TiO_2/CdS composite nanocatalyst was fixed around the light source using a mounting fixture, with a distance of 100 mm from the foam to the center of the LED array.

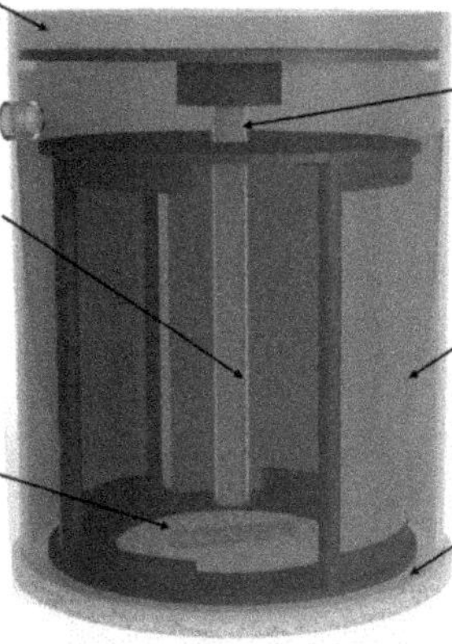

Figure 2. Schematic diagram of the photocatalytic reactor.

Ten liters of a 20 mg/L TCH solution was added to the reactor and stirred for 30 min. In the laboratory, maintain both the room temperature and the solution temperature at 25 °C each time. This helps minimize the impact of temperature fluctuations on the experiment. The pH of the TCH solution was adjusted to the desired value using 10% HCl or NaOH pellets. According to the preliminary experiment, the adsorption–desorption equilibrium between the catalyst and the target pollutant was reached in 30 min. Before the dark reaction, a 10 mL sample was extracted from the reactor for concentration analysis, and the solution was stirred in a dark environment for 30 min to achieve adsorption/desorption equilibrium between the photocatalyst and the TCH molecules. During the light reaction, 10 mL samples were extracted every 20 min from the reactor for concentration analysis.

The concentration of TCH in the solution was measured using an ultraviolet–visible spectrophotometer. Following the methodology of previous studies, the concentration changes in TCH were assessed using this spectrophotometer. The solution reached adsorption–desorption equilibrium within 30 min after adding TiO_2/CdS, after which the degradation stabilized. Therefore, the adsorption equilibrium time was designated as 30 min. The photocatalytic activity of the TiO_2/CdS composite material was evaluated under the illumination of the 24 W visible light LED array. The degradation rate percentage was calculated using Equation (1).

$$D = (1 - \frac{C_t}{C_0}) \times 100\% \tag{1}$$

where D represents the degradation rate of antibiotics at time t in the photocatalytic degradation system. C_0 is the initial concentration of TCH, and C_t is the concentration of TCH at time t.

2.5. Experimental Design and Optimization

Three key parameters were analyzed to evaluate the impact of catalyst dosage, solution stirring speed, and pH on the TCH degradation rate using TiO_2/CdS photocatalysis. Design–Expert (version 13.0.1.0 64-bit) optimized these parameters using the response surface methodology—Box–Behnken design. The BBD can assess the potential interactions between parameters and their effects on the degradation rate. Equation (2) represents the relationship between the variables:

$$Y = \beta_0 + \sum_{i=1}^{k} \beta_i x_i + \sum_{i=1}^{k} \beta_{ii} x_i^2 + \sum_{i=1}^{k} \sum_{j=1}^{k} \beta_{ij} x_i x_j + \varepsilon \quad (2)$$

where β_0 represents the intercept term, β_i represents the linear factor coefficients, β_{ii} represents the quadratic terms, β_{ij} represents the interaction terms, and Y represents the degradation percentage. Table 1 shows the BBD model independent variables and coded values.

Table 1. Independent variables and coded values of the BBD model.

Independent Variables	Symbols	Coded Levels		
		−1	0	1
Catalyst Dose	A	20	30	40
Agitating rate	B	100	300	500
pH	C	6	9	12

2.6. Reusability of TiO_2/CdS

After conducting the photocatalytic degradation reaction on the same nickel foam loaded with TiO_2/CdS, it was slowly rinsed with deionized water and placed in a drying oven at 100 °C for 60 min to ensure complete drying. The photocatalytic degradation reaction was then repeated. This process was performed for five consecutive degradation experiments using TCH at an initial concentration of 20 mg/L to assess the stability of the catalyst. The mass of the nickel foam loaded with TiO_2/CdS was recorded before and after each cycle to calculate the catalyst mass loss during the five cycles.

3. Results and Discussion

3.1. Characterization of the Photocatalyst

XRD analysis compared the crystal structures of TiO_2, CdS, and TiO_2/CdS composite samples. The XRD patterns are shown in Figure 3. The diffraction peaks at 25.30°, 37.80°, 48.02°, 53.89°, 55.02°, 62.66°, and 68.81° correspond to the anatase TiO_2 (JCPDS No. 21-1272) (1 0 1), (0 0 4), (2 0 0), (1 0 5), (2 1 1), and (1 1 6) crystal planes [36]. The diffraction peaks at 26.39°, 43.53°, and 51.69° correspond to the cubic phase of CdS (JCPDS No. 80-0019) (1 1 1), (2 2 0), and (3 1 1) crystal planes [37]. The XRD pattern of the TiO_2/CdS composite retains the characteristic peaks of both TiO_2 and CdS, indicating the successful formation of the composite [38]. Additionally, the peak of TiO_2 at 25.30° overlaps with the (1 1 1) crystal plane of CdS. No significant changes in the characteristic peaks of TiO_2 and CdS were observed, indicating that CdS exists in a separate phase rather than within the lattice of TiO_2.

Figure 4a shows the absorption spectra of all samples in the UV–Vis range. All samples except for TiO_2 exhibit strong light absorption in both the ultraviolet and visible ranges. The absorption edge of TiO_2 is approximately 410 nm, while that of TiO_2/CdS is at 530 nm. The bandgap widths of TiO_2 and CdS were obtained using the extrapolation method, as shown in Figure 4b [39]. The bandgap width of TiO_2 is 3.07 eV, and that of TiO_2/CdS is 2.77 eV, which is lower than the 3.18 eV of TiO_2/Cd prepared by previous researchers [40]. Compared with pure TiO_2, the TiO_2/CdS composite has a significantly narrower bandgap, which enables better visible light absorption.

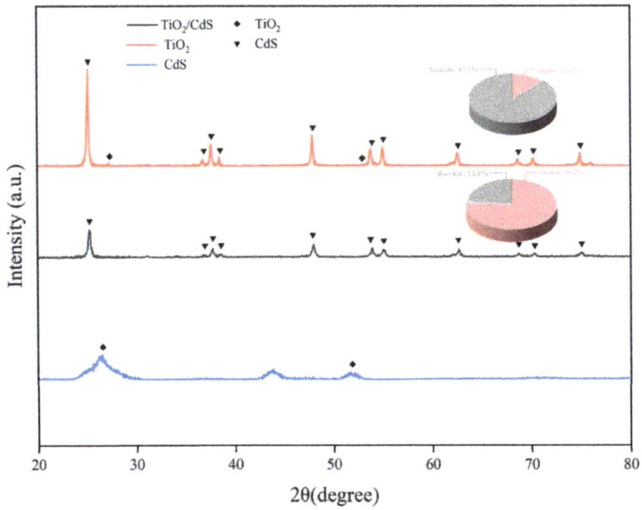

Figure 3. XRD patterns of TiO_2, CdS, and TiO_2/CdS nanocomposites.

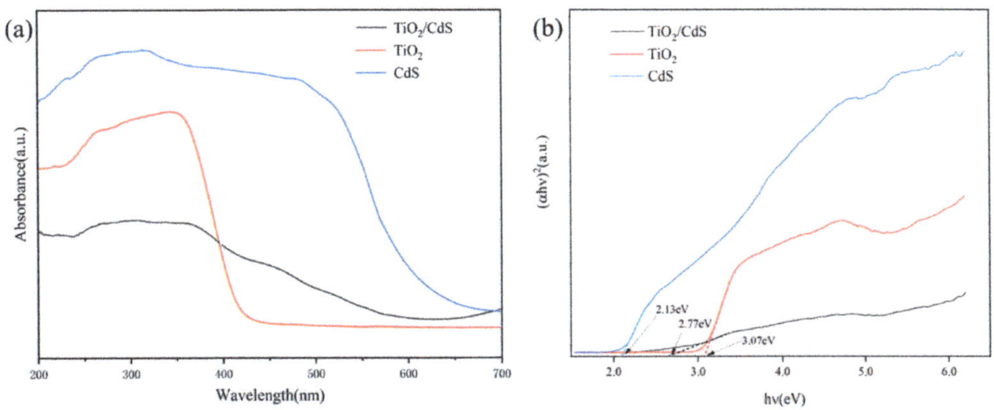

Figure 4. Absorption spectra (**a**) and bandgap values (**b**) of TiO_2, CdS, and TiO_2/CdS nanocomposites.

The nickel foam loaded with different layers of TiO_2/CdS was observed and analyzed using a super depth microscope. The super depth of field microscope, with its large depth of field, can clearly image uneven samples. It has the unique ability to keep different heights in focus simultaneously, making it ideal for observing 3D microstructures such as catalysts. Moreover, it offers high-resolution views, enabling detailed morphological analysis and precise defect detection at the micro level. From Figure 5, it can be seen that the nickel foam has a dense pore structure. As the number of coating layers increases, the coverage of TiO_2/CdS on the nickel foam also increases, and the thickness of the TiO_2/CdS film on the nickel foam grows in Figure 5b–f. The surface area per unit volume was measured, as shown in Figure 5g. The number of TiO_2/CdS plating layers on the nickel foam mesh exhibits a linear correlation with the TiO_2/CdS catalyst dosage applied to the nickel foam mesh. This relationship can be expressed through the equation presented in Figure 5h.

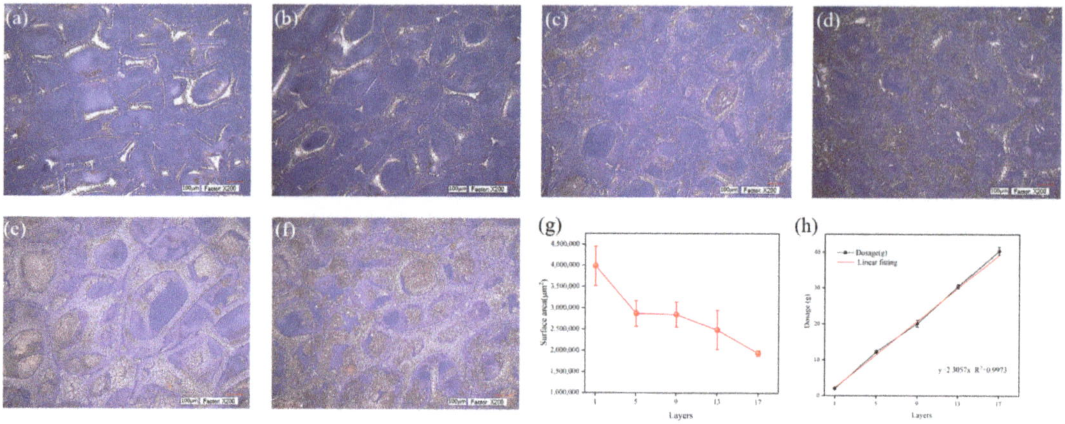

Figure 5. (**a–f**) TiO_2/CdS loaded on nickel foam with 0, 1, 5, 9, 13, and 17 layers in the order of; (**g**) surface area of TiO_2/CdS loaded on nickel foam per unit volume; (**h**) linear relationship between the number of TiO_2/CdS layers and catalyst dose.

The TEM of the samples was observed, as shown in Figure 6a,b, revealing that the TiO_2/CdS loaded on the nickel foam has a high-density particle structure, which facilitates the entry and loading of CdS. TEM images show the lattice fringes of TiO_2/CdS, as depicted in Figure 6e,f. Among them, a lattice fringe of 0.23 nm corresponds to the CdS (2 2 0) plane, while a 0.35 nm lattice fringe corresponds to the TiO_2 (1 0 1) plane [41]. The precise contact interface between CdS and TiO_2 indicates low interface losses during the electron transport process, and the close interface contact is beneficial for the interfacial charge transfer between CdS and TiO_2. In Figure 6d, the X-ray energy spectrum shows that TiO_2/CdS contains elements S, Cd, Ti, and O, indicating that CdS forms a deposit on the surface of TiO_2. Moreover, the Cd and S elements are uniformly and densely distributed within the TiO_2/CdS composite, suggesting good loading and high dispersion of CdS within the TiO_2 particles. Combined with the results from XRD, this confirms the successful synthesis of TiO_2/CdS.

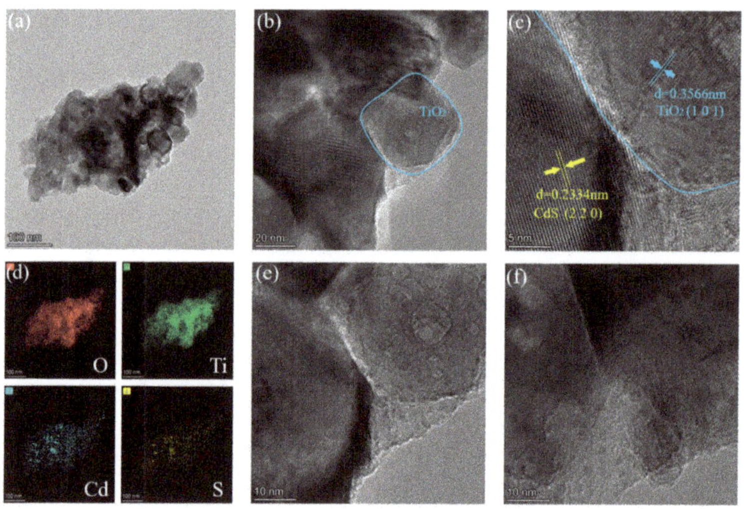

Figure 6. Microstructure characterization of TiO_2/CdS; (**a–c**,**e**,**f**) TEM images of TiO_2/CdS; EDX mapping (**d**) of O, Ti, Cd, and S.

XPS was used to study the chemical states of the surface elements in the composite material, as shown in Figure 7. The full spectrum reveals the presence of Ti, O, Cd, C, and S elements in the TiO_2/CdS composite material, as illustrated in Figure 7a–c, which display the high-resolution spectra of Cd 3d and S 2p, respectively. The characteristic peaks at binding energies of 404.49 eV and 411.30 eV correspond to Cd $3d_{52}$ and $3d_{32}$, indicating that the Cd element in CdS exists in the form of Cd^{2+}. The binding energy difference of 6.8 eV between the Cd $3d_{52}$ and $3d_{32}$ peaks further confirms that the state of Cd is Cd^{2+} [21]. The characteristic peaks at binding energies of 164.18 eV and 168.31 eV correspond to S $2p_{12}$ and S $2p_{32}$, indicating that the S element in CdS exists in the form of S^{2-} [42]. The Ti 2p characteristic peaks near binding energies of 457.87 eV and 463.56 eV correspond to Ti $2p_{32}$ and Ti $2p_{12}$, which are typically attributed to Ti^{4+}, as shown in Figure 7d [43]. Furthermore, the binding energy difference of 5.7 eV between Ti $2p_{32}$ and Ti $2p_{12}$ further indicates that the Ti element exists in the form of Ti^{4+} [44]. Figure 7e shows the characteristic peaks at binding energies of 531.02 eV, which correspond to Ti–O bonds in the TiO_2 lattice, respectively [45].

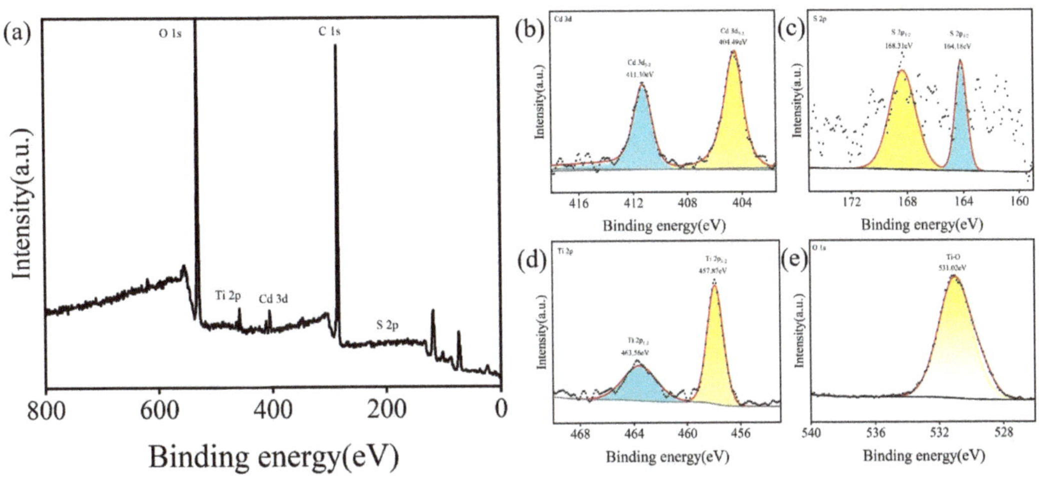

Figure 7. XPS spectra of TiO_2/CdS: (**a**) full spectrum, (**b**) Cd 3d, (**c**) S 2p, (**d**) Ti 2p, and (**e**) O 1s.

3.2. *Quality Ratio*

By adding 1 g, 3 g, 5 g, and 10 g of CdS powder to four identical 50 mL mixtures of tetrabutyl titanate and 200 mL of anhydrous ethanol, mixed solutions with CdS mass ratios of 0.48%, 1.43%, 2.37%, and 4.63% were obtained. A 70 × 36 mm nickel foam was then subjected to a three-time dipping–pull method, followed by heat treatment and other steps, resulting in four pieces of nickel foam loaded with different mass ratios of TiO_2/CdS. These were used for photocatalytic degradation of an 80 mL TCH solution with an initial concentration of 20 mg/L, as shown in Figure 8. The subsequent preparation involved using a mixed solution with a CdS mass ratio of 1.43% to fabricate TiO_2/CdS. In the paper, for the XRD, UV–Vis, TEM, and XPS tests and photocatalytic experiments using the TiO_2/CdS composite, the TiO_2/CdS was prepared from a precursor solution with a CdS mass ratio of 1.43%.

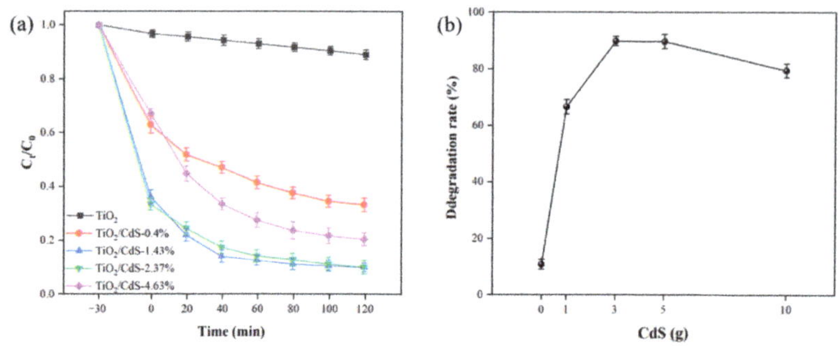

Figure 8. (**a**) Degradation rate curves at different CdS mass ratios, (**b**) effect of different CdS mass ratios on TCH degradation rate.

3.3. RSM–BBD

The results of the variance analysis shown in Table 2 indicate that the model is significant in elucidating the photocatalytic degradation of TCH. The high F-value (115.99) and extremely low *p*-value (<0.0001) indicate that the model is highly effective in explaining the variations in TCH degradation rates. The three factors, including catalyst dosage (A), agitating rate (B), and solution pH (C), are all highly significant, reaffirming their critical roles in influencing TCH degradation rates. The positive coefficients of these factors suggest that increasing the catalyst dosage has a beneficial effect on enhancing removal efficiency, while higher agitating rates and alkaline pH values also improve photocatalytic degradation rates. The analysis of variable interactions (AB, AC, BC) demonstrates their high significance, underscoring the importance of considering these factors' synergistic effects on TCH degradation rates. The significant quadratic terms (A^2, B^2, C^2) indicate a nonlinear relationship between the independent variables and the response. This nonlinear relationship highlights the need to understand further how variations in these parameters collectively affect the degradation rate of TCH. The non-significant goodness-of-fit value ($p = 0.1117$) suggests that the model fits the experimental data well, indicating that the model adequately represents the relationship between the variables and the response. This confirms the reliability of the model regarding TCH degradation rates. The binary polynomial formula (Equation (2)) predicts the response variable while accounting for the independent factors and their interactions. To maximize the degradation rate of TCH, the solver in Design–Expert was used to solve Equation (3) and provide the necessary values. This mathematical formula can predict the values required to achieve the maximum TCH degradation rate.

$$Y = -69.441 + 1.441 \times A + 0.278 \times B + 12.840 \times C + 0.0004 \times AB + 0.052 \times AC - 0.007 \times BC - 0.037A^2 - 0.0004B^2 - 0.667C^2 \quad (3)$$

Figure 9a–f illustrate the relationships among the three factors and the degradation rate of TCH. To simplify the analysis, each figure selects two independent factors while keeping the remaining factor fixed at its midpoint value. The contour plots effectively demonstrate the interactions between the selected factors, where elliptical or saddle-shaped contours indicate significant interactions, while circular contours suggest negligible interactions. Additionally, the three-dimensional plots in Figure 9a–c provide a deeper understanding of these relationships. Figure 10a also presents the optimization results. Figure 10b shows the curve fitting of experimental and predicted TCH degradation rates along with the normal distribution of the residuals. The differences between the experimental and expected results are minimal, and the residual analysis confirms the model's reliability. The residuals in Figure 10c appear as a straight line, indicating that the errors follow a normal distribution.

Table 2. Analysis of variance (ANOVA) for the TCH photodegradation rate regression model.

Source	Sum of Squares	df	Mean Square	F-Value	*p*-Value	
Model	1338.61	9	148.73	115.99	<0.0001	significant
A	20.67	1	20.67	16.12	0.0051	
B	108.71	1	108.71	84.78	<0.0001	
C	7.74	1	7.74	6.04	0.0436	
AB	3.13	1	3.13	2.44	0.1620	
AC	9.86	1	9.86	7.69	0.0276	
BC	68.81	1	68.81	53.66	0.0002	
A^2	56.92	1	56.92	44.39	0.0003	
B^2	830.49	1	830.49	647.66	<0.0001	
C^2	151.79	1	151.79	118.38	<0.0001	
Residual	8.98	7	1.28	1.28		
Lack of Fit	6.68	3	2.23	2.23	3.88	not significant
Pure Error	2.30	4	4	0.5740		
Cor Total	1347.58	16	16			

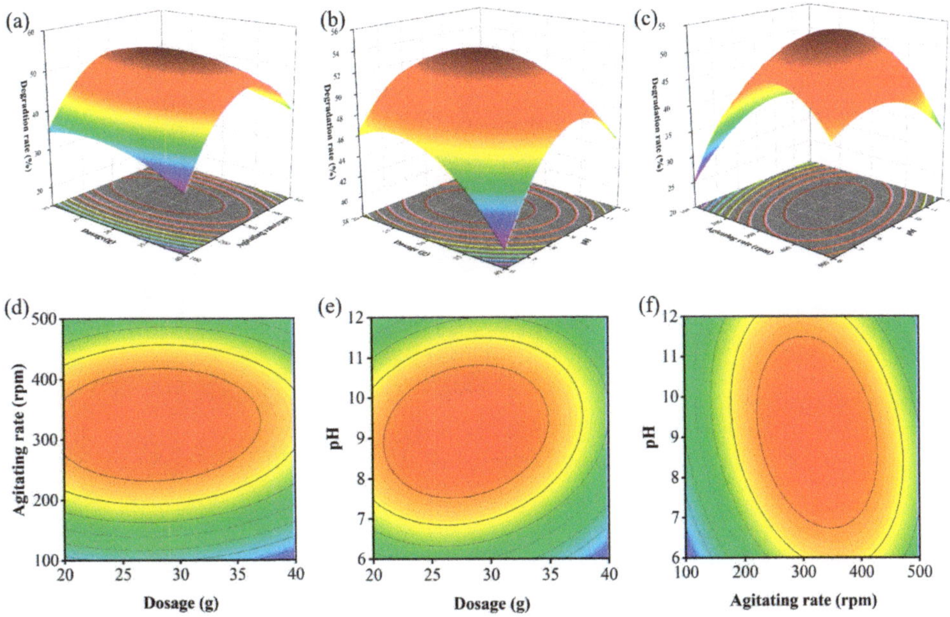

Figure 9. (**a**–**f**) Response surface and contour plots showing the effect of catalyst dosage, agitating rate, and pH on the degradation rate of TCH.

The analysis of this dataset reveals the intricate relationships between various factors and the degradation rate. The catalyst dosage (Factor 1, A) is a crucial parameter affecting photocatalytic degradation, controlled by applying different layers of TiO_2/CdS catalysts onto the foam nickel using the dip-coating method. The range of catalyst dosage (20–40 g) was determined through preliminary experiments and extensive literature review. This dosage range was selected to assess its suitability for degrading TCH. Generally, an increase in catalyst dosage is associated with enhanced degradation capability; for instance, at 29 g (13 layers), the degradation of a 20 mg/L TCH solution reached 53.84%. This phenomenon is expected as a more significant catalyst dosage provides more active sites for the photocatalytic reaction [46–48]. However, this relationship may not be linear. As the number of catalyst layers increases, the pores of the foam nickel can become clogged, hindering sufficient contact between active radicals and the TCH solution. Consequently, an excessive catalyst dosage may not yield better results, indicating an optimal range for catalyst dosage [49]. A possible reason for this is that increasing the catalyst amount may lead to easier detachment of the catalyst from the foam nickel, resulting in an excess of catalyst particles in the solution, which could scatter light and adversely affect the overall enhancement of the photocatalytic degradation rate [50,51].

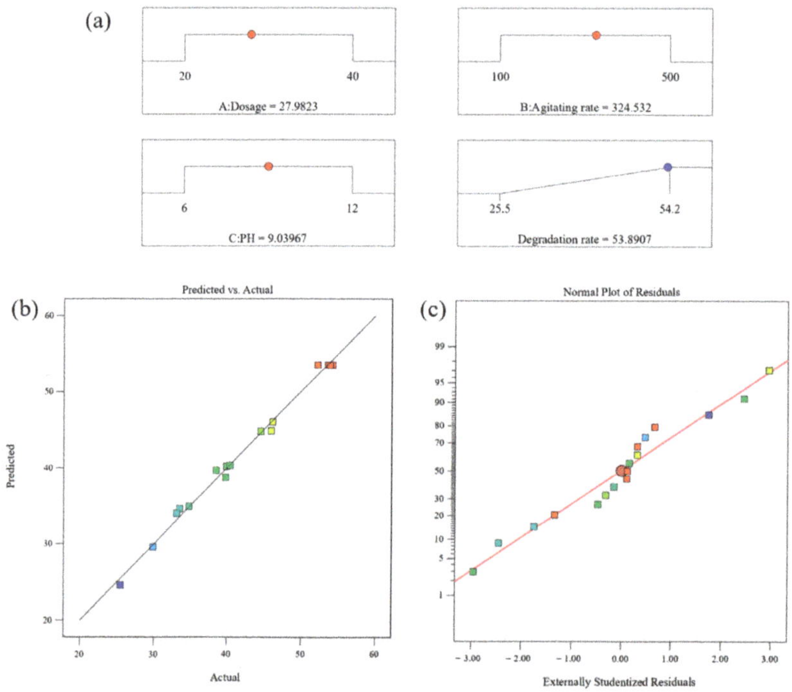

Figure 10. (**a**) Optimal solution obtained from RSM optimization, (**b**) Normal probability plot of residuals, (**c**) Comparison of predicted and actual values.

The agitating rate of the magnetic stirrer (Factor 2, B) significantly impacts the degradation rate of TCH. As the agitating rate increased from 100 rpm to 300 rpm, the degradation rate of TCH rose from 25.5% to 53.61%. This increase in degradation rate can be attributed to the kinetic energy gained by the TCH solution, which improved the intimate interface contact between reactants and facilitated sufficient interaction between active radicals and the TCH solution. The probability of successful collisions between active radicals and TCH during the photocatalytic process increases with the agitating rate [52]. However, it is noteworthy that in this study, the performance slightly declined when the agitating rate was raised to 500 rpm. This may be due to the agitating rate exceeding the optimal value, resulting in ineffective collisions between reactants [53].

The pH value (Factor 3, C) affects the structure and surface properties of TCH in aqueous solution, as well as the surface charge of the photocatalyst, thus influencing the removal efficiency of pollutants [54–56]. The data indicate that the pH value significantly impacts degradation efficiency, with different values yielding varying results. Considering the response variable (degradation rate), the dataset shows a wide range of degradation efficiency, from 25.5% to 54.2%. With 30 g of catalyst, an agitating rate of 300 rpm, and pH = 9, the highest degradation efficiency reached 53.84%. Conversely, with 30 g of catalyst, an agitating rate of 100 rpm, and pH = 6, the lowest degradation efficiency was 25.5%. The dataset suggests that alkaline conditions can enhance photocatalytic removal efficiency. However, with 30 g of catalyst, an agitating rate of 500 rpm, and pH = 12, the degradation efficiency dropped to 33.1%. This decrease of 20.74% when the pH increased from 9 to 12 is attributed to the TCH negative surface charge becoming a double negative charge, significantly increasing repulsion from the photocatalyst particles. Thus, an optimal pH value exists while increasing the pH, which enhances the formation of hydroxyl radicals and the degradation efficiency. The point of zero charge (pzc) of the TiO_2/CdS composite material is 6.25. Although TCH and the catalyst surface carry negative charges at pH = 9, the degradation rate is still higher [57]. A possible reason for this phenomenon is the higher concentration of adsorbed OH^-, which promotes the formation of hydroxyl radicals, thereby enhancing the oxidative effect on TCH [58].

3.4. *Reusability*

The reusability of photocatalytic materials is a critical factor in assessing the performance of photocatalysts. This study investigated the reusability of TiO_2/CdS composite materials by cleaning and drying nickel foam loaded with TiO_2/CdS. This was followed by conducting five consecutive degradation experiments using an initial concentration of 20 mg/L TCH to evaluate the stability of the catalyst.

As shown in Figure 11a, the cyclic experimental results of the TiO_2/CdS composite material indicate that although there is a slight decrease in photocatalytic degradation efficiency after the third cycle, the degradation rate of TCH remains relatively stable during the subsequent two cycles. After five consecutive cycles of degrading the TCH solution, the degradation efficiency decreased to 25.03%. Figure 11b shows the change in mass of the nickel foam loaded with TiO_2/CdS during the continuous cycling experiments. It is evident that the mass significantly decreased after the first cycle, while the mass gradually reduced during the subsequent four cycles. This is attributed to the partial detachment of TiO_2/CdS from the nickel foam during the first cycle. Notably, most TCH degradation occurred within the first 60 min of the first cycle, as loosely adsorbed TiO_2/CdS particles were lost with the flow of the solution, which increased the contact area with the TCH solution, leading to a marked improvement in degradation efficiency. However, as more particles detached, the solution's transparency decreased, resulting in a decline in degradation efficiency after 90 min of the first cycle. In the following four cycles, the concentration of TCH steadily reduced during the photocatalytic reaction, indicating minimal detachment of TiO_2/CdS particles, which correlates with the slow reduction in the mass of the nickel foam loaded with TiO_2/CdS. The observed decrease in photocatalytic performance may be due to surface area loss from nanoparticle agglomeration and particle detachment from the nickel foam. There may also be slight changes in the structure of TiO_2/CdS or accumulation of reaction byproducts on its surface during the photocatalytic process, which could affect the degradation efficiency in multiple cycles.

Nevertheless, the dense structure of TiO_2/CdS loaded on the nickel foam still maintained its effectiveness, as evidenced by the relatively small loss in removal efficiency. There was no significant change in the XRD pattern (Figure 11c), indicating that the TiO_2/CdS composite material is stable even after multiple use cycles.

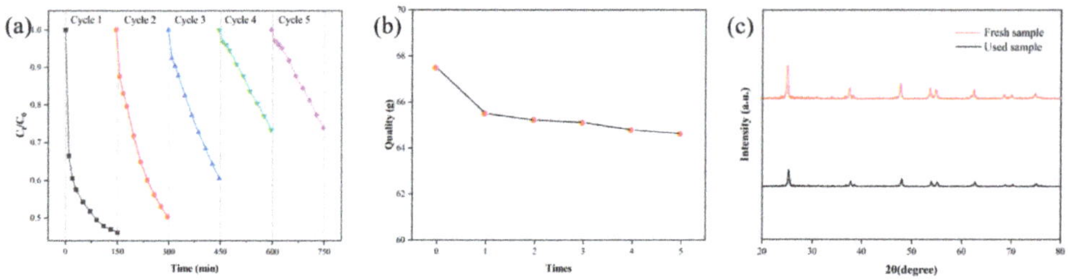

Figure 11. (**a**) Reproducibility test of TiO_2/CdS, (**b**) variation of catalyst mass, (**c**) XRD patterns of fresh sample and after recycling.

The main active species responsible for the degradation of TCH were analyzed through radical scavenging experiments. A total of 1 mM benzoquinone (BQ), isopropanol (IPA), and sodium oxalate (AO) were used as scavengers for $\cdot O_2^-$, $\cdot OH$, and h^+, respectively. As shown in Figure 12a,b, the effects of different scavengers on the photocatalytic degradation capacity of the TiO_2/CdS composite material were investigated. The addition of benzoquinone and ascorbic acid significantly decreased the photocatalytic degradation rate of TCH, indicating that $\cdot O_2^-$ and h^+ are essential for the degradation process. However, the addition of isopropanol did not significantly reduce degradation efficiency, suggesting that the $\cdot OH$ radical has a minimal effect on the degradation of TCH. The degradation rate of TCH without any scavengers was 59.65%, while the degradation rates with the addition of benzoquinone, isopropanol, and ascorbic acid were 34.42%, 56.31%, and 44.06%, respectively. The contributions of the radicals to the photocatalytic degradation of TCH using TiO_2/CdS loaded on nickel foam were calculated. The results indicated that the contribution rates of $\cdot O_2^-$, $\cdot OH$, and h^+ were 50.06%, 16.23%, and 29.99%, respectively. The scavenging experiment results suggest that $\cdot O_2^-$ is the primary active species in the TiO_2/CdS photocatalytic system, while $\cdot OH$ and h^+ play a supporting role.

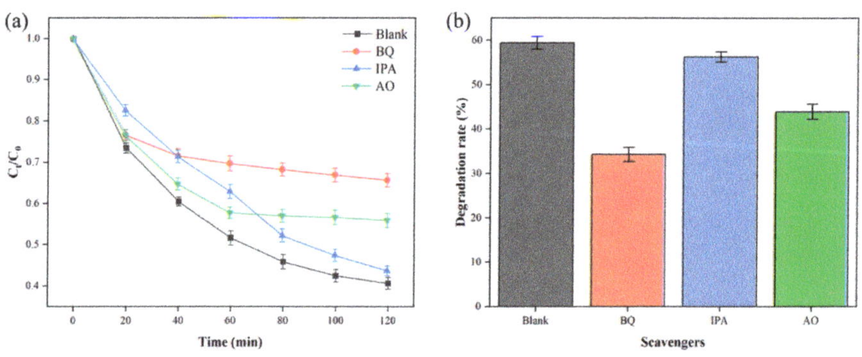

Figure 12. Effect of different capture agents on the degradation rate of tetracycline hydrochloride: (**a**) degradation rate curve, (**b**) degradation rate of TCH at 120 min.

Figure 13 illustrates the photocatalytic reaction mechanism of TiO_2/CdS. Under visible light irradiation, both TiO_2 and CdS are excited to produce photogenerated electrons and holes. The conduction band edge potential of CdS is more negative than that of TiO_2, and the electrostatic field facilitates the transfer of electrons from CdS to the conduction band of TiO_2. Meanwhile, the valence band edge potential of CdS is also more negative than that of TiO_2, allowing holes from the valence band of TiO_2 to transfer to the valence band of CdS. This spatial separation of electrons and holes is one of the essential methods to reduce their recombination. The electrons in the conduction band of TiO_2 react with dissolved oxygen in water to generate superoxide radicals, while the holes in the valence band of CdS react with water to produce ·OH hydroxyl radicals. The results of the radical scavenging experiments indicate that O_2^- plays a major role in the photocatalytic degradation of TCH using TiO_2/CdS. Furthermore, during the photocatalytic reaction, active species decompose TCH into intermediate products and ultimately convert it into CO_2 and H_2O.

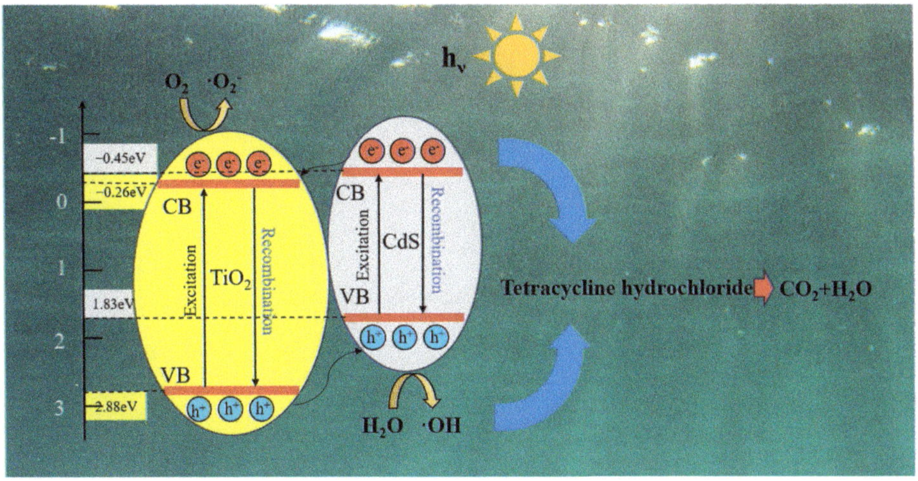

Figure 13. Photocatalytic reaction mechanism of TiO_2/CdS under visible light.

4. Conclusions

This study successfully developed an efficient TiO_2/CdS composite material supported on nickel foam for photocatalytic degradation of tetracycline hydrochloride (TCH). Characterization techniques such as XRD, UV–Vis, TEM, and XPS were employed to comprehensively analyze the TiO_2/CdS composites, revealing that these materials exhibit excellent photocatalytic activity. Under optimized conditions, with a TiO_2 to CdS mass ratio of 3.8:1, the catalyst achieved a maximum degradation rate of 90.03% for 80 mL of 20 mg/L TCH solution within 150 min. The response surface methodology–Box–Behnken design (RSM–BBD) was successfully applied to optimize the operational parameters, with the best degradation rate of 53.89% obtained at a catalyst dosage of 28 g, an agitation rate of 325 rpm, and a pH of 9.04.

Additionally, the TiO_2/CdS composites supported on nickel foam demonstrated good stability and reusability, maintaining a degradation rate of 25.03% after five consecutive cycles with a minimal mass loss of 4.44%. Radical scavenger experiments revealed that superoxide radicals ($\cdot O_2^-$) played the dominant role in the photocatalytic degradation of TCH, with the order of contribution being $\cdot O_2^- > h^+ >$ OH.

Overall, the TiO_2/CdS composite material supported on nickel foam not only exhibited excellent photocatalytic degradation efficiency but also showed promising stability and reusability, indicating its potential for water treatment applications. However, future research could focus on optimizing the synthesis methods, exploring different loading strategies, and investigating the performance of these materials in more complex water matrices. Furthermore, combining this system with other advanced oxidation technologies could enhance the degradation efficiency, providing more effective solutions for practical applications in environmental remediation.

Author Contributions: Conceptualization, Z.F. and Z.Y.; validation, K.Z. and L.M.; formal analysis, Z.F.; investigation, Z.F.; resources, J.D.; data curation, Z.F.; writing—original draft preparation, K.Z.; writing—review and editing, K.Z. and L.M.; visualization, K.Z.; supervision, J.D.; funding acquisition, L.M. and Z.Y. All authors have read and agreed to the published version of the manuscript.

Funding: This research was funded by the China Agriculture Research System of MOF and MARA, grant number CARS-31-11, and the China Postdoctoral Science Foundation, grant number No. 2023M741215.

Data Availability Statement: The data presented in this study are available on request from the corresponding author.

Conflicts of Interest: The authors declare no conflicts of interest.

References

1. Paradina-Fernández, L.; Wünsch, U.; Bro, R.; Murphy, K. Direct Measurement of Organic Micropollutants in Water and Wastewater Using Fluorescence Spectroscopy. *ACS EST Water* **2023**, *3*, 3905–3915. [CrossRef]
2. Pouramini, Z.; Mousavi, S.M.; Babapoor, A.; Hashemi, S.A.; Lai, C.W.; Mazaheri, Y.; Chiang, W.-H. Effect of Metal Atom in Zeolitic Imidazolate Frameworks (ZIF-8 & 67) for Removal of Dyes and Antibiotics from Wastewater: A Review. *Catalysts* **2023**, *13*, 155. [CrossRef]
3. Xuan, J.; Feng, W.; Wang, J.; Wang, R.; Zhang, B.; Bo, L.; Chen, Z.-S.; Yang, H.; Sun, L. Antimicrobial peptides for combating drug-resistant bacterial infections. *Drug Resist. Updat.* **2023**, *68*, 100954. [CrossRef] [PubMed]
4. Fang, S.-Y.; Zhang, P.; Gong, J.-L.; Tang, L.; Zeng, G.-M.; Song, B.; Cao, W.-C.; Li, J.; Ye, J. Construction of highly water-stable metal-organic framework UiO-66 thin-film composite membrane for dyes and antibiotics separation. *Chem. Eng. J.* **2020**, *385*, 123400. [CrossRef]
5. Nasrollahi, N.; Vatanpour, V.; Khataee, A. Removal of antibiotics from wastewaters by membrane technology: Limitations, successes, and future improvements. *Sci. Total Environ.* **2022**, *838*, 156010. [CrossRef] [PubMed]
6. Yao, B.; Luo, Z.; Du, S.; Yang, J.; Zhi, D.; Zhou, Y. Sustainable biochar/$MgFe_2O_4$ adsorbent for levofloxacin removal: Adsorption performances and mechanisms. *Bioresour. Technol.* **2021**, *340*, 125698. [CrossRef] [PubMed]
7. Yu, F.; Sun, S.; Han, S.; Zheng, J.; Ma, J. Adsorption removal of ciprofloxacin by multi-walled carbon nanotubes with different oxygen contents from aqueous solutions. *Chem. Eng. J.* **2016**, *285*, 588–595. [CrossRef]
8. Reddy, C.V.; Kakarla, R.R.; Cheolho, B.; Shim, J.; Aminabhavi, T.M. Heterostructured 2D/2D ZnIn2S4/g-C3N4 nanohybrids for photocatalytic degradation of antibiotic sulfamethoxazole and photoelectrochemical properties. *Environ. Res.* **2023**, *225*, 115585. [CrossRef] [PubMed]
9. Xiao, M.; Li, R.; Yin, J.; Yang, J.; Hu, X.; Xiao, H.; Wang, W.; Yang, T. Enhanced photocatalytic oxidation of As(III) by TiO_2 modified with Fe_3O_4 through Ti–O–Fe interface bonds. *Colloids Surf. Physicochem. Eng. Asp.* **2022**, *651*, 129678. [CrossRef]
10. Mu, R.; Ao, Y.; Wu, T.; Wang, C.; Wang, P. Synthesis of novel ternary heterogeneous anatase-TiO_2 (B) biphase nanowires/$Bi_4O_5I_2$ composite photocatalysts for the highly efficient degradation of acetaminophen under visible light irradiation. *J. Hazard. Mater.* **2020**, *382*, 121083. [CrossRef]
11. Kutuzova, A.; Dontsova, T.; Kwapinski, W. Application of TiO_2-Based Photocatalysts to Antibiotics Degradation: Cases of Sulfamethoxazole, Trimethoprim and Ciprofloxacin. *Catalysts* **2021**, *11*, 728. [CrossRef]
12. Mahmoud, M.E.; Ibrahim, G.A.A. Cr(VI) and doxorubicin adsorptive capture by a novel bionanocomposite of Ti-MOF@TiO_2 incorporated with watermelon biochar and chitosan hydrogel. *Int. J. Biol. Macromol.* **2023**, *253*, 126489. [CrossRef] [PubMed]
13. Abreu-Jaureguí, C.; Andronic, L.; Sepúlveda-Escribano, A.; Silvestre-Albero, J. Improved photocatalytic performance of TiO_2/carbon photocatalysts: Role of carbon additive. *Environ. Res.* **2024**, *251*, 118672. [CrossRef]
14. Anucha, C.B.; Altin, I.; Bacaksiz, E.; Stathopoulos, V.N. Titanium dioxide (TiO_2)-based photocatalyst materials activity enhancement for contaminants of emerging concern (CECs) degradation: In the light of modification strategies. *Chem. Eng. J. Adv.* **2022**, *10*, 100262. [CrossRef]
15. Jiang, M.; Zhang, M.; Wang, L.; Fei, Y.; Wang, S.; Núñez-Delgado, A.; Bokhari, A.; Race, M.; Khataee, A.; Jaromír Klemeš, J.; et al. Photocatalytic degradation of xanthate in flotation plant tailings by TiO2/graphene nanocomposites. *Chem. Eng. J.* **2022**, *431*, 134104. [CrossRef]
16. Etacheri, V.; Di Valentin, C.; Schneider, J.; Bahnemann, D.; Pillai, S.C. Visible-light activation of TiO_2 photocatalysts: Advances in theory and experiments. *J. Photochem. Photobiol. C Photochem. Rev.* **2015**, *25*, 1–29. [CrossRef]
17. Kumari, M.L.A.; Devi, L.G.; Maia, G.; Chen, T.-W.; Al-Zaqri, N.; Ali, M.A. Mechanochemical synthesis of ternary heterojunctions TiO2(A)/TiO2(R)/ZnO and TiO_2(A)/TiO_2(R)/SnO_2 for effective charge separation in semiconductor photocatalysis: A comparative study. *Environ. Res.* **2022**, *203*, 111841. [CrossRef] [PubMed]

18. Acharya, R.; Pani, P. Visible light susceptible doped TiO_2 photocatalytic systems: An overview. *Mater. Today Proc.* **2022**, *67*, 1276–1282. [CrossRef]
19. Li, S.; Liu, Y.; Long, Y.; Mo, L.; Zhang, H.; Liu, J. Facile Synthesis of Bi_2MoO_6 Microspheres Decorated by CdS Nanoparticles with Efficient Photocatalytic Removal of Levfloxacin Antibiotic. *Catalysts* **2018**, *8*, 477. [CrossRef]
20. Guo, X.; Chen, C.; Song, W.; Wang, X.; Di, W.; Qin, W. CdS embedded TiO_2 hybrid nanospheres for visible light photocatalysis. *J. Mol. Catal. Chem.* **2014**, *387*, 1–6. [CrossRef]
21. Sabir, M.; Rafiq, K.; Abid, M.Z.; Quyyum, U.; Shah, S.S.A.; Faizan, M.; Rauf, A.; Iqbal, S.; Hussain, E. Growth of tunable Au-BaO@TiO_2/CdS heterostructures: Acceleration of hydrogen evolution from water splitting. *Fuel* **2023**, *353*, 129196. [CrossRef]
22. Xiao, R.; Zhang, J.; Jiang, T.; Zhou, Y.; Wang, Y.; Xu, W.; Feng, Y. Highly ordered Janus CdS-Au-TiO_2 Z-scheme structure with high efficiency in photocatalysis. *Sci. China Chem.* **2023**, *66*, 1722–1730. [CrossRef]
23. Kaur, A.; Umar, A.; Anderson, W.A.; Kansal, S.K. Facile synthesis of CdS/TiO_2 nanocomposite and their catalytic activity for ofloxacin degradation under visible illumination. *J. Photochem. Photobiol. Chem.* **2018**, *360*, 34–43. [CrossRef]
24. Li, J.; Xia, Z.; Ma, D.; Liu, G.; Song, N.; Xiang, D.; Xin, Y.; Zhang, G.; Chen, Q. Improving photocatalytic activity by construction of immobilized Z-scheme CdS/Au/TiO_2 nanobelt photocatalyst for eliminating norfloxacin from water. *J. Colloid Interface Sci.* **2021**, *586*, 243–256. [CrossRef]
25. Zangeneh, H.; Zinatizadeh, A.A.; Nazari, S.; Joshaghani, M.; Zinadini, S.; Sibali, L.; Feyzi, M. Highly efficient azo dye degradation in a photocatalytic rotating disc reactor with deposited l-histidine-TiO_2-CdS. *Mater. Sci. Semicond. Process.* **2022**, *152*, 107071. [CrossRef]
26. Jin, X.; Huang, Y.; He, S.; Chen, G.; Liu, X.; He, C.; Du, C.; Chen, Q. Preparation of Co-Fe based Prussian blue analogs loaded nickel foams for Fenton-like degradation of tetracycline. *Appl. Catal. Gen.* **2023**, *650*, 118985. [CrossRef]
27. Lu, M.; Sun, J.; Cui, B.; Zhang, J.; Ren, J.; Li, R. Construction of Pd, Ru/2D MXene nanosheets/3D self-supporting nickel foam composite electrode and its electrocatalytic synergistic degradation of antibiotics. *Sep. Purif. Technol.* **2024**, *340*, 126736. [CrossRef]
28. Tang, J.; Cheng, Z.; Li, H.; Xiang, L. Electro-Chemical Degradation of Norfloxacin Using a PbO2-NF Anode Prepared by the Electrodeposition of PbO_2 onto the Substrate of Nickel Foam. *Catalysts* **2022**, *12*, 1297. [CrossRef]
29. Miao, F.; Lu, Y.; Tao, B.; Zhao, M.; Chu, P.K. Nickel foam-loaded Co-MOF@TiO_2/MoS_2 as electrode materials for dual-function devices for glucose detection and hydrogen evolution. *Microchim. Acta* **2024**, *191*, 469. [CrossRef] [PubMed]
30. Jiao, F.-Z.; Wu, J.; Zhang, T.; Pan, R.-J.; Wang, Z.-H.; Yu, Z.-Z.; Qu, J. Simultaneous Solar-Thermal Desalination and Catalytic Degradation of Wastewater Containing Both Salt Ions and Organic Contaminants. *ACS Appl. Mater. Interfaces* **2023**, *15*, 41007–41018. [CrossRef] [PubMed]
31. Yu, J.; Cong, S.; Liu, B.; Teng, W. Construction of MoS_2/NiFe-Ni foam p-n heterojunction as photoanode for tetracycline degradation and simultaneous cathodic hydrogen evolution. *J. Environ. Chem. Eng.* **2022**, *10*, 108437. [CrossRef]
32. Zhu, Y.F.; Zhou, L.; Jiang, Q.S. One-dimensional ZnO nanowires grown on three-dimensional scaffolds for improved photocatalytic activity. *Ceram. Int.* **2020**, *46*, 1158–1163. [CrossRef]
33. Lv, C.; Lan, X.; Wang, L.; Dai, X.; Zhang, M.; Cui, J.; Yuan, S.; Wang, S.; Shi, J. Rapidly and highly efficient degradation of tetracycline hydrochloride in wastewater by 3D IO-TiO_2 -CdS nanocomposite under visible light. *Environ. Technol.* **2021**, *42*, 377–387. [CrossRef] [PubMed]
34. Shen, W.; Xiang, D.; Yang, J.; Tang, Y.; Xin, C.; Guo, Q.; Yu, X. Fabrication of beta zeolite supported Ti_3+-TiO_2/CdS composite for ultrahigh-performance photodegradation of tetracycline under visible-light illumination. *Colloids Surf. Physicochem. Eng. Asp.* **2022**, *653*, 129965. [CrossRef]
35. Zhang, Y.; Shi, Z.; Luo, L.; Liu, Z.; Macharia, D.K.; Duoerkun, G.; Shen, C.; Liu, J.; Zhang, L. Construction of titanium dioxide/cadmium sulfide heterojunction on carbon fibers as weavable photocatalyst for eliminating various contaminants. *J. Colloid Interface Sci.* **2020**, *561*, 307–317. [CrossRef] [PubMed]
36. Bi, X.; Du, G.; Kalam, A.; Sun, D.; Yu, Y.; Su, Q.; Xu, B.; Al-Sehemi, A.G. Tuning oxygen vacancy content in TiO_2 nanoparticles to enhance the photocatalytic performance. *Chem. Eng. Sci.* **2021**, *234*, 116440. [CrossRef]
37. Xu, J.; Cao, X. Characterization and mechanism of MoS_2/CdS composite photocatalyst used for hydrogen production from water splitting under visible light. *Chem. Eng. J.* **2015**, *260*, 642–648. [CrossRef]
38. Gao, B.; Zhao, X.; Liang, Z.; Wu, Z.; Wang, W.; Han, D.; Niu, L. CdS/TiO_2 Nanocomposite-Based Photoelectrochemical Sensor for a Sensitive Determination of Nitrite in Principle of Etching Reaction. *Anal. Chem.* **2021**, *93*, 820–827. [CrossRef] [PubMed]
39. Shi, C.; An, Y.; Gao, G.; Xue, J.; Algadi, H.; Huang, Z.; Guo, Z. Insights into Selective Glucose Photoreforming for Coproduction of Hydrogen and Organic Acid over Biochar-Based Heterojunction Photocatalyst Cadmium Sulfide/Titania/Biochar. *ACS Sustain. Chem. Eng.* **2024**, *12*, 2538–2549. [CrossRef]
40. Ali, H. Facile synthesis of mesoporous TiO_2-CdS-polyaniline ternary system with improved optical properties. *Mater. Res. Express* **2019**, *6*, 115529. [CrossRef]
41. Du, Y.; Niu, X.; He, X.; Hou, K.; Liu, H.; Zhang, C. Synthesis and Photocatalytic Activity of TiO_2/CdS Nanocomposites with Co-Exposed Anatase Highly Reactive Facets. *Molecules* **2021**, *26*, 6031. [CrossRef]

42. Dai, Q.; Li, Y.; Qiu, Z.; Tian, H.; Pu, Y.; Chen, X.; Lv, B.; Wei, J.; Wang, W. CdS/$ZnFe_2O_4$ Core–Shell Nanorod Arrays on Modified TiO_2 Photoanodes for Photoelectrochemical Water Splitting. *ACS Appl. Nano Mater.* **2024**, *7*, 17441–17450. [CrossRef]
43. Rafiq, K.; Sabir, M.; Abid, M.Z.; Jalil, M.; Nadeem, M.A.; Iqbal, S.; Rauf, A.; Hussain, E. Tuning of TiO_2 /CdS Hybrid Semiconductor with Au Cocatalysts: State-of-the-Art Design for Sunlight-Driven H $_2$ Generation from Water Splitting. *Energy Fuels* **2024**, *38*, 4625–4636. [CrossRef]
44. Li, J.; Wang, L.; Wang, W.; Jia, X.; Zhang, Y.; Yang, H.; Li, Y.; Zhou, Q. Cooperative effects of surface plasmon resonance and type-II band alignment to significantly boost photoelectrochemical H_2 generation of TiO_2/CdS/TiN nanorod array photoanode. *Appl. Catal. B Environ.* **2023**, *334*, 122833. [CrossRef]
45. Luo, T.; Sun, X.; Ma, D.; Wang, G.; Yang, F.; Zhang, Y.; Huang, J.; Zhang, H.; Wang, J.; Peng, F. Fabrication of TiO_2 /CdS Heterostructure by Soluble Solid-State Titanium-oxo-Clusters for Fast Photocatalytic Degradation of Tetracycline. *J. Phys. Chem. C* **2023**, *127*, 1372–1380. [CrossRef]
46. Sinar Mashuri, S.I.; Ibrahim, M.L.; Kasim, M.F.; Mastuli, M.S.; Rashid, U.; Abdullah, A.H.; Islam, A.; Asikin Mijan, N.; Tan, Y.H.; Mansir, N.; et al. Photocatalysis for Organic Wastewater Treatment: From the Basis to Current Challenges for Society. *Catalysts* **2020**, *10*, 1260. [CrossRef]
47. Hassan, A.F.; El-Naggar, G.A.; Braish, A.G.; Abd El-Latif, M.M.; Shaltout, W.A.; Elsayed, M.S. Fabrication of titania/calcium alginate nanocomposite matrix for efficient adsorption and photocatalytic degradation of malachite green. *Int. J. Biol. Macromol.* **2023**, *249*, 126075. [CrossRef] [PubMed]
48. Silerio-Vázquez, F.D.J.; González-Burciaga, L.A.; Antileo, C.; Núñez-Núñez, C.M.; Proal-Nájera, J.B. Photocatalytic degradation of antibiotics in water via TiO_2-x: Research needs for technological advancements. *J. Hazard. Mater. Adv.* **2024**, *16*, 100506. [CrossRef]
49. Zhang, T.; Liu, Y.; Rao, Y.; Li, X.; Yuan, D.; Tang, S.; Zhao, Q. Enhanced photocatalytic activity of TiO_2 with acetylene black and persulfate for degradation of tetracycline hydrochloride under visible light. *Chem. Eng. J.* **2020**, *384*, 123350. [CrossRef]
50. Ebrahimi-Koodehi, S.; Ghodsi, F.E.; Mazloom, J. Ni/Mn metal–organic framework decorated bacterial cellulose (Ni/Mn-MOF@BC) and nickel foam (Ni/Mn-MOF@NF) as a visible-light photocatalyst and supercapacitive electrode. *Sci. Rep.* **2023**, *13*, 19260. [CrossRef] [PubMed]
51. Sadegh Jafari Zadegan, M.; Sabbaghi, S.; Rasouli, K.; Moosaei, R.; Mahdi Zerafat, M. Innovatively-synthesized α-Fe_2O_3/Ti_3C_2Tx MXene nanocomposite by dry-impregnation method: Photocatalyst characterization and influence of operational parameters on highly efficient tetracycline degradation. *Inorg. Chem. Commun.* **2024**, *163*, 112314. [CrossRef]
52. Chakachaka, V.M.; Tshangana, C.S.; Mamba, B.B.; Muleja, A.A. CFD-Assisted Process Optimization of an Integrated Photocatalytic Membrane System for Water Treatment. *Membranes* **2023**, *13*, 827. [CrossRef] [PubMed]
53. Fu, Y.; Ren, Z.; Wu, J.; Li, Y.; Liu, W.; Li, P.; Xing, L.; Ma, J.; Wang, H.; Xue, X. Direct Z-scheme heterojunction of ZnO/MoS_2 nanoarrays realized by flowing-induced piezoelectric field for enhanced sunlight photocatalytic performances. *Appl. Catal. B Environ.* **2021**, *285*, 119785. [CrossRef]
54. Cheikh, S.; Imessaoudene, A.; Bollinger, J.-C.; Hadadi, A.; Manseri, A.; Bouzaza, A.; Assadi, A.; Amrane, A.; Zamouche, M.; El Jery, A.; et al. Complete Elimination of the Ciprofloxacin Antibiotic from Water by the Combination of Adsorption–Photocatalysis Process Using Natural Hydroxyapatite and TiO_2. *Catalysts* **2023**, *13*, 336. [CrossRef]
55. Santoso, S.P.; Angkawijaya, A.E.; Bundjaja, V.; Hsieh, C.-W.; Go, A.W.; Yuliana, M.; Hsu, H.-Y.; Tran-Nguyen, P.L.; Soetaredjo, F.E.; Ismadji, S. TiO_2/guar gum hydrogel composite for adsorption and photodegradation of methylene blue. *Int. J. Biol. Macromol.* **2021**, *193*, 721–733. [CrossRef]
56. Zhang, B.; He, X.; Yu, C.; Liu, G.; Ma, D.; Cui, C.; Yan, Q.; Zhang, Y.; Zhang, G.; Ma, J.; et al. Degradation of tetracycline hydrochloride by ultrafine TiO_2 nanoparticles modified g-C_3N_4 heterojunction photocatalyst: Influencing factors, products and mechanism insight. *Chin. Chem. Lett.* **2022**, *33*, 1337–1342. [CrossRef]
57. Zangeneh, H.; Zinatizadeh, A.A.; Zinadini, S.; Feyzi, M.; Rafiee, E.; Bahnemann, D.W. A novel L-Histidine (C, N) codoped-TiO_2-CdS nanocomposite for efficient visible photo-degradation of recalcitrant compounds from wastewater. *J. Hazard. Mater.* **2019**, *369*, 384–397. [CrossRef]
58. Zeinali Heris, S.; Etemadi, M.; Mousavi, S.B.; Mohammadpourfard, M.; Ramavandi, B. Preparation and characterizations of TiO_2/ZnO nanohybrid and its application in photocatalytic degradation of tetracycline in wastewater. *J. Photochem. Photobiol. Chem.* **2023**, *443*, 114893. [CrossRef]

catalysts

MDPI

Review

Modified Zeolites for the Removal of Emerging Bio-Resistive Pollutants in Water Resources

Fatin Samara [1,2], Amer A. Al Abdel Hamid [3], Venkatesh Gopal [1], Lara Dronjak [1], Fares Feghaly [1] and Sofian Kanan [1,2,*]

[1] Department of Biology, Chemistry & Environmental Sciences, American University of Sharjah, Sharjah P.O. Box 26666, United Arab Emirates
[2] Energy Water and Sustainable Environments Research Center (EWSERC), American University of Sharjah, Sharjah P.O. Box 26666, United Arab Emirates
[3] Department of Chemistry, Yarmouk University, Irbid 21163, Jordan
* Correspondence: skanan@aus.edu

Abstract: The increasing presence of pollutants, including pharmaceuticals and pesticides, in water resources necessitates the development of effective remediation technologies. Zeolites are promising agents for pollutant removal due to their high surface area, ion-exchange capacity, natural abundance, and diverse tailorable porous structures. This review focuses on the efficient application of modified zeolites and mesoporous materials as photocatalysts and adsorbents for removing contaminants from water bodies. The adsorption and photodegradation of pesticides and selected non-steroidal anti-inflammatory drugs and antibiotics on various zeolites reveal optimal adsorption and degradation conditions for each pollutant. In most reported studies, higher SiO_2/Al_2O_3 ratio zeolites exhibited improved adsorption, and thus photodegradation activities, due to increased hydrophobicity and lower negative charge. For example, SBA-15 demonstrated high efficiency in removing diclofenac, ibuprofen, and ketoprofen from water in acidic conditions. Metal doped into the zeolite framework was found to be a very active catalyst for the photodegradation of organic pollutants, including pesticides, pharmaceuticals, and industrial wastes. It is shown that the photocatalytic activity depends on the zeolite-type, metal dopant, metal content, zeolite pore structure, and the energy of the irradiation source. Faujasite-type Y zeolites combined with ozone achieved up to 95% micropollutant degradation. Bentonite modified with cellulosic biopolymers effectively removed pesticides such as atrazine and chlorpyrifos, while titanium and/or silver-doped zeolites showed strong catalytic activity in degrading carbamates, highlighting their environmental application potential.

Keywords: zeolite; photodegradation; adsorption; pesticides; pharmaceuticals; pollution; water treatment

1. Introduction

The vast growth of the world's population has imposed a significant need for sufficient food production over the past two decades. Consequently, the extensive use of pesticides, combined with the expansion of specialized food and pharmaceutical industries, increases the discharge of associated waste into water supplies, leading to adverse effects on human health [1–4]. Despite the strict efforts by various environmental agencies to limit and regulate the use of pesticides and pharmaceutical products, they continue to be detected in groundwater and municipal water treatment plants worldwide [5]. The discharge of organic micropollutants from diffuse sources, such as pesticides used in agriculture, industrial wastewater effluent, and municipal wastewater effluent, poses significant threats to public health and raises major toxicological concerns for aquatic environments, including drinking water sources [6–10].

Pesticides are mainly characterized by their chemical classes and target organisms. Humans are exposed to pesticides through several routes of exposure, such as inhalation, ingestion, and dermal contact [11]. Long-term exposure to pesticides has been linked to neurobehavioral disorders, allergies, autoimmune diseases, reproductive abnormalities, leukemia, malignant lymphoma, nausea, vomiting, sweating, hyperventilation, pain, fasciculation, myotonia, weakness, and myoglobinuria [12–14].

The high concentrations of pesticides, pharmaceuticals, and industrial waste deposited in water resources have made their removal from water bodies a significant environmental concern in recent times. Given the wide range of pesticides with various functionalities currently in use, it is extremely difficult to identify a single method for the removal of these pesticides from water resources. A variety of methods have been designed for the removal of organic pollutants from the environment. Some of these methods include adsorption, photocatalytic degradation, advanced oxidation, nanofiltration membranes, ozonation, extraction, and photocatalytic degradation [15–22]. Photocatalysis is one of the alternative methods, recognized for its efficiency in water treatment. The main advantages of this technique are the utilization of light as an energy source and the ability to enhance versatility by incorporating dopants into the semiconductor to improve its properties. Additionally, photocatalysis does not produce harmful toxins or toxic byproducts, which is also considered an advantage. However, it also faces several challenges, including limited absorption of the light spectrum, issues with catalyst stability and durability, restricted reaction rates, low selectivity, high production costs, specific working conditions, dependence on light availability, and low quantum efficiency. An example of such a catalyst includes titanium dioxide (TiO_2), a chemical with numerous applications across various industries, as well as in environmental remediation for areas affected by persistent pollution [23,24]. In addition, aluminosilicate complexes, also known as zeolites, have been reported to provide a beneficial platform for TiO_2 and other metal complexes due to their porous configuration, which directs the interaction between the catalyst and pollutants [25].

Zeolites are microporous crystalline solids made through the linking of aluminate (AlO_4) and silicate (SiO_4) moieties in tetrahedral units with various Si/Al ratios that are used to describe the hydrophobicity of the zeolite surface. The framework type mainly defines the pore opening, cage, and channel structure of these zeolites [22,26]. Both aluminate and silicate units are linked through oxygen atoms, which form regular crystalline cavities and channels of various dimensions [27,28]. Figure 1 shows the framework structure, along with the associated cell units obtained from the International Zeolite Association network (iza-structure.org). Based on the obtained pore diameters, zeolites can be classified as small-pore zeolites, with diameters ranging from 3 to 4.5 Å, intermediate-pore zeolites, with diameters ranging from 4.5 to 6 Å, and large-pore zeolites, with diameters above 6 Å [29]. Since the pore diameter depends on the silicon content, another classification is based on the Si/Al content [30]. Because of their crystalline structure, porosity, the shape and size of the cavities, and their hardness, zeolites have unique features for water treatment applicability, including a high specific surface area associated with their pore structure, biocompatibility, ion-exchange capacity, natural abundance, and diverse, tailorable porous structures [31–39].

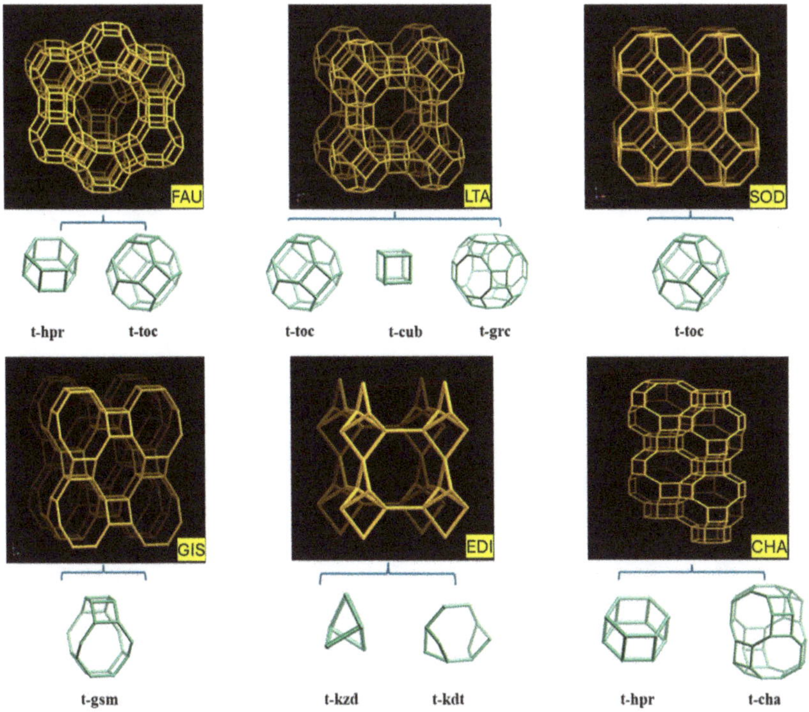

Figure 1. Zeolite framework type along, with their associated composite building units (retrieved from https://www.iza-structure.org/databases/, accessed on 11 June 2024).

Zeolites have been identified as strong adsorbents for wastewater treatment due to their potential to adsorb metal ions, as well as organic pollutants. The surface of these zeolites can be modified to enhance their adsorption potential for pollutants, resulting in changes in surface properties and pore structure [40]. The selectivity of adsorption may be assessed by the polarity, shape, and size of the diffusing molecules relative to the size and geometry of the pores of the zeolites, along with the presence of exchangeable cations in the zeolite structure. Consequently, extensive research efforts are focused on exploring the applicability of zeolites in water treatment, particularly for the decontamination of organic waste. Figure 2 illustrates the number of articles published on Scopus that use "zeolite-adsorption" and "zeolite-catalysts" as keywords. The data presented in Figure 2, show a significant increase during the past five years, indicating the benefit of modified zeolites in environmental management and waste treatment [41–44].

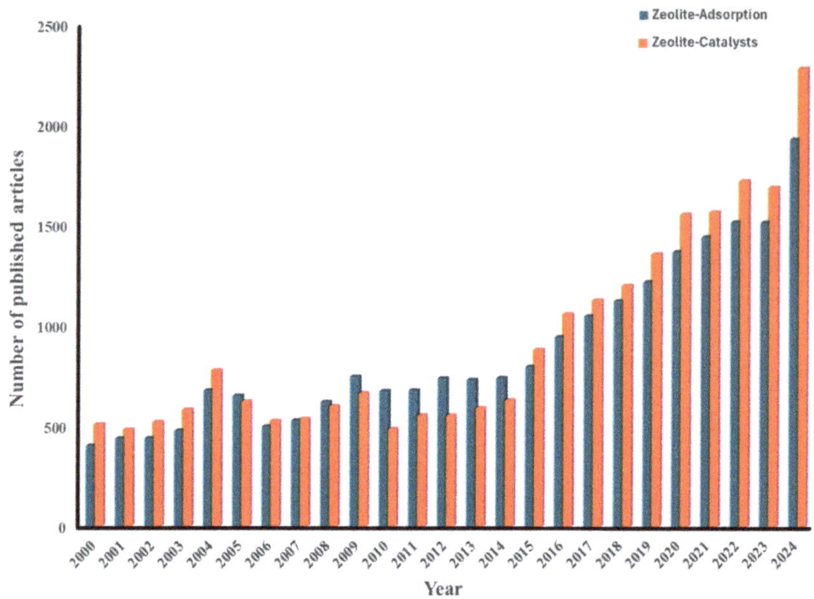

Figure 2. Statistics on the use of zeolites in water treatment, based on core publications from the Scopus database, using the keywords "zeolite adsorption" and "zeolite catalysts".

2. The Use of Modified Zeolites for the Removal of Pharmaceutical Products

Zeolite composition is summarized by the formula $M_{x/n}\cdot[Al_xSi_yO_{2(x+y)}]\cdot pH_2O$, where M is (Na, K, Li) and/or (Ca, Mg, Ba, Sr), while n is the cation charge; y/x = 1–6, p/x = 1–4. The [SiO_4] and [AlO_4] tetrahedral units are the basic structural building units of a zeolite framework, with uniformly sized pores of molecular dimensions [45]. The widespread use of zeolites in environmental protection is attributed to their tremendous adsorption capability—primarily for substances potentially classified as harmful to the environment, especially to aquatic bodies and their ecosystems. Pharmaceuticals and other industrial pollutants are representative examples of the applicability of zeolites in reducing their toxic levels in water resources.

The extensive use of pharmaceuticals (defined as substances used in the diagnosis, treatment, and prevention of disease, in addition to the restoration, correction, and modification of organic functions) has significantly contributed to improvements in human health, including the extension of life expectancy [46]. A recent example is the critical role of pharmaceuticals during the SARS-CoV-2 pandemic, where their application in the treatment, prevention, and diagnosis was unprecedented on a global scale. However, despite the many medicinal benefits of these products, their usage still poses unintended consequences, as a significant percentage of their concentrations persist in various environmental compartments, potentially affecting ecosystems and human health. Excretion is the primary route by which pharmaceuticals enter the environment, either in their parent form or as metabolites. The resulting pharmacologically active compounds, formed through biotransformation processes, may produce unpredictable byproducts [47]. While pharmaceuticals are generally susceptible to diverse transformation reactions, some resulting products are very stable. Such transformed hydrophilic compounds easily pass through sewage treatment plants, resulting in a considerable impact on the aquatic environment [48–50]. Thus, the presence of pharmaceutical contaminants—particularly in aquatic environments—

presents a complex challenge, making it an increasingly important field of study [51], with a focus on surface water samples, sewage treatment plants, and drinking water [52–54].

2.1. The Effective Removal of Pharmaceuticals with Zeolites

There are numerous known methods for the removal of pharmaceuticals from the aquatic environment, including biological processes, hydrodynamic cavitation, and ultraviolet light treatment [55]. Upon reviewing various conventional procedures, advanced oxidation processes (AOPs) stand out as the most promising method for treating industrial wastewater. Integrating biological processes with AOPs can facilitate the sustainable degradation of toxic pollutants in wastewater [56]. AOPs can achieve either complete or partial degradation of emerging toxic pollutants through the action of free radicals, such as hydroxyl, superoxide, hydroperoxyl, and sulfate radicals. Additionally, the presence of bio-enzymes and microorganisms contributes to the sustainable degradation of pollutants, offering an economical and environmentally friendly solution. Thus, they are an efficient method for removing pharmaceutical pollutants from highly contaminated waters, hospital waste effluents, or pharmaceutical industries. However, they could also cause major environmental and health risks [57–59].

Due to their high surface area, high chemical and mechanical stability, ease of regeneration, and reuse with little performance loss and low cost, zeolites are ideal photocatalysts for large-scale wastewater treatment and cleanup. They can be modified for various heterogeneous catalytic applications. They are promising materials for sustainable development and energy savings, as well as effective photocatalysts for degrading dye molecules. Adding transition metal ions or semiconductor nanoparticles to the zeolite framework can also increase photocatalytic activity by extending light absorption into the visible range, which tunes the catalytic process to sunlight, as a natural source.

As part of AOPs, photocatalysis is the process in which a photon of light catalyzes a reaction. Upon irradiation, electrons (e−) are excited from the valance band (VB) to the conduction band (CB), leaving positively charged holes (h+) that develop the active oxidizing species, which then cause organic pollutants, such as dyes in wastewater, to degrade when the irradiation energy is equivalent to its band-gap energy (EBG), as presented in Figure 3.

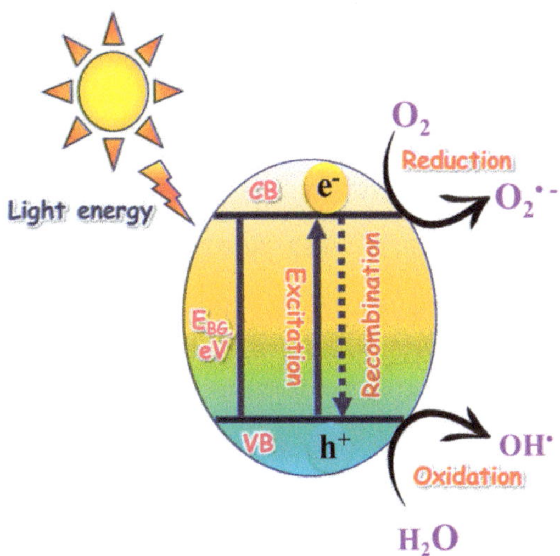

Figure 3. Mechanism of heterogeneous photocatalysis.

Pure zeolites are not considered efficient photocatalysts due to their relatively high band-gap energy (EBG), which necessitates an irradiation source in the UV range. Therefore, modified zeolites were highly investigated for the degradation of various pollutants, especially when they are incorporated with semiconducting metal oxides. These doped materials tend to lower the energy bandgap, allowing them to absorb visible light and provide sufficient active surface sites that capture pollutants, enhancing their photocatalytic activity, as illustrated in Figure 4.

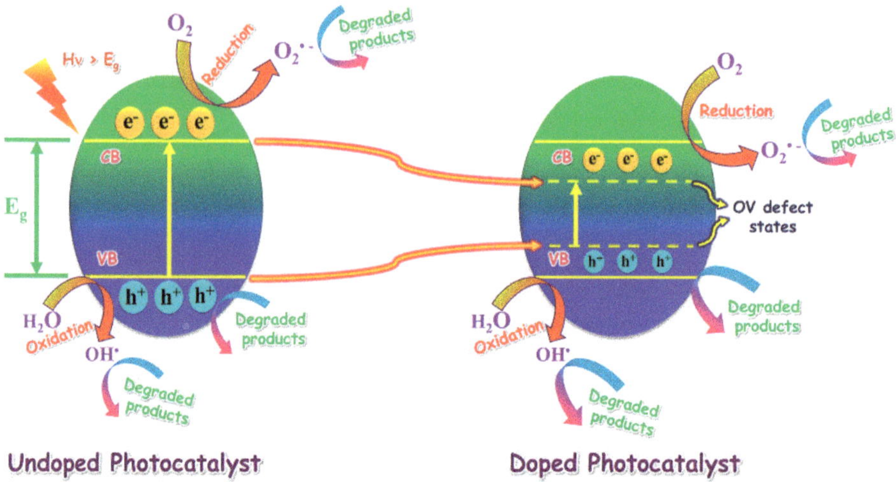

Figure 4. An illustration of the difference between normal photocatalysts and semiconducting modified photocatalysts for potential pharmaceutical degradation using a visible light source.

Bezu et al. reported that using ZnO/Ag_3PO_4 with g-C_3N_4 support over natural zeolite enhanced the degradation of methyl orange (MO) dye from an aqueous stream, with the degradation efficiency reaching 98.2% in 180 min of solar irradiation [59]. The sol-gel method was used to incorporate the Au–TiO_2/HZSM-5 nanocomposite as a potential photocatalyst for the degradation of MO [60]. While 50% of the MO was degraded in 30 min without light, the photodegradation efficiency enhanced to 96.6% with reusability of up to four cycles, during which the efficiency drops by 10.1% [60]. Cr-doped TiO_2 photocatalysts supported on natural zeolite showed an increase in the Cr-O-Ti bond formation upon calcination [61]. The percentage degradation of MO by the calcined 10% Cr/TiO_2/zeolite reaches 41.73% after 5 h of illumination, which is 17.9% higher than the degradation efficiency of undoped TiO_2/zeolite [61]. In addition, deposited g-C_3N_4 with zeolite-Y was shown to enhance the degradation of methylene blue (MB) dye by using the pore-size structure and high surface area of the zeolite, as well as enhancing the use of the photoinduced carriers in the zeolite-Y/g-C_3N_4 composite to prevent the charge recombination rate efficiently [62]. Aguiñaga et al. reported the use of natural zeolite (mordenite) as an efficient self-photocatalyst, with the results obtained indicating that the complete degradation of caffeine under similar conditions on both zeolite and TiO_2 requires the same amount of time [63]. This study showed that, when such natural material is incorporated with active metal oxides, the cavities of zeolites behave as a support for metal oxides, increasing the surface area of metal oxides, thereby improving adsorption and photocatalytic properties, as illustrated in Figure 5.

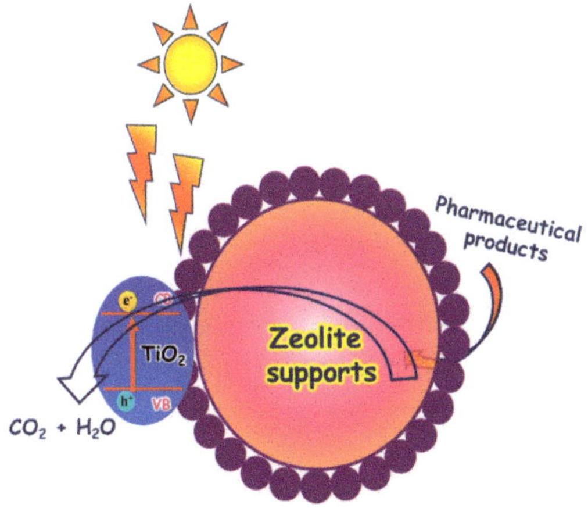

Figure 5. The potential use of zeolite pore structure to capture pollutants to enhance the photocatalytic activity of the Ti-doped catalyst.

Sacco et al. integrated semiconducting ZnO into zeolite (ZnO/ZEO) cavities by wet impregnation as adsorption-assisted photocatalysis for caffeine. Applying both the adsorption and the photocatalysis approach, it was shown that UV irradiation resulted in a significantly higher total removal of caffeine, with 100% of caffeine being removed within 120 min of reaction time by using the ZnO/ZeO composite, compared to the removal by adsorption only, which reached only 69% removal in 120 min [64]. Liu et al. evaluated the TiO_2/zeolite composite for the removal of sulfadiazine (SDZ) via adsorption and photocatalysis under UV light [65]. The prepared TiO_2/zeolite material removed 93.31% of SDZ in the presence of UV light in 120 min, while the uncatalyzed degradation of SDZ under UV irradiation results in 32.76% removal within 120 min [65].

Clinoptilolite, a surface-modified natural zeolite, was used as a carrier for the sustained release of diclofenac, a common anti-inflammatory drug [66]. The study concluded that clinoptilolite serves as a versatile drug carrier, potentially capable of loading molecules with various chemical properties onto or within the zeolite structure. Its easily modifiable surface broadens the range of its application, which has led research to explore its pharmaceutical potential. However, it was unclear whether the drugs were adsorbed on the zeolite's surface or within its internal pores. The study also led researchers to propose that additional degradation products of Na-ibuprofen might also be formed, and clinoptilolite might catalyze this degradation upon adsorption. Thus, clinoptilolite is believed to form complexes with degradation products due to its larger pore volume. By identifying the degradation products, it would be possible to determine a degradation mechanism, improving the understanding of clinoptilolite's role as a carrier and degradation catalyst [66].

Ali et al. investigated the zeolite/Fe_3O_4/CuS/$CuWO_4$ nanocomposite solar photocatalytic degradation of levofloxacin and acetaminophen pharmaceutical pollutants from aqueous solutions [67]. Under optimized conditions, the maximum removal was 84.77% for acetaminophen and 73.26% for LEVO at 220 and 200 min, respectively [67]. MoS_2-zeolite was synthesized via ultrasonic and hydrothermal methods as a photocatalyst for the degradation of tetracycline [68]. The catalytic efficiency of the catalyst for tetracycline reached up to 87.23%, which is 3.58 times higher than alkali-modified zeolite and 1.80 times more than pure MoS_2 [68].

Zeolites have emerged as a simple, efficient, low-cost, and environmentally friendly solution to treat various forms of wastewater. They are extensively known for the adsorption of pesticides, either in pristine or modified forms [69–73]. Research efforts in this field have focused on exploring and developing materials with enhanced adsorption characteristics, with zeolites particularly found to consistently satisfy these requirements, hence becoming regularly utilized in the field, as shown in Figure 6.

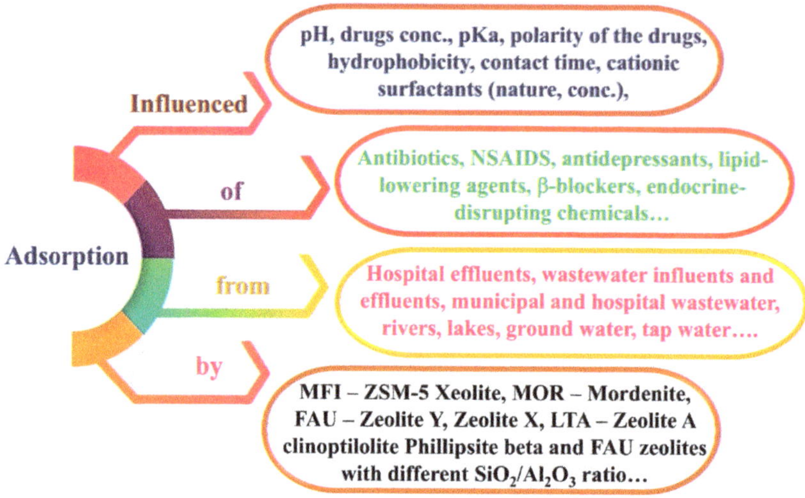

Figure 6. Mode of adsorption of pharmaceutical products by zeolites from aquatic media.

As an example, dexamethasone—A glucocorticosteroid used to treat inflammation, allergies, and various organ disorders, which was used extensively during the SARS-CoV-2 pandemic was investigated for its removal from an aqueous solution using Clinoptilolite zeolite (CP) [74]. When introduced to the environment, dexamethasone has proven to pose several risks to aquatic ecosystems, including the potential increase in algal blooms, which could catalyze eutrophication [75] and alter fish populations within water bodies due to reproductive impairment and hormonal effects [76]. Additionally, studies have shown that dexamethasone exposure during pregnancy may lead to short- and long-term health concerns to fetuses, particularly to structural and functional changes in neurons, which may impact essential brain functions [77]. CP is described as a natural, versatile, and inexpensive mineral, which has been investigated and applied in the last few decades [78]. Due to the resistant structure of dexamethasone against biological degradation and its low concentration in hospital wastewater, its decomposition by conventional wastewater treatment plants is challenging. Thus, the use of modified CP as a natural adsorbent to remove dexamethasone from aqueous solutions was investigated under various conditions [78]. The results revealed that the removal efficiency depends on the pH of the solution, the initial concentration of dexamethasone, the adsorbent dose, and contact time. With that in mind, the maximum removal efficiency of 78% of dexamethasone occurred at pH 4, with 0.5 g CP. When considering the natural, inexpensive nature of CP and its lack of sludge production, these factors highlight the advantages of this adsorbent.

In each adsorption process, the pH of the solution is an influential factor, as it directly affects the surface properties of adsorbents and the ionic forms of contaminants. The type and ionic state of functional groups on both adsorbent and adsorbate surfaces play a critical role in determining the impact of pH on the adsorption process. To assess the effect of pH on adsorption efficiency, the adsorption process was performed under acidic (pH 4), neutral (pH 7), and alkaline (pH 9) conditions [79]. The results showed that the removal efficiency of dexamethasone using modified zeolite increases with decreasing pH, with the maximum removal efficiency achieved at pH 4. When the pH is low, the electrostatic attraction between the positively charged adsorbent surface and the negatively charged dexamethasone ions increases. In neutral and alkaline conditions (pH 7 and higher, respectively), the number of H^+ ions is low. In these conditions, the OH^- ions from the solution, and the contaminant anions compete for placement on the adsorbent surface, resulting in decreased removal efficiency [78].

In addition to pH, the initial concentration of the contaminant has a notable impact on removal efficiency and adsorption capacity. As the initial concentration increases, so does the adsorption capacity, while the removal efficiency decreases [80]. At low concentrations, the ratio of the available surface area to the initial concentration is large; therefore, the removal efficiency is high. However, with increased adsorbate concentration, this ratio declines, resulting in a decreased removal efficiency [81,82]. The effect of initial dexamethasone concentration on removal efficiency was examined using various concentrations of 5, 10, 20, and 40 mg/L, which confirmed the trend that as the initial concentration increased, removal efficiency decreased, while the adsorption capacity increased [80].

The effect of adsorbent dose on removal efficiency was also investigated by varying doses (0.1, 0.3, and 0.5 g/50 mL) in aqueous solutions. The results demonstrated that an increase in adsorbent dose increased removal efficiency while decreasing adsorption capacity [83,84]. The increase in removal efficiency was attributed to a rise in the number of exchangeable sites and the surface area when the adsorbent dose increased. In contrast, the decrease in adsorption capacity may be due to the overlapping of adsorption sites resulting from the overcrowding of adsorbent particles [85].

2.2. *Potential Applications of Zeolites in Water Treatment*

Zeolites have a three-dimensional pore structure, with crystalline voids in the form of chambers and channels [86]. Their properties vary depending on the Si/Al ratio, and their hydrophobic nature is beneficial for facilitating the adsorption of pharmaceuticals from aqueous solutions [87]. Commercially available zeolites include mordenite (MOR), faujasite (FAU), and MFI, all of which have been tested for the removal of antibiotics and non-steroidal drugs from aqueous solutions.

Environmental antibiotic pollution creates a complex challenge due to the development of microbial antibiotic resistance from prolonged exposure in the environment [88]. Almost 5 million deaths were associated with antimicrobial resistance (AMR) in 2019, with AMR being directly responsible for 1.27 million deaths [89]. To effectively address this issue, the sorption of four antibiotics, namely chlortetracycline (CTC), oxytetracycline (OTC), ofloxacin (OFL), and enrofloxacin (ENR), on natural zeolite was investigated. The study examined the effects of pH and natural organic matter (NOM) on the sorption process, using the Langmuir–Freundlich sorption model to calculate the maximum sorption capacity. The gathered results showed that the maximum sorption capacity for the antibiotics was up to 8 mg/g at a neutral pH, with increased sorption capacity as the pH of the solution decreased. Additionally, NOM reduced the sorption of OTC—presumably due to the formation of ion pairs between OTC and NOM—but unexpectedly enhanced the sorption of the remaining antibiotics. Nevertheless, the study demonstrated that sorption by natural zeolite appears to be an economically feasible method for removing antibiotics [88–90].

In a study focused on the adsorption of acetaminophen and Na-ibuprofen by clinoptilolite and zeolite BEA, researchers examined the effectiveness of these zeolites in removing the pharmaceuticals from water [91]. Given the frequent use of these drugs for both human and livestock pain relief and the lack of adequate removal procedures within conventional wastewater treatment plants, both acetaminophen and ibuprofen are regularly introduced into the environment [92]. Their persistence in the environment causes a wide range of developmental and functional impacts on aquatic organisms. Ibuprofen poses a variety of risks, including genetic and cellular damage, oxidative stress, dystrophy, and inflammation, as well as a decrease in egg production across various species of aquatic animals [93]. Similarly, acetaminophen may cause genetic damage, protein denaturation in cells, and the oxidative degradation of lipids [94]. Thus, the removal of both is critical to minimize their impact on aquatic ecosystems.

The adsorption capacity was affected by the cation-exchange and the altered elemental composition of natural clinoptilolite obtained from different geographical locations. The zeolites clinoptilolite and beta (BEA) were used as the absorbents in both their natural form and after ion exchange with Cu(II), Fe(III), and NH_4^+, aiming to remove acetaminophen and ibuprofen from water. It was found that both zeolites were able to adsorb the pharmaceuticals in question, and the results showed that clinoptilolite was the better adsorbent. The best uptake was exhibited by Cu-exchanged clinoptilolite, which showed the highest adsorption affinity for ibuprofen. Researchers hypothesized that ibuprofen degraded during the adsorption experiments to produce ions that form stable complexes with the cations [91,95,96]. The zeolites modified by cation-exchange were found to exhibit different adsorption capacities compared to the initial zeolite, suggesting that the modifications could be tailored to better fit certain applications or target specific pollutants. Notably, the addition of Cu(II) cations to natural clinoptilolite specifically enhances the adsorption of Na-ibuprofen. Furthermore, the results also demonstrated that drug adsorption can vary with the use of a natural zeolite obtained from different geographical locations due to differences in elemental compositions. Overall, the results showed that both clinoptilolite and BEA can effectively remove analgesics through adsorption. Since clinoptilolite achieved the highest adsorption results, it would be of interest to investigate if it exhibits similar adsorption affinities for other drugs, such as naproxen and diclofenac [91,95,96].

The efficacy of adsorption of selected non-steroidal medicinal products and antibiotics by zeolites and mesoporous silica adsorbents was assessed to explore their potential for use in water treatment [91]. The adsorption capacity of sorbents with high silica content was found to depend on their surface hydrophobicity (or hydrophilicity), structural features (such as micropore volume and pore size), as well as the properties of the studied medicinal products [94]. Therein, the focus was on non-steroidal anti-inflammatory drugs (NSAIDs) and antibiotics, as they are considered the most widely used medicinal products. NSAIDs (diclofenac, ibuprofen, ketoprofen, and naproxen) are widely distributed in surface waters and wastewater [95,96]. Moreover, these drugs have exhibited a tendency to bioaccumulate in the bodies of aquatic organisms upon their release in the environment—particularly in the effluent from wastewater treatment plants. A previous study showed that the quantity of NAIDS detected in the bile of wild fish species exceeded the concentrations of those drugs in the surrounding water by roughly 1000 times [97]. In addition, antibiotics (erythromycin, sulfamethoxazole, tetracycline, and trimethoprim) were selected, as they are included in the World Health Organization's List of Essential Medicines, and owing to their wide use as antimicrobial substances [95,98,99]. It was shown that antibiotic concentrations were highest in hospital effluents, wastewater effluents, and river water. In the three types of water (tap water, hospital wastewater, surface water (lakes)), erythromycin and trimethoprim were not detected. The highest concentrations of diclofenac, ibuprofen, and ketoprofen were found in wastewater influents, municipal wastewater, and hospital effluents. Thus, it was concluded that the drug concentrations in different types of waters and wastewater follow this order: hospital effluents > wastewater influents > municipal wastewater > secondary wastewater > river water > wastewater effluents > groundwater > surface water > seawater > tap water > hospital wastewater > surface water (lakes). Furthermore, due to human activities, pharmaceuticals are detected in various types of water and wastewater across each continent [100].

It has been demonstrated that at similar concentrations, removal strategies for these pharmaceuticals are largely consistent across different locations. Non-steroidal analgesics and anti-inflammatories, including ibuprofen, ketoprofen, and naproxen, are among the most readily removable pharmaceuticals. However, the removal of drugs such as diclofenac from wastewater remains challenging. Although the transformations of pharmaceuticals in the aquatic environment have not been thoroughly studied [56,101,102], these substances are known to undergo various reactions and changes upon entering surface water and water reservoirs, with dilution being among the least impactful factors [103]. The byproducts resulting from such transformations are sometimes more stable than the parent compounds and may exhibit varying levels of toxicity, influenced by both biotic and abiotic changes in the environment [104,105]. This is most observed with lipophilic and difficult-to-degrade products. Conversely, hydrophilic products may pass through sewage treatment plants and enter flowing and stagnant surface waters [48]. Moreover, it has been shown that pharmaceuticals can exhibit various removal routes with variable rates, which do not necessarily follow predictable patterns, even if they belong to the same therapeutic groups [49]. Figure 7 presents the approximate non-metabolized fractions of selected pharmaceuticals/antibiotics from the NSAID group as they enter wastewater after ingestion and human metabolism.

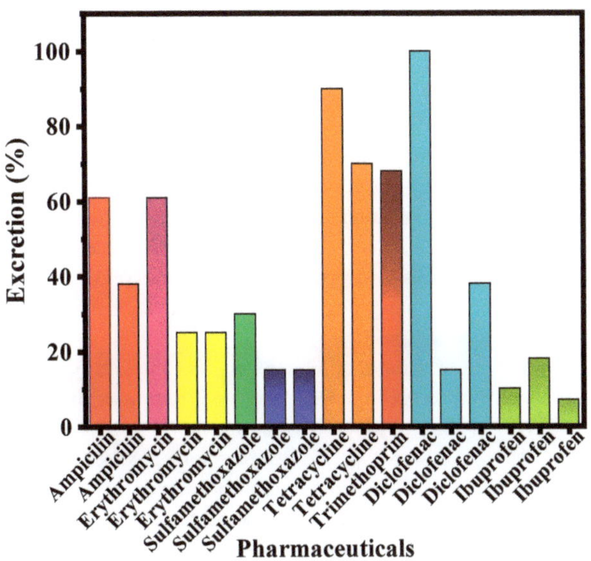

Figure 7. Typical pharmaceuticals and their approximate nonmetabolized fractions entering sewage after ingestion and human metabolism. Note: the bars showing excretion percentages are based on reported values extracted from ref. [91].

Adsorption is recognized as the preferred method among the physicochemical methods for removing pharmaceuticals from wastewater [95]. This preference stems from the ability of adsorption to produce high-quality treated wastewater, its ease of operation, cost-effectiveness, and the lack of undesirable byproducts [106,107]. Additionally, adsorption can be employed for treating various types of water and wastewater, including those with a high organic compound content that cannot be effectively removed by other methods [108]. Adsorption has some drawbacks, including the gradual loss of adsorbent efficiency over time, which can reduce its effectiveness in long-term applications. Additionally, the material required for water treatment, especially in large-scale operations such as mining, can be substantial. The proper disposal of spent sorbent and regeneration solutions poses another challenge, as these processes can be complex and environmentally taxing. Furthermore, the desorption of pollutants may not always be successful, potentially limiting the reuse of adsorbents. Lastly, while novel nanomaterials offer promising advancements, they are still in the developmental stage, which may delay their widespread adoption and application.

While adsorption using porous materials (activated carbon) is effective for removing these pharmaceuticals, limitations include regeneration issues and low selectivity for organic adsorbents [108,109]. Activated charcoal, characterized by its well-developed pore structure, large surface area, and high fragmentation, adsorbs a wide spectrum of medicines, especially hydrophobic compounds [110–112]. However, its disadvantage lies in its limited capacity and the adverse effects of regenerative processes on its pore structure and the chemical properties of its functional groups, reducing its adsorption efficiency for pharmaceutical removal.

As an alternative, zeolites and mesoporous silica materials offer advantages, including shorter contact time, lower desorption percentages, and better structural stability (enabling regeneration at high temperature). Herein, the general characteristics of zeolites and mesoporous silica materials, along with a review of relevant data, are presented. While individual publications describe the removal efficiency of selected pharmaceuticals using zeolite sorbents or mesoporous materials, there is a need for studies that provide comparative summaries of their removal efficiencies. Therefore, this review aims to collect and introduce examples of the successful and efficient use of zeolites and mesoporous materials in removing pharmaceuticals from water.

Challenges of Using Modified Zeolites for Wastewater and Water Treatment

Modified zeolites hold significant potential for wastewater and water treatment; however, several challenges must be addressed to optimize their application. The size of zeolite particles impacts their performance, as larger particles slow ammonium diffusion rates, reducing ion-exchange efficiency, while powdered zeolites enhance diffusion but are impractical for industrial use due to high pressure drops and material losses [113]. The efficiency of surface-modified zeolites is also influenced by solution pH, with higher pH levels reducing anion adsorption, as OH^- ions compete with target anions for active sites [114]. Furthermore, assessing the cost benefit of using modified zeolites is complex, given the variability in raw material availability, processing needs, and modification methods, although they remain cost-effective and efficient for pollutant removal [115]. Maintaining adsorption efficiency over multiple cycles is critical, and regeneration processes, such as chemical treatments, have proven effective in restoring their performance [116]. Scaling up zeolite use for industrial applications faces hurdles, including the impracticality of powdered forms in treatment columns and the need for consistent performance across varying sizes and modifications [113]. Despite these challenges, modified zeolites derived from materials such as fly ash demonstrate promising results in nutrient removal, such as ammonium and phosphorus, though their efficiency is dependent on specific modifications and environmental conditions [117]. Addressing these limitations is essential for realizing the full potential of modified zeolites in large-scale water treatment applications.

2.3. *Adsorption of Selected NSAIDs (Sodium Diclofenac, Ibuprofen, and Naproxen) and Selected Antibiotics (Erythromycin and Sulfamethoxazole)*

The sorption of diclofenac sodium is influenced by both the geometry of the pores in the zeolite and the molecules of the zeolite (such as hierarchical ZSM-5), as well as the interaction with active sites [118]. In a study using natural Jordanate zeolite (intermediate silica), the optimal sorption of diclofenac sodium occurred at pH 6. This improvement was attributed to the penetration of diclofenac sodium cation into the zeolite's pores. More efficient removal of diclofenac sodium was noted at higher initial concentrations, where stronger interactions occurred between the natural zeolite and diclofenac sodium [119].

Additionally, the modification of MOR using TiO_2 with SiO_2/Al_2O_3 ratios of 18 and 240—denoted as TMOR18 and TMOR240—demonstrated more effective adsorption of sodium diclofenac on zeolites of higher SiO_2/Al_2O_3 ratio, due to the high hydrophobicity of the zeolite [120], along with lower negative charges compared to those with a lower SiO_2/Al_2O_3 ratio [121]. On the other hand, sorption of diclofenac sodium onto magnetic nanoparticle-coated zeolite (MNCZ) was found to agree with earlier findings, in which removal was observed to be more selective in solutions with an acidic pH. See Figure 8 [122].

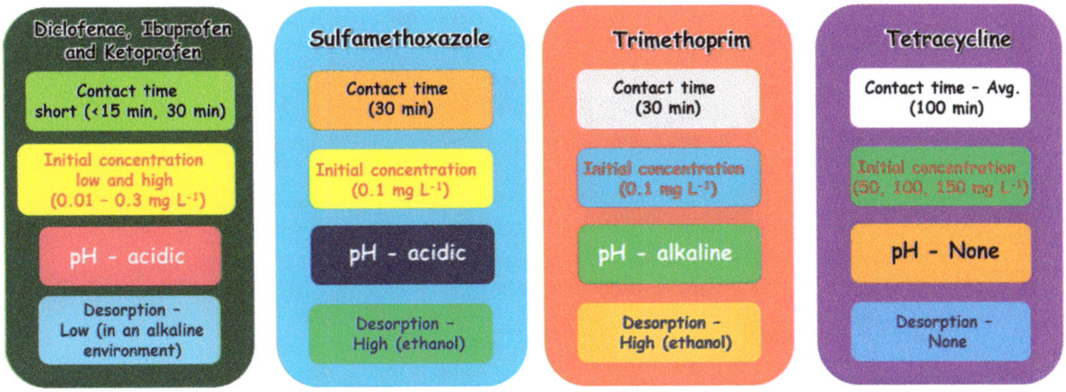

Figure 8. Optimized factors influencing the removal efficiency of selected pollutants.

Jordanate natural zeolite was found to adsorb ibuprofen effectively at an acidic pH, with optimal results at pH 2 [119]. Similar results were observed with MNCZ, where complete adsorption occurred after 10 min [118]. At higher initial concentrations, the resistance to mass transfer of ibuprofen molecules between the aqueous and solid phases of Jordanate natural zeolite was quickly overcome [119]. However, with MNCZ, the efficiency of ibuprofen removal decreases at higher concentrations and improves at lower concentrations [119]. For naproxen, adsorption was less effective under alkaline conditions due to the formation of water complexes that delayed the sorption because of the increased number of hydroxyl ions. As for acidic conditions, sorption of naproxen occurs rapidly, requiring only 10 min of contact time for effective removal. In addition, removal occurs better with zeolites having a higher SiO_2/Al_2O_3 ratio and a low initial concentration [118,123].

As for erythromycin, three zeolites were tested for their efficiency in its removal, including the two zeolites, *MOR-400* and *Y-zeolites*, and natural Slovak zeolite from Zeocem [124]. The removal of erythromycin is particularly critical, as a study revealed that even low concentrations of erythromycin in the environment could create the necessary pressure for microbes to develop resistance—especially with chronic exposure to the pharmaceutical. The analysis indicates that adsorption of erythromycin involved both the entire zeolite structure and its micropores [124]. Using three fractions of different concentrations of erythromycin, its removal from wastewater was tested using Zeocem natural zeolite. Over 90% removal was achieved within a contact time of only 30 min, the high efficiency being attributed to the optimization of pH (*the pH of the wastewater must be lower than the pKa of erythromycin*, i.e., *8.88*) [125]. Water samples collected at the outlet of a wastewater treatment plant in Ferrara (northern Italy) were used to investigate the adsorption efficiency of zeolite Y. The concentration of erythromycin in these samples was about 1.10 $\mu g\ g^{-1}$, and the results showed that 100% removal occurred using this zeolite [124].

The effectiveness of sulfamethoxazole removal was improved when using high-silicon zeolites [126]. Studies carried on MOR and ZSM-5 zeolites confirmed a temperature dependency on sorption efficacy. Sorption with MOR was found to be more effective at room temperature, while ZSM-5 showed higher removal efficacy at 65 °C. Research findings also showed that sorption efficiency of individual zeolites was affected by the initial concentration [127]. For examples, at low concentrations (20 μM) of contaminants, the best results were noted with the use of zeolite Y. In contrast, for high concentrations (180 μM), using ZSM-5 showed more satisfactory results. Another advantageous aspect for using zeolite Y, MOR, and ZSM-5 is that contaminants were adsorbed irreversibly [128]. Among the three sorbents, zeolite Y was found to be the most effective. Moreover, the modification of MOR18 zeolite with TiO_2 was observed to significantly enhance the adsorption of sulfamethoxazole [129].

The removal of sodium diclofenac and ibuprofen using Jordanate natural zeolite (*intermediate silica*) was investigated. The optimum pH for ibuprofen was 2, whereas a pH of 6 was shown to be more effective for sodium diclofenac. In addition, based on the chemical structure of zeolites (*at a low ibuprofen concentration*), the drug was found to interact strongly with zeolites through the carboxylic group and oxygen atoms combined with silicon and aluminum elements. At pH 6, sodium diclofenac exhibited similar interactions with zeolite molecules through sodium cations, which penetrate the pores of the zeolite and are involved in an electrostatic attraction with the amine group of diclofenac [119].

The maximum removal percentage of both diclofenac and ibuprofen was reached after 10 min, though the optimum contact time was extended up to 80 min to ensure satisfactory contact. Peak removal was observed at sodium diclofenac and ibuprofen concentrations of 2.0 and 1.0 g L^{-1}, respectively. Tests within a concentration range of 10–50 mg L^{-1} showed that removal efficiency increased with concentration. This trend can be explained by a higher initial concentration overcoming the resistance to mass transfer of pharmaceutical molecules between the aqueous and solid phases, promoting a stronger interaction between the natural zeolite and medicines. It was shown that the adsorption capacity of tested zeolite for sodium diclofenac was 4.8 mg g^{-1} (*based on the Langmuir isotherm*), while for ibuprofen it was 1.23 mg g^{-1}(*based on the Freundlich isotherm*) [119].

Modifying specific zeolites with SiO_2/Al_2O_3 or TiO_2 powders (*such as TMOR-18* or *TMOR-240*) has been explored for enhancing the removal of pharmaceuticals such as diclofenac and sulfamethoxazole. While modification with TiO_2 powder decreases the specific surface area, it allows for the formation of new mesopores and macropores [119]; however, these changes did not have a significant impact on the adsorption of medicines. Similarly, adjusting SiO_2/Al_2O_3 ratios showed little improvement in adsorption, with the zeolites having higher SiO_2/Al_2O_3 ratios becoming more hydrophobic and exhibiting lower negative charges compared to those with lower SiO_2/Al_2O_3 ratios. For example, on TMOR-240, diclofenac was removed only by 20–30. This is due to diclofenac's negative charge, which may cause electrostatic repulsion between MOR240 and the drug [119].

Another form of zeolite modification is coating the zeolite surface with magnetic nanoparticles (e.g., *MNCZ zeolite*). Optimization studies for sorption parameters (*contact time, solution pH, and initial concentration*) showed a high dependence on the pH of the solution, with pH changes from acidic to alkaline conditions reducing MNCZ removal efficiency. The lower sorption efficiency at an alkaline pH was attributed to the increased number of hydroxyl ions and thus the formation of aqueous complexes, which delayed the sorption. For instance, the adsorption values recorded at a pH of 2 were 99.6% for diclofenac sodium, 98.8% for ibuprofen, and 99.8% for naproxen solutions, while the values determined for the same solutions at a pH of 11 were 93.9%, 90.8%, and 90.7%, respectively. The study also showed that MNCZ has a very wide workable pH range (2–9), suggesting that MNCZ could be used for drinking water applications and/or wastewater treatment from medicines. In another study [122], contact time with MNCZ had no significant effect on the adsorption of diclofenac sodium, and the removal efficiency was estimated over 95% within 10 min contact time [122]. Moreover, the initial concentration had no significant effect on the removal efficiency of the tested pharmaceuticals at low concentrations. Nevertheless, a significant reduction in the removal efficiency of the investigated compounds (e.g., *ibuprofen*) was noticed upon increasing the concentration [122].

Inorganic mesoporous silica materials SBA-15 and MCM-41 were alternatively employed as adsorbents for water treatment [87]. These materials have been characterized by their high porosity, uniform and narrow pore sizes, ordered arrangement of pore structures, large pore volume, and large surface area, which increase their adsorption capacity. Silica-based porous materials have already been used by other researchers to remove pharmaceutical residues [130]. Two such materials—MCM-41 and SBA-15—were tested for the selected NSAIDs and antibiotics. SBA-15 was tested in two independent studies for the removal of diclofenac, ibuprofen, and ketoprofen. The contact time needed to remove the three mentioned NSAIDs was less than 15 min, whereas in the other two studies, the required contact time was 30 min [94]. Furthermore, SBA-15 exhibited better adsorption efficiency than MCM-41. Additionally, the enhancement of the adsorption process observed at low pH values indicates that the interaction between drugs and the surface of mesoporous silica is hydrophilic in nature. In addition, the low desorption rate of these pharmaceuticals in an alkaline environment demonstrates that adsorption onto SBA-15 and MCM-41 was very strong.

Similarly, sulfamethoxazole and trimethoprim were removed using trimethylsilylated mesoporous SBA-15 (TMS-SBA-15) material. The initial concentrations of both antibiotics were high, but the contact time was short (30 min), and complete desorption was achieved with the use of ethanol. Sulfamethoxazole, an anionic compound, was efficiently removed at low pH values. On the other hand, trimethoprim is a cationic compound, and its sorption was more effective as the pH of the solution increased [131]. The mesoporous material MCM-41, impregnated with zeolite-A, was used for tetracycline removal [132]. Upon testing high concentrations of the antibiotic with this material, the results showed that tetracycline adsorption was very effective at high concentrations [132]; however, the removal process was very slow (*contact time reached 100 min*), with MCM-41 reactive sites becoming saturated because of tetracycline adsorption.

In summary, SBA-15 showed promise as an adsorbent for the removal of diclofenac, ibuprofen, and ketoprofen from surface water and wastewater from industrial pharmaceuticals, both of which have high concentrations of these drugs. For an SBA-15 dose of 2.0 g L^{-1}, the removal efficiencies of diclofenac, ibuprofen, and ketoprofen were 66.7%, 95.1%, and 91.2%, respectively. Within the dose range of 0.1–1.0 g L^{-1}, the adsorption of the three pharmaceuticals was noticed to increase linearly with the increase in the concentration of the adsorbent. The increase in the adsorbent dose from 1.0 to 2.0 g L^{-1} has not shown any significant effects. The pH adsorption dependence of these drugs suggests a hydrophilic interaction with the mesoporous silica surface. Moreover, the lower desorption from the silica surface in alkaline media suggests that these pharmaceuticals were strongly adsorbed onto SBA-15. The advantage of using SBA-15 is the ease of regeneration after sorption by combustion, without any material losses, attributed to the stability of the mesoporous silica structure at high temperatures (up to 850 °C) [87].

3. Modified Zeolites for the Removal of Pesticides from Water Resources

Zeolite modified with various loadings of the surfactant stearyl dimethyl-benzyl ammonium chloride (SDBAC) was used for pesticides removal from water bodies [133].

Batch reactors using 2.5 g of modified zeolite to 100 mL of different concentrations of three pesticides including atrazine, lindane, and diazinone. Significant removal capacities of 2.01 μmol/kg for atrazine, 3.4 μmol/kg for lindane, and 4 μmol/kg for diazinone were obtained [133]. Metal-anchored zeolites have been presented in the literature as efficient and active catalysts for various organic pollutants. Several studies have shown that silver-doped zeolites exhibit strong catalytic activity in aqueous solutions. We investigated various dopants in zeolite A in several studies. For example, three dopants in zeolite 5A, including Ag, Pt, and mixed Ag-Pt, were tested for the degradation of naptalam [134]. While Ag-doped 5A showed high affinity for adsorbing naptalam from the solution upon mixing, the mixed silver–platinum–5A catalyst showed strong activity in the degradation of naptalam [134].

Doped metals into zeolites play a major role in the decontamination of water resources from organophosphorus pesticides. In another study, Ag(I)-anchored A-type zeolites with various compositions and colors were investigated. All modified samples enhanced the degradation of the malathion pesticide by 18 to 35 folds compared to malathion alone in solution. The rate variation was dependent on the size and the charge of the Ag clusters. The large size of malathion prevented its entrapment in the zeolite channels; therefore, the photodecomposition and catalytic behavior of Ag clusters occurred on the Ag/zeolite A surface [135]. Silver exchanged into zeolite X and mordenite provides good heterogeneous catalysts for the decomposition of carbamate pesticides. Various silver loadings into X and mordenite zeolites were tested for the degradation of carbofuran and carbaryl pesticides. The catalytic activity aligned with the high Ag loadings, with an observed enhancement in the carbofuran degradation by 184 folds compared to the uncatalyzed reaction under optimized conditions [136]. The catalytic activities were directly aligned to the metal loading, which enhances the possibility of making large $[Ag^+]_n$ oligomers and thus improves the photocatalytic properties of the doped zeolites [136,137].

Prada-Vásquez et al., investigated the use of two faujasite-type Y zeolite with 60 and 12 Si/Al ratios. The modified materials were tested for the removal of 25 pharmaceutical compounds (PhCs) present in real effluents from two municipal wastewater treatment plants located in the Mediterranean coast of Spain. It was shown that the use of ozone in the presence of NaY zeolites significantly improved the micropollutants degradation rate, able to degrade 95 % of the total mixture of PhCs within the early 9- and 12-min using NaY-12 and NaY-60, respectively. Environmental assessment showed that the total risk quotients of pharmaceuticals were reduced between 87–99% after ozonation in the presence of NaY zeolite indicating that the catalyst is a promising route for eliminating micropollutants from wastewater matrices [138]. We recently reported Ag doped zeolite Y as photocatalyst towards the photodecomposition of 2,3,7,8-tetrachlorodibenzo-p-dioxin (TCDD). The catalyst was found to effectively decompose TCDD in a methanol/tetrahydrofuran solution to reach 86% after 5 h with various products were characterized under various catalytic experimental conditions as illustrated in Figure 9 [139]. Natural zeolites used as coating material over porous ceramic substrates produced filters of various thicknesses. Using a flow rate of 240 L/h, various micropollutant adsorption was tested in distilled water, surface water, and wastewater treatment plant effluents. The adsorption of ten tested micropollutants showed an overall efficiency of 41.5% and 43% on the surface modified zeolite filters. While, high adsorption capacity for metoprolol, citalopram, venlafaxine, and clarithromycin over natural zeolite filters, the surface modified zeolite filters showed high adsorption affinity for hydrochlorothiazide, diclofenac, candesartan, mecoprop and irbesartan [140].

The potential of Fe doped ZSM-5 zeolite immobilized with P25-commercial TiO_2 (TiO_2-FeZ) was used in the photocatalytic treatment of diclofenac (DCF) from water solution [141].

It was found that the biodegradability increased when there was a decrease in the aromatic DCF by-products and enhanced dechlorination of organic structures. The degradation occurred via DCF adsorption on the catalyst surface followed by hydroxylation and ring cleavage, which illustrates the photodegradation pathways and all products formed [141].

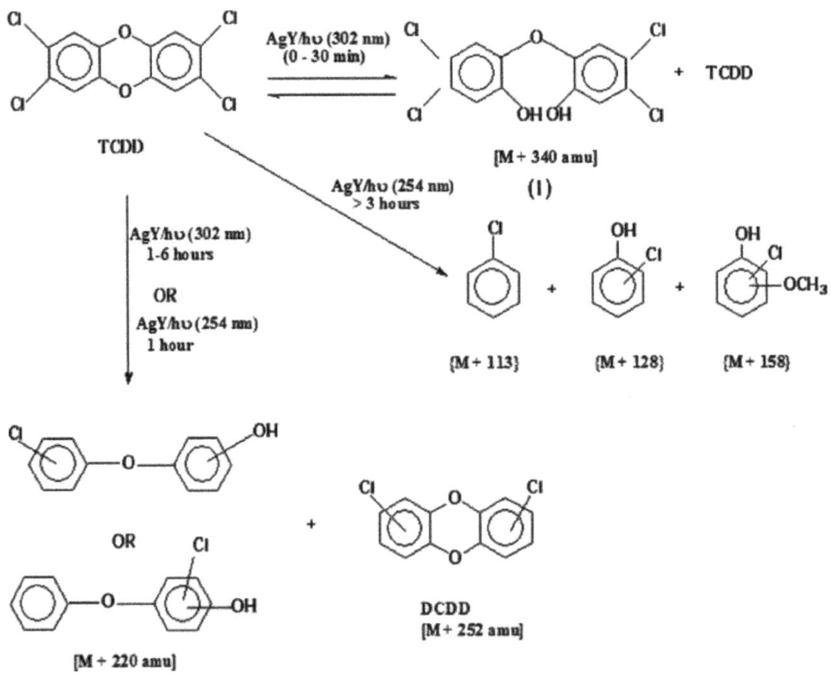

Figure 9. Possible pathway for degradation of TCDD by Ag/zeolite nanocatalyst [139].

TiO_2 is one of the most common photocatalysts studied in the literature; however, the separation and reuse of the catalyst from solution is one of the associated limitations for its industrial use [142,143]. Various supports were implemented with TiO_2 to enhance catalyst separation. Particularly, TiO_2-incorporated HZSM-11 zeolite was used for the degradation of the insecticide dichlorvos in water, resulting in a degradation efficiency similar to that of commercial TiO_2 P25 [142]. In another study, TiO_2 doped into H_β exhibited better activity for both the degradation and mineralization of monocrotophos pesticides (MCPs), where MCP and its intermediates were found to be strongly adsorbed onto supported TiO_2 [143]. Similarly, TiO_2-supported photocatalyst decontaminate dichlorvos pesticides present in indoor air samples, with one identified by-product in the gas phase posing potential toxicity [144]. Table 1 summarizes the major catalytic studies for various pollutants that were addressed in this study.

Zeolite X along with its carbon composite, obtained through the chemical transformation of fly, enhanced the adsorption capacity towards 2,4-dichlorophenoxyacetic acid (2,4-D), 2-methyl-4-chlorophenoxyacetic acid (MCPA), carbendazim, and simazine. The pesticide adsorption occurred within 60 s [145]. It has been shown that cationic surfactants significantly enhance the adsorption of 2,4-D and MCPA, whereas the unmodified zeolite-carbon composite has high affinity toward carbendazim and simazine sorption efficiency [145]. Polyaniline (PANI)-modified zeolite NaY was investigated as a solid-phase extraction sorbent of commonly used pesticides including carbamate, organophosphate, sulfonylurea, pyrethroid and neonicotinoid. It was found that a sorption capacity of

833 mg/kg sorbent and a preconcentration factor of 42 in the mixture of twenty pesticides. The use of the new zeolite-based sorbent along with HPLC-PDA detection provides an effective pesticide detection method for food and environmental samples with various matrices. This method provided LODs and LOQs in the ranges of 0.001–1.00 mg/L and 0.005−2.50 mg/L, respectively with recoveries of the target pesticides in the range of 64−128% [146].

Table 1. Modified zeolites and their catalytic applications for various pollutants.

Catalyst	Target Pollutant	Removal Efficiency/Comments	Energy Source (Reaction Time)	Ref.
g-C_3N4:ZnO/Ag_3PO_4 in pristine and natural zeolite	MO	98.2% removal using zeolite support	Sunlight (180 min)	[59]
Au–TiO_2/HZSM-5 nanocomposite	MO	96.6% with high stability for 4 cycles	365 nm mercury lamp (30 min)	[60]
Cr/TiO_2/zeolite	MO	41.73% efficiency, which is 17.9% higher than that of TiO_2/zeolite alone	254 nm UV lamp (5 h)	[61]
Modified clinoptilolite–mordenite zeolite	Caffeine	99% decomposition	254 nm UV light (4 h)	[63]
ZnO/zeolite pellets	Caffeine	100% removal in two hours	UV-LEDs strip 365–370 nm (2 h)	[64]
TiO_2/zeolite	Sulfadiazine	93.31%	UV light (120 min)	[65]
Zeolite/Fe_3O_4/CuS/$CuWO_4$ nanocomposite	Levofloxacin and acetaminophen	total organic carbon removal efficiency of 57.43%	solar	[67]
MoS_2-zeolite	Tetracycline	87.23% removal efficiency	A 300 W Xe-lamp with a UV light filter (420 nm)	[68]
Ag clusters–A-zeolite	Malathion	Removal enhancement of up to 35-fold that of malathion alone	UV irradiation	[135]
Ag-X & Ag-Mordenite	Carbofuran	Removal enhanced by 184 fold compared to uncatalyzed s	UV light	[136]
NaY-12 and NaY-60	Mixture pharmaceuticals	Total removal of 87–99% after ozonation	(9–12 min)	[138]
Fe doped ZSM-5 zeolite (TiO_2-FeZ)	Diclofenac	Up to 99% removal in presence of chloride and sulfates	Xe lamp of 450 W	[141]
TiO_2 incorporated HZSM-11	Dichlorvos	Similar activity as P-25 but it provides an easy separation and reuse	365 nm	[142]

Mustafa et al., reported the potential of a chitosan/zeolite-A composite (CS/ZA) to enhance the adsorption of three species of organophosphorus pesticides (acephate (AC), omthosate (OM), and methyl parathion (MP)). The material possesses high capacity towards AC, OM, and MP pesticides even in the presence of various metal ions with adsorption levels of 650.7 mg/g, 506.5 mg/g, and 560.8 mg/g, respectively [147]. The use of CS/ZA particles for the treatment of 7.5 L of the polluted solutions in fixed-beds continuous column revealed a removal of AC, OM, and MP to reach 78%, 57.6%, and 74.3%, respectively [147]. Chitosan-MgO/Zeolite (CS-MgO/Zeolite) composite was made and used as an adsorbent for the removal of chlorpyrifos from water bodies [148]. It is reported that using an initial dosage of 0.4 g/L solution, around 80.9% of chlorpyrifos was removed over CS-MgO/Zeolite at a neutral conditions pH of 7 after 40 min, with good fit to Freundlich adsorption isotherm model [148].

Silver-dodecatungstophosphoric acid and HBEA zeolite-modified composites showed superior adsorption potentials for aqueous pesticide removal. The study evaluated the amount of glyphosate adsorbed from a water suspension of 378 mg per gram of composite material with 25% AgWPA content. This high affinity is due to the modified active sites for glyphosate adsorption, and the radical-scavenging activity, based on the adsorption data, fits the Krishnamurti isotherm [149]. Composites of BEA zeolite and silver tungstophosphate were prepared by ion-exchange, and as physical mixtures with varying compositions. Batch adsorption testing of the silver-zeolite composite was optimized to eliminate nicosulfuron from an aqueous solution. Under optimized conditions, it was shown that the composite with a 2:1 mass ratio of BEA/silver tungstophosphate showed the highest removal efficiency for nicosulfuron, with 38.2 mg/g of the composite being reached [70]. In another study, tungstophosphoric acid and a BEA zeolite composite were synthesized by wet impregnation. The prepared composite exhibited higher nicosulfuron adsorption efficiency than the parent zeolite. The pesticide removal from all modified composites was within the range of 12.1–25.8 mg of nicosulfuron per gram of the investigated sample [150]. Further, potassium tungstophosphate-doped BEA zeolite showed high affinity for the adsorption of glyphosate, with adsorption capacities in the range of 45.2 and 92.2 mg of glyphosate per gram of the investigated composite [69]. Modification with potassium tungstophosphate affected the active sites present in BEA zeolite, which enhanced the amount of adsorbed pesticide in comparison to BEA zeolite, indicating their potential environmental application [69].

Modified natural clay minerals provide high adsorption capacity for organic pollutants [151,152]. Narayanan et al. reported that the use of a cellulosic biopolymer with bentonite provides 90% removal of pesticides, including pendimethalin, thiophanate methyl, and butachlor from aqueous systems, but the adsorption capacity tends to decline after two successive dosages [152]. In another study, modified bentonite with octadecyl amine (ODA-M) and combined ODA with amino propyl triethoxysilane (ODAAPS-M) demonstrated high affinity for pesticide removal from aqueous solutions via both adsorption and coagulation–flocculation [153]. Specifically, using unmodified bentonite removed various pesticides with efficiencies of 47 and 66%, whereas the modified forms ODA-M and ODAAPS-M exhibited an improved performance, with removal percentage ranges of 59–74% and 69–82%, respectively. In particular, combined PAC ODA-M-clay, provided 100% removal of atrazine, chlorpyriphos, and cypermethrin, with the minimum removal efficiency observed for endosulphan sulphate (82%) [153].

Functionalized zeolites such as cationic surfactant-modified zeolite adsorbents showed hydrophobic interactions between the zeolite framework and the methamidophos molecule in pesticides. In-depth analysis revealed that the mode of interaction was due to electrostatic and the van Der Waals force of attraction induced by the hydrophobic phase created by the alkyl chain of the zeolite composite [154]. As one of the most used pesticides, diazinon has been used in various absorption studies. In two separate studies, Amani et al. used MCM-41, MCM-48, and MCM-41 functionalized with propyl methacrylate (MPS-MCM-41) as adsorbates for P-based pesticides such as diazinon and fenitrothion [44,155,156]. The modified composite exhibited better adsorption capacity than the bare MCM-41 composite. The adsorption capacity tended to vary with the initial concentration of the pesticide and the solution's pH which, was mainly attributed to the increase in hydrogen bonding between the pesticide and the functional active sites on the surface of the adsorbents [155]. The adsorption of diazinon and fenitrothion onto MCM-41 and MCM-48 was also investigated and found to occur immediately (max adsorption within 5 min), with a higher adsorption rate occurring on MCM-41 [156]. It was found that there was a clear reduction in the specific surface area of each composite upon adsorption by the filling of pores and channels, indicating that the pores and voids were occupied after adsorption [156].

Chen et al. [157] analyzed the adsorption of seven different pesticides onto SBA-15, carbonized SBA-15 (SBA-15-c), SBA-15 monolith (SBA-15-m), MgO-modified SBA-15 (MgO-SBA-15), MCM-41, carbonized MCM-41 (MCM-41c), MCM-41 monolith (MCM-41m), and MCM-48. While pure MCM-41 revealed better sorption capacity for the selected pesticides, the removal efficiency of SBA-15 was found to be poor for all analyzed pesticides. This is because it has a greater surface area and different pore sizes compared to SBA-15. MgO/SBA-15 showed a higher immobilization rate compared to pure SBA-15, with 97% of dipterex being immobilized, indicating the importance of the alkali sites in the entrapping of organophosphorus pesticides [157]. Table 2 reveals the various types of modified zeolites and their adsorption capacities for various pesticides.

Table 2. Modified zeolites and their adsorption capacities for various pesticides.

Adsorbate	Adsorbents	Adsorption Capacity mg/g OR (%)	Surface Area m^2/g (Particle Size mm)	Ref.
Atrazine	Organo-zeolite	0.43	(0.4–0.8)	[133]
Lindane		0.99	(0.4–0.8)	
Diazinone		1.34	(0.4–0.8)	
Acephate	Chitosan-zeolite A	650.7	---	[147]
Omthosate		506.5	---	
methylparathion		560.8	---	
gylphosphate	AgWPA-HBEA zeolite	378	---	[149]
Organochlorine	ZSM-5	(98%)	---	[140]
chloridazon	Kerolite 400	1.568	248	[158]
chloridazon	Kerolite 600	2.253	224	
chloridazon	Kerolite 200	0.636	256	[159]
Ethion	Zif-8	279.3	>800	[160]
Ethion	Zif-67	210.8	~593.6	
prothiofos	Zif-8	366.7	>800	
prothiofos	Zif-67	261.1	~593.6	
paraquat	NaY (si/Al = 2.2)	210–240	789–856	[161]
2,4-D	HY (Si/Al = 100)	92.95	---	[162]
2,4-D	HY	11.39	744	
2,4-D	NaY	12.15	797	
2,4-D	30SNaY	33.15	466	

Table 2. *Cont.*

Adsorbate	Adsorbents	Adsorption Capacity mg/g OR (%)	Surface Area m^2/g (Particle Size mm)	Ref.
bentazone	CHA	(15.5%)	11 (external)	[163]
bentazone	FAU	(100%)	661	
Bentazone	LTA	(1.59%)	1 (external)	
bentazone	MOR	(22.05%)	34	
Imidacloprid	CHA	(8.37%)	11 (external)	
Imidacloprid	FAU	(99.72%)	661	
Imidacloprid	LTA	(7.26%)	1 (external)	
Imidacloprid	MOR	(79.48%)	34	
Isoproturon	CHA	(3.7%)	11 (external)	
Isoproturon	FAU	(100%)	661	
Isoproturon	LTA	(0.00%)	1 (external)	
Isoproturon	MOR	(94.71%)	34	
Metalaxyl-M	CHA	(24.06%)	11 (external)	
Metalaxyl-M	FAU	(99.34%)	661	
Metalaxyl-M	LTA	(46.53%)	1 (external)	
Metalaxyl-M	MOR	(31.23%)	34	
Phoxim	NaY	(5.5%)	672	[157]
	FeY	(19.1%)	459	
	SBA-15c	(70.9%)	885	
	MgO/SBA-15	(29.2%)	481	
Chlorpyrifos	NaY	(11.0%)	672	
	FeY	(19.5%)	459	
	SBA-15c	(84.6%)	885	
	MgO/SBA-15	(22.7%)	481	
Dipeterex	NaY	(15.1%)	672	
	FeY	(31.9%)	459	
	SBA-15c	(7.4%)	885	
	MgO/SBA-15	(97.4%)	481	
Carbendazim	NaY	(4.5%)	672	
	FeY	(20.2%)	459	
	SBA-15c	(98.0%)	885	
	MgO/SBA-15	(8.7%)	481	
imidacloprid	NaY	(0.0%)	672	
	FeY	(21.5%)	459	
	SBA-15c	(23.0%)	885	
	MgO/SBA-15	(1.4%)	481	
atrazine	Octadecylamine–bentonite	(100%)	7.15 mm	[153]
diazinone	Octadecylamine–bentonite	(100%)	7.15 mm	

The hydrophobicity of high silica zeolites makes them suitable for organic micropollutant (OMP) adsorption in aqueous solutions [164]. Despite their potential to remove OMPs, including pharmaceuticals, personal care products, and industrial chemicals [165,166], zeolites with an aluminum content are stable under oxidative conditions, enhancing their potential for oxidation without compromising their surface properties and pore structure [167–169]. The variations in the Si/Al ratio change the acidic site strength, with the adsorbed amount of organic and inorganic pollutants being affected by the effect of modified zeolitic materials. As an example, by adjusting the temperature, initial concentration, and pH of the methylene blue solution, the adsorbed amount may increase from the order of 28 to 660 mg/g in a modified zeolite. Similar results were obtained for cadmium adsorption on zeolitic materials, which exceeded 190 mg/g, compared to 50 mg/g for the non-modified zeolite [170].

Studies have shown that the efficiency of zeolite ZSM-5 for polar molecules such as atrazine was better than that of zeolite Y [44,171,172]. In addition, three types of zeolites with various Si/Al ratios—3.61 (Y), 8.61 (Y-10), and 111.35 (Y-100)—along with the corresponding HDTMA-modified composites, were also tested for the removal of atrazine and linuron [172]. It was found that linuron showed higher adsorption capacity than atrazine, given the zeolite pore size of 7.4 Å, which can accommodate linuron (thickness 6.12 Å) but not atrazine (thickness 9.6 Å) [172]. In addition, zeolite Y modified with HDTMA surfactant showed a reduction in adsorption efficiency for linuron and atrazine [172]. The adsorption of atrazine was faster on X-zeolite compared to A-type zeolites, with equilibrium being established within 60 min on X compared to that obtained over zeolite A, which took 6 h. While the surface area of X-zeolite is double of that for A-zeolite, atrazine is too large in size to fit the zeolite pore, so it is believed that the cation acts as a strong localized positive charge to attract atrazine [173].

4. Degradation Stability and Production Costs of Modified Zeolites

Zeolites exhibit a remarkable photochemical stability, making them ideal for repeated use in photocatalytic applications [174,175]. The research indicates that zeolites retain their structure and morphology even after multiple degradation cycles, demonstrating exceptional stability and reusability [176]. When combined with other materials, such as ZnO/SOD, zeolites have shown stable degradation rates across multiple cycles, although slight reductions in efficiency may occur due to mass loss or the adsorption of intermediate products [177]. However, some structural degradation has been observed under specific conditions, highlighting the importance of using milder conditions to preserve stability during adsorption processes [178]. Their high surface area, efficient electron transfer, and durable porous structure make zeolites particularly effective for environmental applications.

The cost of producing modified zeolites varies widely, depending on factors such as the type of zeolite, the raw materials utilized, and the production process. Zeolite production costs typically range from USD 85 to USD 160 per ton, depending on the synthesis methods and materials employed [179]. Utilizing inexpensive raw materials such as fly ash, clay, bentonite, lithium slag, paper sludge, and coal fly ash can significantly lower production expenses. The production cost of zeolite 13X, synthesized from fly ash, is approximately one-fifth of its commercial price [180]. The manufacturing cost for NaX, NaA, or NaY zeolites, derived from rice husk, is estimated at around USD 1.4 per 100 g [181].

5. Summary and Perspectives

5.1. Pharmaceuticals

Pharmaceuticals, which have been present in the environment for two to three decades, are an environmental concern. They have been detected by various investigations, and their dangerous effects on the flora, fauna, and humans have been proven. The migration of pharmacological contaminants takes place in surface waters because of their incomplete removal from sewage treatment plants. Therefore, the constant searching for advanced and effective processes for removing these pollutants is in high demand. Adsorption is advantageous over AOPs (advanced oxidation processes), as they do not produce new harmful products. However, the requirement of a large amount of adsorbent, the need for regeneration or disposal after use, as well as consequences related to the adsorbed pharmaceuticals and their derivatives, are all considered shortages in adsorption methods that may be encountered when applied for their removal. Zeolites and mesoporous silica materials have emerged as cost-effective and readily available adsorbents, widely used for the removal of NSAIDs and antibiotics in both laboratory experiments and small-scale implementation studies.

- This review has provided a comprehensive analysis of research findings related to pharmaceutical disposal, focusing on the application of these materials for retaining selected medicines.
- New technologies for the effective removal of pharmaceuticals from wastewater must be adopted. These removal technologies should be affordable and easy to apply in the field, as well as at a large scale, and at a low cost.
- Standards for defining the maximum allowable concentration and amounts of pharmaceuticals entering the environment (*aquatic ecosystems*) should be implemented to reduce the disposal amounts into wastewater treatment plants.
- Advanced methods for the continuous detection of drug disposal processes should be developed and applied, especially in rapidly developing industrial countries.
- Research should be conducted on environmental systems continually to determine the pharmaceuticals' effect on aquatic life. NSAIDs and antibiotics may be the best recommended to accomplish this purpose.

The development and implementation of sorption-based techniques (using zeolites or mesoporous materials) for the elimination of NSAIDs and antibiotics may also be applied to clean environmental systems from other classes of drugs (such as hormones, antidepressants, anticancer substances, or beta-blockers).

5.2. *Pesticides*

Over the past two decades, the widespread use of pesticides in agriculture for increased food production, along with the expansion of specialized food, has led to a significant rise in pesticide-related waste entering water supplies, posing serious health risks. This review discussed the use of zeolite-based materials for pesticide adsorption, highlighting their effectiveness and their enhanced adsorption properties, achieved through surface functionalization by AOP, for the elimination of harmful products. Despite promising results, the variability in experimental conditions and functionalization methods complicates the determination of the optimal sorbent for specific pesticides, underscoring the need for further research. The potential of zeolite-based composites, which include metal oxides, polymers, and carbon-based materials, is also explored, as they exhibit higher removal efficiencies and photocatalytic capabilities.

Finally, this review article calls for detailed assessments of adsorbent regeneration and cost analysis to ensure the economic viability of the adsorption process, aiming to advance the application of zeolite-based materials in sustainable water treatment solutions.

- The different conditions and methods used in various studies make it difficult to identify the most suitable adsorbent for specific pesticides. Standardizing experimental protocols could help draw more definitive conclusions.
- Composites of zeolites with materials such as metal oxides, polymers, and carbon-based materials show superior removal efficiencies and photocatalytic properties, suggesting a promising direction for future research and application.

- The synthesis of zeolites faces significant challenges, such as high costs, complex processes, and the generation of alkaline wastewater. Research should aim at simplifying synthesis methods, reducing costs, and achieving homogenous properties for industrial scalability.
- The potential of zeolite-based materials in photocatalysis for contaminant degradation is underexplored. More studies are needed to understand the mechanisms and enhance the efficiency of these materials for practical use in wastewater treatment.
- Adsorption stands out as an effective, easy-to-handle, and non-toxic method for pesticide removal. However, it requires careful consideration of adsorbent properties, modification processes, and potential secondary pollution.
- The economic viability of adsorption processes depends on the reuse and regeneration of adsorbents. Detailed studies on regeneration methods and cost analyses are crucial for the practical application of adsorption techniques.
- Future research should focus on developing cost-effective, easy-to-synthesize adsorbents that combine photocatalysis and adsorption. Comprehensive studies on adsorption mechanisms, kinetic and thermodynamic modeling, and the management of secondary pollution are essential for advancing the field.

By addressing these perspectives, researchers can advance the understanding and application of mesoporous silica materials and synthetic zeolites for pesticide removal, contributing to more effective and sustainable water treatment solutions.

5.3. Future Perspectives of Modified Zeolites for Wastewater and Water Treatment

Modified zeolites present promising future advancements for wastewater and water treatment, driven by their enhanced efficiency and broad applicability. Emerging technologies, such as nanoparticle-doped zeolites, are expected to revolutionize water treatment, offering sustainable solutions to global contamination challenges [182]. Optimizing regeneration processes and improving permeation flux and contaminant removal rates will enhance their cost-effectiveness and environmental benefits [179,183]. Zeolites also demonstrate great potential for nutrient removal, including ammonium and phosphorus, addressing eutrophication concerns [117]. Future innovations may focus on utilizing waste materials to create affordable modified adsorbents, which are particularly beneficial for rural and developing areas [183]. Additionally, functionalized zeolites offer improved selectivity and activity, expanding their role in environmental and wastewater treatment applications [184].

Funding: This work was funded by the Research Office at the American University of Sharjah [FRG22-C-S77].

Acknowledgments: This work was funded by the Research Office at the American University of Sharjah [FRG22-C-S77]. This work was also partially supported by the Open Access Program at the American University of Sharjah. This paper represents the opinions of the authors and does not mean to represent the position or opinions of the American University of Sharjah.

Conflicts of Interest: The authors declare that they have no known competing financial interests or personal relationships that could have appeared to influence the work reported in this paper.

References

1. Hassan, M.; Zhao, Y.; Xie, B. Employing TiO2 Photocatalysis to Deal with Landfill Leachate: Current Status and Development. *Chem. Eng. J.* **2016**, *285*, 264–275. [CrossRef]
2. Liu, Y.; Liu, F.; Pan, X.; Li, J. Protecting the Environment and Public Health from Pesticides. *Environ. Sci. Technol.* **2012**, *46*, 5658–5659. [CrossRef]

3. Yu, C.; Zhou, W.; Liu, H.; Liu, Y.; Dionysiou, D.D. Design and Fabrication of Microsphere Photocatalysts for Environmental Purification and Energy Conversion. *Chem. Eng. J.* **2016**, *287*, 117–129. [CrossRef]
4. Kanan, S.; Moyet, M.A.; Arthur, R.B.; Patterson, H.H. Recent Advances on TiO 2 -Based Photocatalysts toward the Degradation of Pesticides and Major Organic Pollutants from Water Bodies. *Catal. Rev.* **2020**, *62*, 1–65. [CrossRef]
5. Snyder, S.A.; Wert, E.C.; Rexing, D.J.; Zegers, R.E.; Drury, D.D. Ozone Oxidation of Endocrine Disruptors and Pharmaceuticals in Surface Water and Wastewater. *Ozone Sci. Eng.* **2006**, *28*, 445–460. [CrossRef]
6. Eggen, R.I.L.; Hollender, J.; Joss, A.; Schärer, M.; Stamm, C. Reducing the Discharge of Micropollutants in the Aquatic Environment: The Benefits of Upgrading Wastewater Treatment Plants. *Environ. Sci. Technol.* **2014**, *48*, 7683–7689. [CrossRef] [PubMed]
7. Michael, I.; Rizzo, L.; McArdell, C.S.; Manaia, C.M.; Merlin, C.; Schwartz, T.; Dagot, C.; Fatta-Kassinos, D. Urban Wastewater Treatment Plants as Hotspots for the Release of Antibiotics in the Environment: A Review. *Water Res.* **2013**, *47*, 957–995. [CrossRef]
8. Reemtsma, T.; Berger, U.; Arp, H.P.H.; Gallard, H.; Knepper, T.P.; Neumann, M.; Quintana, J.B.; de Voogt, P. Mind the Gap: Persistent and Mobile Organic Compounds—Water Contaminants That Slip Through. *Environ. Sci. Technol.* **2016**, *50*, 10308–10315. [CrossRef]
9. Petrie, B.; Barden, R.; Kasprzyk-Hordern, B. A Review on Emerging Contaminants in Wastewaters and the Environment: Current Knowledge, Understudied Areas and Recommendations for Future Monitoring. *Water Res.* **2015**, *72*, 3–27. [CrossRef] [PubMed]
10. Hughes, S.R.; Kay, P.; Brown, L.E. Global Synthesis and Critical Evaluation of Pharmaceutical Data Sets Collected from River Systems. *Environ. Sci. Technol.* **2013**, *47*, 661–677. [CrossRef]
11. Rekha; Naik, S.N.; Prasad, R. Pesticide Residue in Organic and Conventional Food-Risk Analysis. *J. Chem. Health Saf.* **2006**, *13*, 12–19. [CrossRef]
12. Corsini, E.; Liesivuori, J.; Vergieva, T.; Van Loveren, H.; Colosio, C. Effects of Pesticide Exposure on the Human Immune System. *Hum. Exp. Toxicol.* **2008**, *27*, 671–680. [CrossRef] [PubMed]
13. Foo, K.Y.; Hameed, B.H. Detoxification of Pesticide Waste via Activated Carbon Adsorption Process. *J. Hazard. Mater.* **2010**, *175*, 1–11. [CrossRef] [PubMed]
14. Bolognesi, C. Genotoxicity of Pesticides: A Review of Human Biomonitoring Studies. *Mutat. Res. Mutat. Res.* **2003**, *543*, 251–272. [CrossRef]
15. Aungpradit, T.; Sutthivaiyakit, P.; Martens, D.; Sutthivaiyakit, S.; Kettrup, A.A.F. Photocatalytic Degradation of Triazophos in Aqueous Titanium Dioxide Suspension: Identification of Intermediates and Degradation Pathways. *J. Hazard. Mater.* **2007**, *146*, 204–213. [CrossRef]
16. Lagadec, A.J.M.; Miller, D.J.; Lilke, A.V.; Hawthorne, S.B. Pilot-Scale Subcritical Water Remediation of Polycyclic Aromatic Hydrocarbon- and Pesticide-Contaminated Soil. *Environ. Sci. Technol.* **2000**, *34*, 1542–1548. [CrossRef]
17. Maldonado, M.I.; Malato, S.; Pérez-Estrada, L.A.; Gernjak, W.; Oller, I.; Doménech, X.; Peral, J. Partial Degradation of Five Pesticides and an Industrial Pollutant by Ozonation in a Pilot-Plant Scale Reactor. *J. Hazard. Mater.* **2006**, *138*, 363–369. [CrossRef]
18. Mahalakshmi, M.; Arabindoo, B.; Palanichamy, M.; Murugesan, V. Photocatalytic Degradation of Carbofuran Using Semiconductor Oxides. *J. Hazard. Mater.* **2007**, *143*, 240–245. [CrossRef]
19. Saritha, P.; Aparna, C.; Himabindu, V.; Anjaneyulu, Y. Comparison of Various Advanced Oxidation Processes for the Degradation of 4-Chloro-2 Nitrophenol. *J. Hazard. Mater.* **2007**, *149*, 609–614. [CrossRef]
20. Ahmad, A.L.; Tan, L.S.; Shukor, S.R.A. Dimethoate and Atrazine Retention from Aqueous Solution by Nanofiltration Membranes. *J. Hazard. Mater.* **2008**, *151*, 71–77. [CrossRef] [PubMed]
21. Rafatullah, M.; Sulaiman, O.; Hashim, R.; Ahmad, A. Adsorption of Methylene Blue on Low-Cost Adsorbents: A Review. *J. Hazard. Mater.* **2010**, *177*, 70–80. [CrossRef] [PubMed]
22. Villa, C.C.; Valencia, G.A.; López Córdoba, A.; Ortega-Toro, R.; Ahmed, S.; Gutiérrez, T.J. Zeolites for Food Applications: A Review. *Food Biosci.* **2022**, *46*, 101577. [CrossRef]
23. Zhang, X.; Wang, Y.; Liu, B.; Sang, Y.; Liu, H. Heterostructures Construction on TiO2 Nanobelts: A Powerful Tool for Building High-Performance Photocatalysts. *Appl. Catal. B Environ.* **2017**, *202*, 620–641. [CrossRef]
24. Fujishima, A.; Zhang, X.; Tryk, D. TiO2 Photocatalysis and Related Surface Phenomena. *Surf. Sci. Rep.* **2008**, *63*, 515–582. [CrossRef]
25. Pan, Z.; Stemmler, E.A.; Cho, H.J.; Fan, W.; LeBlanc, L.A.; Patterson, H.H.; Amirbahman, A. Photocatalytic Degradation of 17α-Ethinylestradiol (EE2) in the Presence of TiO2-Doped Zeolite. *J. Hazard. Mater.* **2014**, *279*, 17–25. [CrossRef] [PubMed]
26. Huwei, S.; Asghari, M.; Zahedipour-Sheshglani, P.; Alizadeh, M. Modeling and Optimizing the Changes in Physical and Biochemical Properties of Table Grapes in Response to Natural Zeolite Treatment. *LWT* **2021**, *141*, 110854. [CrossRef]
27. Magri, A.; Petriccione, M.; Gutiérrez, T.J. Metal-Organic Frameworks for Food Applications: A Review. *Food Chem.* **2021**, *354*, 129533. [CrossRef] [PubMed]
28. Amani, S.; Bagheri Garmarudi, A.; Rahmani, N.; Khanmohammadi, M. The β-Cyclodextrin-Modified Nanosized ZSM-5 Zeolite as a Carrier for Curcumin. *RSC Adv.* **2019**, *9*, 32348–32356. [CrossRef] [PubMed]

29. Zarrintaj, P.; Mahmodi, G.; Manouchehri, S.; Mashhadzadeh, A.H.; Khodadadi, M.; Servatan, M.; Ganjali, M.R.; Azambre, B.; Kim, S.; Ramsey, J.D.; et al. Zeolite in Tissue Engineering: Opportunities and Challenges. *MedComm* **2020**, *1*, 5–34. [CrossRef] [PubMed]
30. Ramesh, K.; Reddy, D.D. Zeolites and Their Potential Uses in Agriculture. *Adv. Agron.* **2011**, *113*, 219–241. [CrossRef]
31. Alonso, A.; Moral-Vico, J.; Abo Markeb, A.; Busquets-Fité, M.; Komilis, D.; Puntes, V.; Sánchez, A.; Font, X. Critical Review of Existing Nanomaterial Adsorbents to Capture Carbon Dioxide and Methane. *Sci. Total Environ.* **2017**, *595*, 51–62. [CrossRef] [PubMed]
32. Ennaert, T.; Van Aelst, J.; Dijkmans, J.; De Clercq, R.; Schutyser, W.; Dusselier, M.; Verboekend, D.; Sels, B.F. Potential and Challenges of Zeolite Chemistry in the Catalytic Conversion of Biomass. *Chem. Soc. Rev.* **2016**, *45*, 584–611. [CrossRef]
33. Li, Y.; Li, L.; Yu, J. Applications of Zeolites in Sustainable Chemistry. *Chem* **2017**, *3*, 928–949. [CrossRef]
34. Perego, C.; Bosetti, A.; Ricci, M.; Millini, R. Zeolite Materials for Biomass Conversion to Biofuel. *Energy Fuels* **2017**, *31*, 7721–7733. [CrossRef]
35. Shi, Y.; Xing, E.; Wu, K.; Wang, J.; Yang, M.; Wu, Y. Recent Progress on Upgrading of Bio-Oil to Hydrocarbons over Metal/zeolite Bifunctional Catalysts. *Catal. Sci. Technol.* **2017**, *7*, 2385–2415. [CrossRef]
36. Zhang, L.-L.; Xu, Y.-L.; Tang, Z.-H.; Xu, X.-H.; Chen, X.; Li, T.; Ding, C.-Y.; Huang, M.-Q.; Chen, X.-P.; Wang, Y.-T.; et al. Effects of Alisol B 23-Acetate on Ovarian Cancer Cells: G1 Phase Cell Cycle Arrest, Apoptosis, Migration and Invasion Inhibition. *Phytomedicine* **2016**, *23*, 800–809. [CrossRef] [PubMed]
37. Ghasemi, Z.; Sourinejad, I.; Kazemian, H.; Rohani, S. Application of Zeolites in Aquaculture Industry: A Review. *Rev. Aquac.* **2018**, *10*, 75–95. [CrossRef]
38. Eroglu, N.; Emekci, M.; Athanassiou, C.G. Applications of Natural Zeolites on Agriculture and Food Production. *J. Sci. Food Agric.* **2017**, *97*, 3487–3499. [CrossRef]
39. Noviello, M.; Gattullo, C.E.; Allegretta, I.; Terzano, R.; Gambacorta, G.; Paradiso, V.M. Synthetic Zeolite Materials from Recycled Glass and Aluminium Food Packaging as Potential Oenological Adjuvant. *Food Packag. Shelf Life* **2020**, *26*, 100572. [CrossRef]
40. Shi, J.; Yang, Z.; Dai, H.; Lu, X.; Peng, L.; Tan, X.; Shi, L.; Fahim, R. Preparation and Application of Modified Zeolites as Adsorbents in Wastewater Treatment. *Water Sci. Technol.* **2018**, *2017*, 621–635. [CrossRef] [PubMed]
41. Rathi, B.S.; Kumar, P.S.; Show, P.-L. A Review on Effective Removal of Emerging Contaminants from Aquatic Systems: Current Trends and Scope for Further Research. *J. Hazard. Mater.* **2021**, *409*, 124413. [CrossRef]
42. Suo, F.; Liu, X.; Li, C.; Yuan, M.; Zhang, B.; Wang, J.; Ma, Y.; Lai, Z.; Ji, M. Mesoporous Activated Carbon from Starch for Superior Rapid Pesticides Removal. *Int. J. Biol. Macromol.* **2019**, *121*, 806–813. [CrossRef]
43. Suo, F.; You, X.; Ma, Y.; Li, Y. Rapid Removal of Triazine Pesticides by P Doped Biochar and the Adsorption Mechanism. *Chemosphere* **2019**, *235*, 918–925. [CrossRef]
44. Andrunik, M.; Bajda, T. Removal of Pesticides from Waters by Adsorption: Comparison between Synthetic Zeolites and Mesoporous Silica Materials. A Review. *Materials* **2021**, *14*, 3532. [CrossRef]
45. Wang, S.; Peng, Y. Natural Zeolites as Effective Adsorbents in Water and Wastewater Treatment. *Chem. Eng. J.* **2010**, *156*, 11–24. [CrossRef]
46. Lichtenberg, F.R. How Many Life-Years Have New Drugs Saved? A Three-Way Fixed-Effects Analysis of 66 Diseases in 27 Countries, 2000–2013. *Int. Health* **2019**, *11*, 403–416. [CrossRef]
47. Heberer, T. Occurrence, Fate, and Removal of Pharmaceutical Residues in the Aquatic Environment: A Review of Recent Research Data. *Toxicol. Lett.* **2002**, *131*, 5–17. [CrossRef]
48. Halling-Sørensen, B.; Nors Nielsen, S.; Lanzky, P.F.; Ingerslev, F.; Holten Lützhøft, H.C.; Jørgensen, S.E. Occurrence, Fate and Effects of Pharmaceutical Substances in the Environment—A Review. *Chemosphere* **1998**, *36*, 357–393. [CrossRef] [PubMed]
49. Jelic, A.; Gros, M.; Ginebreda, A.; Cespedes-Sánchez, R.; Ventura, F.; Petrovic, M.; Barcelo, D. Occurrence, Partition and Removal of Pharmaceuticals in Sewage Water and Sludge during Wastewater Treatment. *Water Res.* **2011**, *45*, 1165–1176. [CrossRef]
50. Kanan, S.; Moyet, M.; Obeideen, K.; El-Sayed, Y.; Mohamed, A.A. Occurrence, Analysis and Removal of Pesticides, Hormones, Pharmaceuticals, and Other Contaminants in Soil and Water Streams for the Past Two Decades: A Review. *Res. Chem. Intermed.* **2022**, *48*, 3633–3683. [CrossRef]
51. Khetan, S.K.; Collins, T.J. Human Pharmaceuticals in the Aquatic Environment: A Challenge to Green Chemistry. *Chem. Rev.* **2007**, *107*, 2319–2364. [CrossRef] [PubMed]
52. Nikolaou, A.; Meric, S.; Fatta, D. Occurrence Patterns of Pharmaceuticals in Water and Wastewater Environments. *Anal. Bioanal. Chem.* **2007**, *387*, 1225–1234. [CrossRef]
53. aus der Beek, T.; Weber, F.; Bergmann, A.; Hickmann, S.; Ebert, I.; Hein, A.; Küster, A. Pharmaceuticals in the environment—Global Occurrences and Perspectives. *Environ. Toxicol. Chem.* **2016**, *35*, 823–835. [CrossRef]
54. Heberer, T. Tracking Persistent Pharmaceutical Residues from Municipal Sewage to Drinking Water. *J. Hydrol.* **2002**, *266*, 175–189. [CrossRef]

55. Zupanc, M.; Kosjek, T.; Petkovšek, M.; Dular, M.; Kompare, B.; Širok, B.; Blažeka, Ž.; Heath, E. Removal of Pharmaceuticals from Wastewater by Biological Processes, Hydrodynamic Cavitation and UV Treatment. *Ultrason. Sonochem.* **2013**, *20*, 1104–1112. [CrossRef]
56. Li, W.C. Occurrence, Sources, and Fate of Pharmaceuticals in Aquatic Environment and Soil. *Environ. Pollut.* **2014**, *187*, 193–201. [CrossRef] [PubMed]
57. Dzekashu, L.; Akoachere, J.; Mbacham, W. Pharmaceutical Waste Management Practices in Pharmacies and Medicine Stores: Evidence from Bamenda Health District, Cameroon. *Eur. J. Health Sci.* **2021**, *6*, 14–32. [CrossRef]
58. Rojas, M.R.; Leung, C.; Bonk, F.; Zhu, Y.; Edwards, L.; Arnold, R.G.; Sáez, A.E.; Klečka, G. Assessment of the Effectiveness of Secondary Wastewater Treatment Technologies to Remove Trace Chemicals of Emerging Concern. *Crit. Rev. Environ. Sci. Technol.* **2013**, *43*, 1281–1314. [CrossRef]
59. Bezu, Z.; Taddesse, A.M.; Diaz, I. Natural Zeolite Supported G-C3N4/ZnO/Ag3PO4 Composite: A Tandem N-N Heterojunction for Simultaneous Photodegradation of Dyes under Visible and Solar Irradiation. *J. Photochem. Photobiol. A Chem.* **2024**, *449*, 115369. [CrossRef]
60. Sun, J.; Liu, N.; Zhai, S.; Xiao, Z.; An, Q.; Huang, D. Gold-Titania/protonated Zeolite Nanocomposite Photocatalysts for Methyl Orange Degradation under Ultraviolet and Visible Irradiation. *Mater. Sci. Semicond. Process.* **2014**, *25*, 286–293. [CrossRef]
61. Wang, C.; Shi, H.; Li, Y. Synthesis and Characterization of Natural Zeolite Supported Cr-Doped TiO2 Photocatalysts. *Appl. Surf. Sci.* **2012**, *258*, 4328–4333. [CrossRef]
62. Ma, X.; Li, S.-Y.; Liu, S.; Li, H.-D.; Chen, H.-L.; Chen, Q.-L. Zeolite-Y/g-C3N4 Composite with Enhanced Photocatalytic Activity for Dye Degradation and Nitrogen Fixation. *Res. Chem. Intermed.* **2023**, *49*, 1147–1164. [CrossRef]
63. Alvarez-Aguiñaga, E.; Elizalde-González, M.; Sabinas-Hernández, S. Unpredicted photocatalytic activity of clinoptilolite–mordenite natural zeolite. *RSC Adv.* **2020**, *10*, 39251–39260. [CrossRef]
64. Sacco, O.; Vaiano, V.; Matarangolo, M. ZnO Supported on Zeolite Pellets as Efficient Catalytic System for the Removal of Caffeine by Adsorption and Photocatalysis. *Sep. Purif. Technol.* **2018**, *193*, 303–310. [CrossRef]
65. Liu, X.; Liu, Y.; Lu, S.; Guo, W.; Xi, B. Performance and Mechanism into TiO2/Zeolite Composites for Sulfadiazine Adsorption and Photodegradation. *Chem. Eng. J.* **2018**, *350*, 131–147. [CrossRef]
66. de Gennaro, B.; Catalanotti, L.; Cappelletti, P.; Langella, A.; Mercurio, M.; Serri, C.; Biondi, M.; Mayol, L. Surface Modified Natural Zeolite as a Carrier for Sustained Diclofenac Release: A Preliminary Feasibility Study. *Colloids Surf. B Biointerfaces* **2015**, *130*, 101–109. [CrossRef] [PubMed]
67. Ali, A.H.; Alwared, A.I. Construction of Ternary Heterostructure of zeolite/Fe 3 O 4 /CuS/CuWO 4 as a Reusable: Characterization Studies. *Asia-Pac. J. Chem. Eng.* **2024**, *19*, e3125. [CrossRef]
68. Liu, J.; Lin, H.; Dong, Y.; He, Y.; Liu, C. MoS2 Nanosheets Loaded on Collapsed Structure Zeolite as a Hydrophilic and Efficient Photocatalyst for Tetracycline Degradation and Synergistic Mechanism. *Chemosphere* **2022**, *287*, 132211. [CrossRef]
69. Vasiljević, B.N.; Obradović, M.; Bajuk-Bogdanović, D.; Milojević-Rakić, M.; Jovanović, Z.; Gavrilov, N.; Holclajtner-Antunović, I. In Situ Synthesis of Potassium Tungstophosphate Supported on BEA Zeolite and Perspective Application for Pesticide Removal. *J. Environ. Sci.* **2019**, *81*, 136–147. [CrossRef]
70. Janićijević, D.; Uskoković-Marković, S.; Ranković, D.; Milenković, M.; Jevremović, A.; Nedić Vasiljević, B.; Milojević-Rakić, M.; Bajuk-Bogdanović, D. Double Active BEA Zeolite/silver Tungstophosphates—Antimicrobial Effects and Pesticide Removal. *Sci. Total Environ.* **2020**, *735*, 139530. [CrossRef] [PubMed]
71. Jevremović, A.; Nedić Vasiljević, B.; Popa, A.; Uskoković-Marković, S.; Ignjatović, L.; Bajuk-Bogdanović, D.; Milojević-Rakić, M. The Environmental Impact of Potassium tungstophosphate/ZSM-5 Zeolite: Insight into Catalysis and Adsorption Processes. *Microporous Mesoporous Mater.* **2021**, *315*, 110925. [CrossRef]
72. Jevremović, A.; Božinović, N.; Arsenijević, D.; Marmakov, S.; Nedić Vasiljević, B.; Uskoković-Marković, S.; Bajuk-Bogdanović, D.; Milojević-Rakić, M. Modulation of Cytotoxicity by Consecutive Adsorption of Tannic Acid and Pesticides on Surfactant Functionalized Zeolites. *Environ. Sci. Process. Impacts* **2020**, *22*, 2199–2211. [CrossRef]
73. Milojević-Rakić, M.; Bajuk-Bogdanović, D.; Nedić Vasiljević, B.; Rakić, A.; Škrivanj, S.; Ignjatović, L.; Dondur, V.; Mentus, S.; Ćirić-Marjanović, G. Polyaniline/FeZSM-5 Composites—Synthesis, Characterization and Their High Catalytic Activity for the Oxidative Degradation of Herbicide Glyphosate. *Microporous Mesoporous Mater.* **2018**, *267*, 68–79. [CrossRef]
74. Johnson, D.; Lopez, M.; Kelley, B. Dexamethasone. [Updated 2023 May 2]. In *StatPearls [Internet]*; StatPearls Publishing: Treasure Island, FL, USA, 2025. Available online: https://www.ncbi.nlm.nih.gov/books/NBK482130/ (accessed on 30 June 2024).
75. Dutta, V.; Dubey, D.; Kumar, S. Cleaning the River Ganga: Impact of lockdown on water quality and future implications on river rejuvenation strategies. *Sci Total Environ.* **2020**, *743*, 140756. [CrossRef]
76. Guiloski, I.; Ribas, J.; Pereira, L.; Neves, A.; de Assis, H. Effects of trophic exposure to dexamethasone and diclofenac in freshwater fish. *Ecotoxicol. Environ. Saf.* **2015**, *114*, 204–211. [CrossRef] [PubMed]

77. Dai, C.; Zhang, Y.; Gong, Y.; Bradley, A.; Tang, Z.; Sellick, K.; Shrestha, S.; Spears, E.; Covington, B.A.; Stanley, J.; et al. Hyperaminoacidemia from interrupted glucagon signaling increases pancreatic acinar cell proliferation and size via mTORC1 and YAP pathways. *iScience* **2024**, *27*, 111447. [CrossRef] [PubMed]
78. Khalfaoui, A.; Meniai, A.H. Application of Chemically Modified Orange Peels for Removal of copper(II) from Aqueous Solutions. *Theor. Found. Chem. Eng.* **2012**, *46*, 732–739. [CrossRef]
79. Mohseni, S.N.; Amooey, A.A.; Tashakkorian, H.; Amouei, A.I. Removal of Dexamethasone from Aqueous Solutions Using Modified Clinoptilolite Zeolite (Equilibrium and Kinetic). *Int. J. Environ. Sci. Technol.* **2016**, *13*, 2261–2268. [CrossRef]
80. Jain, M.; Garg, V.K.; Kadirvelu, K. Chromium(VI) Removal from Aqueous System Using Helianthus Annuus (Sunflower) Stem Waste. *J. Hazard. Mater.* **2009**, *162*, 365–372. [CrossRef]
81. Jain, M.; Garg, V.K.; Kadirvelu, K. Equilibrium and Kinetic Studies for Sequestration of Cr(VI) from Simulated Wastewater Using Sunflower Waste Biomass. *J. Hazard. Mater.* **2009**, *171*, 328–334. [CrossRef]
82. Jalali, M.; Aboulghazi, F. Sunflower Stalk, an Agricultural Waste, as an Adsorbent for the Removal of Lead and Cadmium from Aqueous Solutions. *J. Mater. Cycles Waste Manag.* **2013**, *15*, 548–555. [CrossRef]
83. Bhaumik, M.; Leswifi, T.Y.; Maity, A.; Srinivasu, V.V.; Onyango, M.S. Removal of Fluoride from Aqueous Solution by polypyrrole/Fe3O4 Magnetic Nanocomposite. *J. Hazard. Mater.* **2011**, *186*, 150–159. [CrossRef]
84. Ai, L.; Luo, X.; Lin, X.; Zhang, S. Biosorption Behaviors of Uranium (VI) from Aqueous Solution by Sunflower Straw and Insights of Binding Mechanism. *J. Radioanal. Nucl. Chem.* **2013**, *298*, 1823–1834. [CrossRef]
85. Garg, V.K.; Gupta, R.; Kumar, R.; Gupta, R.K. Adsorption of Chromium from Aqueous Solution on Treated Sawdust. *Bioresour. Technol.* **2004**, *92*, 79–81. [CrossRef]
86. Li, Y.; Yu, J. New Stories of Zeolite Structures: Their Descriptions, Determinations, Predictions, and Evaluations. *Chem. Rev.* **2014**, *114*, 7268–7316. [CrossRef] [PubMed]
87. Bui, T.X.; Choi, H. Adsorptive Removal of Selected Pharmaceuticals by Mesoporous Silica SBA-15. *J. Hazard. Mater.* **2009**, *168*, 602–608. [CrossRef] [PubMed]
88. Larsson, D.; Flach, C. Antibiotic resistance in the environment. *Nat Rev Microbiol.* **2022**, *20*, 257–269. [CrossRef]
89. Antimicrobial Resistance Collaborators. Global burden of bacterial antimicrobial resistance in 2019: A systematic analysis. *Lancet* **2019**, *399*, 629–655.
90. Aga, D.S.; Samara, F.; Dronjak, L.; Kanan, S.; Mortula, M.M.; Vahapoglu, L. Rising Water, Rising Risks: The Hidden Dangers of Emerging Contaminants in Climate-Intensified Storms. *ACS EST Water* **2024**, *4*, 2785–2788. [CrossRef]
91. Grela, A.; Kuc, J.; Bajda, T. A Review on the Application of Zeolites and Mesoporous Silica Materials in the Removal of Non-Steroidal Anti-Inflammatory Drugs and Antibiotics from Water. *Materials* **2021**, *14*, 4994. [CrossRef]
92. Żur, J.; Piński, A.; Marchlewicz, A.; Hupert-Kocurek, K.; Wojcieszyńska, D.; Guzik, U. Organic micropollutants paracetamol and ibuprofen-toxicity, biodegradation, and genetic background of their utilization by bacteria. *Environ. Sci. Pollut. Res. Int.* **2018**, *25*, 21498–21524. [CrossRef] [PubMed] [PubMed Central]
93. Jan-Roblero, J.; Cruz-Maya, J.A. Ibuprofen: Toxicology and Biodegradation of an Emerging Contaminant. *Molecules* **2023**, *28*, 2097. [CrossRef] [PubMed]
94. Vo, H.; Koottatep, T.; Chapagain, S.; Panuvatvanich, A.; Polprasert, C.; Nguyen, T.; Chaiwong, C.; Nguyen, N. Removal and monitoring acetaminophen-contaminated hospital wastewater by vertical flow constructed wetland and peroxidase enzymes. *J. Environ. Manag.* **2019**, *250*, 109526. [CrossRef] [PubMed]
95. Benotti, M.J.; Trenholm, R.A.; Vanderford, B.J.; Holady, J.C.; Stanford, B.D.; Snyder, S.A. Pharmaceuticals and Endocrine Disrupting Compounds in U.S. Drinking Water. *Environ. Sci. Technol.* **2009**, *43*, 597–603. [CrossRef]
96. Mhuka, V.; Dube, S.; Nindi, M.M. Occurrence of Pharmaceutical and Personal Care Products (PPCPs) in Wastewater and Receiving Waters in South Africa Using LC-Orbitrap^{TM} MS. *Emerg. Contam.* **2020**, *6*, 250–258. [CrossRef]
97. Brozinski, J.; Lahti, M.; Meierjohann, A.; Oikari, A.; Kronberg, A. The Anti-Inflammatory Drugs Diclofenac, Naproxen and Ibuprofen are found in the Bile of Wild Fish Caught Downstream of a Wastewater Treatment Plant. *Environ. Sci. Technol.* **2013**, *47*, 342–348. [CrossRef]
98. Yang, S.; Carlson, K. Evolution of Antibiotic Occurrence in a River through Pristine, Urban and Agricultural Landscapes. *Water Res.* **2003**, *37*, 4645–4656. [CrossRef] [PubMed]
99. Kulkarni, P.; Olson, N.; Raspanti, G.; Rosenberg Goldstein, R.; Gibbs, S.; Sapkota, A.; Sapkota, A. Antibiotic Concentrations Decrease during Wastewater Treatment but Persist at Low Levels in Reclaimed Water. *Int. J. Environ. Res. Public Health* **2017**, *14*, 668. [CrossRef]
100. Kallenborn, R.; Fick, J.; Lindberg, R.; Moe, M.; Nielsen, K.M.; Tysklind, M.; Vasskog, T. Pharmaceutical Residues in Northern European Environments: Consequences and Perspectives. In *Pharmaceuticals in the Environment*; Springer: Berlin/Heidelberg, Germany, 2008; pp. 61–74.
101. Lapworth, D.J.; Baran, N.; Stuart, M.E.; Ward, R.S. Emerging Organic Contaminants in Groundwater: A Review of Sources, Fate and Occurrence. *Environ. Pollut.* **2012**, *163*, 287–303. [CrossRef]

102. Pal, A.; Gin, K.Y.-H.; Lin, A.Y.-C.; Reinhard, M. Impacts of Emerging Organic Contaminants on Freshwater Resources: Review of Recent Occurrences, Sources, Fate and Effects. *Sci. Total Environ.* **2010**, *408*, 6062–6069. [CrossRef] [PubMed]
103. Mompelat, S.; Le Bot, B.; Thomas, O. Occurrence and Fate of Pharmaceutical Products and by-Products, from Resource to Drinking Water. *Environ. Int.* **2009**, *35*, 803–814. [CrossRef] [PubMed]
104. Joss, A.; Keller, E.; Alder, A.C.; Göbel, A.; McArdell, C.S.; Ternes, T.; Siegrist, H. Removal of Pharmaceuticals and Fragrances in Biological Wastewater Treatment. *Water Res.* **2005**, *39*, 3139–3152. [CrossRef] [PubMed]
105. Castiglioni, S.; Bagnati, R.; Fanelli, R.; Pomati, F.; Calamari, D.; Zuccato, E. Removal of Pharmaceuticals in Sewage Treatment Plants in Italy. *Environ. Sci. Technol.* **2006**, *40*, 357–363. [CrossRef] [PubMed]
106. Brown, P.A.; Brown, J.M.; Allen, S.J. The Application of Kudzu as a Medium for the Adsorption of Heavy Metals from Dilute Aqueous Wastestreams. *Bioresour. Technol.* **2001**, *78*, 195–201. [CrossRef] [PubMed]
107. Homem, V.; Santos, L. Degradation and Removal Methods of Antibiotics from Aqueous Matrices—A Review. *J. Environ. Manag.* **2011**, *92*, 2304–2347. [CrossRef] [PubMed]
108. El Qada, E.N.; Allen, S.J.; Walker, G.M. Adsorption of Basic Dyes from Aqueous Solution onto Activated Carbons. *Chem. Eng. J.* **2008**, *135*, 174–184. [CrossRef]
109. El Qada, E.N.; Allen, S.J.; Walker, G.M. Adsorption of Methylene Blue onto Activated Carbon Produced from Steam Activated Bituminous Coal: A Study of Equilibrium Adsorption Isotherm. *Chem. Eng. J.* **2006**, *124*, 103–110. [CrossRef]
110. Ternes, T.A.; Meisenheimer, M.; McDowell, D.; Sacher, F.; Brauch, H.-J.; Haist-Gulde, B.; Preuss, G.; Wilme, U.; Zulei-Seibert, N. Removal of Pharmaceuticals during Drinking Water Treatment. *Environ. Sci. Technol.* **2002**, *36*, 3855–3863. [CrossRef] [PubMed]
111. Westerhoff, P.; Yoon, Y.; Snyder, S.; Wert, E. Fate of Endocrine-Disruptor, Pharmaceutical, and Personal Care Product Chemicals during Simulated Drinking Water Treatment Processes. *Environ. Sci. Technol.* **2005**, *39*, 6649–6663. [CrossRef]
112. Vieno, N.M.; Härkki, H.; Tuhkanen, T.; Kronberg, L. Occurrence of Pharmaceuticals in River Water and Their Elimination in a Pilot-Scale Drinking Water Treatment Plant. *Environ. Sci. Technol.* **2007**, *41*, 5077–5084. [CrossRef]
113. Currie, L.; Millar, G.J.; Gray, V.; Moghaddam, L. Process development of PVC/natural zeolite beads for the removal of ammonium ions from water. *J. Water Process Eng.* **2024**, *64*, 105714. [CrossRef]
114. Wen, J.; Dong, H.; Zeng, G. application of zeolite in removing salinity/sodicity from wastewater: A review of mechanisms, challenges and opportunities. *J. Clean. Prod.* **2018**, *197*, 1435–1446. [CrossRef]
115. Kumari, S.; Chowdhry, J.; Kumar, M.; Garg, M. Zeolites in wastewater treatment: A comprehensive review on scientometric analysis, adsorption mechanisms, and future prospects. *Environ. Res.* **2024**, *260*, 119782. [CrossRef]
116. Feisal, N.; Kamaludin, N.; Ahmad, M.; Ibrahim, T. A comprehensive review of nanomaterials for efficient heavy metal ions removal in water treatment. *J. Water Process Eng.* **2024**, *64*, 105566. [CrossRef]
117. Liang, Z.; Liu, Z.; Yu, L.; Wang, W. Fly Ash-Based Zeolites: From Waste to Value–A Comprehensive Overview of Synthesis, Properties, and Applications. *Chem. Eng. Res. Des.* **2024**, *212*, 240–260. [CrossRef]
118. Rac, V.; Rakić, V.; Stošić, D.; Pavlović, V.; Bosnar, S.; Auroux, A. Enhanced Accessibility of Active Sites in Hierarchical ZSM-5 Zeolite for Removal of Pharmaceutically Active Substances: Adsorption and Microcalorimetric Study. *Arab. J. Chem.* **2020**, *13*, 1945–1954. [CrossRef]
119. Al-rimawi, F.; Daana, M.; Khamis, M.; Karaman, R.; Khoury, H.; Qurie, M. Removal of Selected Pharmaceuticals from Aqueous Solutions Using Natural Jordanian Zeolite. *Arab. J. Sci. Eng.* **2019**, *44*, 209–215. [CrossRef]
120. Kuwahara, Y.; Aoyama, J.; Miyakubo, K.; Eguchi, T.; Kamegawa, T.; Mori, K.; Yamashita, H. TiO2 Photocatalyst for Degradation of Organic Compounds in Water and Air Supported on Highly Hydrophobic FAU Zeolite: Structural, Sorptive, and Photocatalytic Studies. *J. Catal.* **2012**, *285*, 223–234. [CrossRef]
121. Duke, M.C.; O'Brien-Abraham, J.; Milne, N.; Zhu, B.; Lin, J.Y.S.; Diniz da Costa, J.C. Seawater Desalination Performance of MFI Type Membranes Made by Secondary Growth. *Sep. Purif. Technol.* **2009**, *68*, 343–350. [CrossRef]
122. Salem Attia, T.M.; Hu, X.L.; Yin, D.Q. Synthesized Magnetic Nanoparticles Coated Zeolite for the Adsorption of Pharmaceutical Compounds from Aqueous Solution Using Batch and Column Studies. *Chemosphere* **2013**, *93*, 2076–2085. [CrossRef]
123. de Ridder, D.J.; Verberk, J.Q.J.C.; Heijman, S.G.J.; Amy, G.L.; van Dijk, J.C. Zeolites for Nitrosamine and Pharmaceutical Removal from Demineralised and Surface Water: Mechanisms and Efficacy. *Sep. Purif. Technol.* **2012**, *89*, 71–77. [CrossRef]
124. Martucci, A.; Pasti, L.; Marchetti, N.; Cavazzini, A.; Dondi, F.; Alberti, A. Adsorption of Pharmaceuticals from Aqueous Solutions on Synthetic Zeolites. *Microporous Mesoporous Mater.* **2012**, *148*, 174–183. [CrossRef]
125. Szabová, P.; Plekancová, M.; Gróf, N.; Bodík, I. Slovak Natural Zeolites as a Suitable Medium for Antibiotics Elimination from Wastewater. *Acta Chim. Slovaca* **2019**, *12*, 163–167. [CrossRef]
126. Braschi, I.; Martucci, A.; Blasioli, S.; Mzini, L.L.; Ciavatta, C.; Cossi, M. Effect of Humic Monomers on the Adsorption of Sulfamethoxazole Sulfonamide Antibiotic into a High Silica Zeolite Y: An Interdisciplinary Study. *Chemosphere* **2016**, *155*, 444–452. [CrossRef] [PubMed]
127. de Sousa, D.N.R.; Insa, S.; Mozeto, A.A.; Petrovic, M.; Chaves, T.F.; Fadini, P.S. Equilibrium and Kinetic Studies of the Adsorption of Antibiotics from Aqueous Solutions onto Powdered Zeolites. *Chemosphere* **2018**, *205*, 137–146. [CrossRef]

128. Blasioli, S.; Martucci, A.; Paul, G.; Gigli, L.; Cossi, M.; Johnston, C.T.; Marchese, L.; Braschi, I. Removal of Sulfamethoxazole Sulfonamide Antibiotic from Water by High Silica Zeolites: A Study of the Involved Host–guest Interactions by a Combined Structural, Spectroscopic, and Computational Approach. *J. Colloid Interface Sci.* **2014**, *419*, 148–159. [CrossRef] [PubMed]
129. An, Y.; de Ridder, D.J.; Zhao, C.; Schoutteten, K.; Bussche, J.V.; Zheng, H.; Chen, G.; Vanhaecke, L. Adsorption and Photocatalytic Degradation of Pharmaceuticals and Pesticides by Carbon Doped-TiO2 Coated on Zeolites under Solar Light Irradiation. *Water Sci. Technol.* **2016**, *73*, 2868–2881. [CrossRef] [PubMed]
130. Wang, C.-F.; Li, J.-S.; Wang, L.-J.; Sun, X.-Y. Influence of NaOH Concentrations on Synthesis of Pure-Form Zeolite A from Fly Ash Using Two-Stage Method. *J. Hazard. Mater.* **2008**, *155*, 58–64. [CrossRef]
131. Bui, T.X.; Pham, V.H.; Le, S.T.; Choi, H. Adsorption of Pharmaceuticals onto Trimethylsilylated Mesoporous SBA-15. *J. Hazard. Mater.* **2013**, *254–255*, 345–353. [CrossRef]
132. Liu, M.; Hou, L.; Yu, S.; Xi, B.; Zhao, Y.; Xia, X. MCM-41 Impregnated with A Zeolite Precursor: Synthesis, Characterization and Tetracycline Antibiotics Removal from Aqueous Solution. *Chem. Eng. J.* **2013**, *223*, 678–687. [CrossRef] [PubMed]
133. Lemić, J.; Kovačević, D.; Tomašević-Čanović, M.; Kovačević, D.; Stanić, T.; Pfend, R. Removal of Atrazine, Lindane and Diazinone from Water by Organo-Zeolites. *Water Res.* **2006**, *40*, 1079–1085. [CrossRef] [PubMed]
134. Kanan, S.M.; Nusri, S.E. The Effect of Silver and Silver-Platinum Doped Into 5A Zeolite on the Degradation of Naptalam. *Adv. Mater. Res.* **2013**, *856*, 43–47. [CrossRef]
135. Kanan, S.M.; Kanan, M.C.; Patterson, H.H. Photophysical Properties of Ag(I)-Exchanged Zeolite A and the Photoassisted Degradation of Malathion. *J. Phys. Chem. B* **2001**, *105*, 7508–7516. [CrossRef]
136. Kanan, S.M.; Kanan, M.C.; Patterson, H.H. Silver Nanoclusters Doped in X and Mordenite Zeolites as Heterogeneous Catalysts for the Decomposition of Carbamate Pesticides in Solution. *Res. Chem. Intermed.* **2006**, *32*, 871–885. [CrossRef]
137. Kanan, S.M.; Moyet, M.A. Fabricated Metal Zeolites as Photocatalysts for the Degradation of Organic Pollutants. *Res. Chem. Intermed.* **2021**, *47*, 433–458. [CrossRef]
138. Prada-Vásquez, M.A.; Simarro-Gimeno, C.; Vidal-Barreiro, I.; Cardona-Gallo, S.A.; Pitarch, E.; Hernández, F.; Torres-Palma, R.A.; Chica, A.; Navarro-Laboulais, J. Application of Catalytic Ozonation Using Y Zeolite in the Elimination of Pharmaceuticals in Effluents from Municipal Wastewater Treatment Plants. *Sci. Total Environ.* **2024**, *925*, 171625. [CrossRef]
139. Samara, F.; Jermani, E.; Kanan, S. Photocatalytic UV-degradation of 2,3,7,8-tetrachlorodibenzo-p-dioxin (TCDD) in the presence of silver doped zeolite. *Arab. J. Chem.* **2019**, *12*, 1870–1878. [CrossRef]
140. König, R.; Spaggiari, M.; Santoliquido, O.; Principi, P.; Bianchi, G.; Ortona, A. Micropollutant Adsorption from Water with Engineered Porous Ceramic Architectures Produced by Additive Manufacturing and Coated with Natural Zeolite. *J. Clean. Prod.* **2020**, *258*, 120500. [CrossRef]
141. Salaeh, S.; Juretic Perisic, D.; Biosic, M.; Kusic, H.; Babic, S.; Lavrencic Stangar, U.; Dionysiou, D.D.; Loncaric Bozic, A. Diclofenac Removal by Simulated Solar Assisted Photocatalysis Using TiO2-Based Zeolite Catalyst; Mechanisms, Pathways and Environmental Aspects. *Chem. Eng. J.* **2016**, *304*, 289–302. [CrossRef]
142. Gomez, S.; Marchena, C.L.; Pizzio, L.; Pierella, L. Preparation and Characterization of TiO2/HZSM-11 Zeolite for Photodegradation of Dichlorvos in Aqueous Solution. *J. Hazard. Mater.* **2013**, *258–259*, 19–26. [CrossRef]
143. Shankar, M.V.; Cheralathan, K.K.; Arabindoo, B.; Palanichamy, M.; Murugesan, V. Enhanced Photocatalytic Activity for the Destruction of Monocrotophos Pesticide by TiO2/Hβ. *J. Mol. Catal. A Chem.* **2004**, *223*, 195–200. [CrossRef]
144. Sleiman, M.; Ferronato, C.; Chovelon, J.-M. Photocatalytic Removal of Pesticide Dichlorvos from Indoor Air: A Study of Reaction Parameters, Intermediates and Mineralization. *Environ. Sci. Technol.* **2008**, *42*, 3018–3024. [CrossRef]
145. Andrunik, M.; Skalny, M.; Gajewska, M.; Marzec, M.; Bajda, T. Comparison of Pesticide Adsorption Efficiencies of Zeolites and Zeolite-Carbon Composites and Their Regeneration Possibilities. *Heliyon* **2023**, *9*, e20572. [CrossRef] [PubMed]
146. Arnnok, P.; Patdhanagul, N.; Burakham, R. Dispersive Solid-Phase Extraction Using Polyaniline-Modified Zeolite NaY as a New Sorbent for Multiresidue Analysis of Pesticides in Food and Environmental Samples. *Talanta* **2017**, *164*, 651–661. [CrossRef] [PubMed]
147. Mostafa, M.; Bin Jumah, M.N.; Othman, S.I.; Alruhaimi, R.S.; Salama, Y.F.; Allam, A.A.; Abukhadra, M.R. Effective Removal of Different Species of Organophosphorus Pesticides (Acephate, Omthosate, and Methyl Parathion) Using chitosan/Zeolite-A as Multifunctional Adsorbent. *Environ. Technol. Innov.* **2021**, *24*, 101875. [CrossRef]
148. Hamad, M.T.M.H.; Mahran, B.N.A. Biosynthesized CS-MgO/zeolite Hybrid Material: An Efficient Adsorbent for Chlorpyrifos Removal—Kinetic Studies and Response Surface Methodology. *Emerg. Contam.* **2024**, *10*, 100324. [CrossRef]
149. Janićijević, D.; Jevremović, A.; Janošević Ležaić, A.; Nedić Vasiljević, B.; Usković-Marković, S.; Bajuk-Bogdanović, D.; Milojević-Rakić, M. Comparative Assessment of Pesticide Adsorption Capacity and Antioxidant Activity of Silver Dodecatungstophosphate/HBEA Zeolite Composites. *J. Environ. Chem. Eng.* **2021**, *9*, 106341. [CrossRef]
150. Bajuk-Bogdanović, D.; Jović, A.; Nedić Vasiljević, B.; Milojević-Rakić, M.; Kragović, M.; Krajišnik, D.; Holclajtner-Antunović, I.; Dondur, V. 12-Tungstophosphoric acid/BEA Zeolite Composites—Characterization and Application for Pesticide Removal. *Mater. Sci. Eng. B* **2017**, *225*, 60–67. [CrossRef]

151. Ponnuchamy, M.; Kapoor, A.; Senthil Kumar, P.; Vo, D.-V.N.; Balakrishnan, A.; Mariam Jacob, M.; Sivaraman, P. Sustainable Adsorbents for the Removal of Pesticides from Water: A Review. *Environ. Chem. Lett.* **2021**, *19*, 2425–2463. [CrossRef]
152. Narayanan, N.; Gupta, S.; Gajbhiye, V.T. Decontamination of Pesticide Industrial Effluent by Adsorption–coagulation–flocculation Process Using Biopolymer-Nanoorganoclay Composite. *Int. J. Environ. Sci. Technol.* **2020**, *17*, 4775–4786. [CrossRef]
153. Shabeer, T.P.A.; Saha, A.; Gajbhiye, V.T.; Gupta, S.; Manjaiah, K.M.; Varghese, E. Simultaneous Removal of Multiple Pesticides from Water: Effect of Organically Modified Clays as Coagulant Aid and Adsorbent in Coagulation–flocculation Process. *Environ. Technol.* **2014**, *35*, 2619–2627. [CrossRef]
154. Alvarez-García, S.; Ramírez-García, J.J.; Granados-Correa, F.; Sánchez-Meza, J.C. Determination of Kinetic, Isotherm, and Thermodynamic Parameters of the Methamidophos Adsorption onto Cationic Surfactant-Modified Zeolitic Materials. *Water Air Soil Pollut.* **2018**, *229*, 347. [CrossRef]
155. Amani, M.A.; Latifi, A.M.; Tahvildari, K.; Karimian, R. Removal of Diazinon Pesticide from Aqueous Solutions Using MCM-41 Type Materials: Isotherms, Kinetics and Thermodynamics. *Int. J. Environ. Sci. Technol.* **2018**, *15*, 1301–1312. [CrossRef]
156. Armaghan, M.; Amini, M.M. Adsorption of Diazinon and Fenitothion on MCM-41 and MCM-48 Mesoporous Silicas from Non-Polar Solvent. *Colloid J.* **2009**, *71*, 583–588. [CrossRef]
157. Chen, J.L.; Gao, L.; Jiang, Q.; Hou, Q.; Hong, Y.; Shen, W.; Wang, Y.; Zhu, J.H. Fabricating Efficient Porous Sorbents to Capture Organophosphorus Pesticide in Solution. *Microporous Mesoporous Mater.* **2020**, *294*, 109911. [CrossRef]
158. Ureña-Amate, M.D.; Socías-Viciana, M.M.; González-Pradas, E.; Cantos-Molina, A.; Villafranca-Sánchez, M.; López-Teruel, C. Adsorption of Chloridazon from Aqueous Solution on Modified Kerolite-Rich Materials. *J. Environ. Sci. Heal. Part B* **2008**, *43*, 141–150. [CrossRef]
159. Bell, J.P.; Tsezos, M. Removal of Hazardous Organic Pollutants by Adsorption on Microbial Biomass. *Water Sci. Technol.* **1987**, *19*, 409–416. [CrossRef]
160. Abdelhameed, R.M.; Taha, M.; Abdel-Gawad, H.; Mahdy, F.; Hegazi, B. Zeolitic Imidazolate Frameworks: Experimental and Molecular Simulation Studies for Efficient Capture of Pesticides from Wastewater. *J. Environ. Chem. Eng.* **2019**, *7*, 103499. [CrossRef]
161. Keawkumay, C.; Rongchapo, W.; Sosa, N.; Suthirakun, S.; Koleva, I.Z.; Aleksandrov, H.A.; Vayssilov, G.N.; Wittayakun, J. Paraquat Adsorption on NaY Zeolite at Various Si/Al Ratios: A Combined Experimental and Computational Study. *Mater. Chem. Phys.* **2019**, *238*, 121824. [CrossRef]
162. Pukcothanung, Y.; Siritanon, T.; Rangsriwatananon, K. The Efficiency of Zeolite Y and Surfactant-Modified Zeolite Y for Removal of 2,4-Dichlorophenoxyacetic Acid and 1,1′-Dimethyl-4,4′-Bipyridinium Ion. *Microporous Mesoporous Mater.* **2018**, *258*, 131–140. [CrossRef]
163. De Smedt, C.; Ferrer, F.; Leus, K.; Spanoghe, P. Removal of Pesticides from Aqueous Solutions by Adsorption on Zeolites as Solid Adsorbents. *Adsorpt. Sci. Technol.* **2015**, *33*, 457–485. [CrossRef]
164. Maesen, T. The Zeolite Scene—An Overview. *Stud. Surf. Sci. Catal.* **2001**, *137*, 1–9. [CrossRef]
165. Damjanović, L.; Rakić, V.; Rac, V.; Stošić, D.; Auroux, A. The Investigation of Phenol Removal from Aqueous Solutions by Zeolites as Solid Adsorbents. *J. Hazard. Mater.* **2010**, *184*, 477–484. [CrossRef] [PubMed]
166. Rakić, V.; Damjanović, L.; Rac, V.; Stošić, D.; Dondur, V.; Auroux, A. The Adsorption of Nicotine from Aqueous Solutions on Different Zeolite Structures. *Water Res.* **2010**, *44*, 2047–2057. [CrossRef] [PubMed]
167. Liu, S.; Lim, M.; Amal, R. TiO2-Coated Natural Zeolite: Rapid Humic Acid Adsorption and Effective Photocatalytic Regeneration. *Chem. Eng. Sci.* **2014**, *105*, 46–52. [CrossRef]
168. Zhang, Y.; Mancke, R.G.; Sabelfeld, M.; Geißen, S.-U. Adsorption of Trichlorophenol on Zeolite and Adsorbent Regeneration with Ozone. *J. Hazard. Mater.* **2014**, *271*, 178–184. [CrossRef]
169. Jiang, N.; Shang, R.; Heijman, S.G.J.; Rietveld, L.C. High-Silica Zeolites for Adsorption of Organic Micro-Pollutants in Water Treatment: A Review. *Water Res.* **2018**, *144*, 145–161. [CrossRef] [PubMed]
170. Dehmani, Y.; Ba Mohammed, B.; Oukhrib, R.; Dehbi, A.; Lamhasni, T.; Brahmi, Y.; El-Kordy, A.; Franco, D.S.P.; Georgin, J.; Lima, E.C.; et al. Adsorption of Various Inorganic and Organic Pollutants by Natural and Synthetic Zeolites: A Critical Review. *Arab. J. Chem.* **2024**, *17*, 105474. [CrossRef]
171. Bottero, J.Y.; Khatib, K.; Thomas, F.; Jucker, K.; Bersillon, J.L.; Mallevialle, J. Adsorption of Atrazine onto Zeolites and Organoclays, in the Presence of Background Organics. *Water Res.* **1994**, *28*, 483–490. [CrossRef]
172. Nutcharin, S. The Difference of Si/Al Ratio on Organo-Zeolite in the Adsorption of Atrazine and Linuron. *Appl. Mech. Mater.* **2015**, *804*, 295–298. [CrossRef]
173. Jamil, T.S.; Gad-Allah, T.A.; Ibrahim, H.S.; Saleh, T.S. Adsorption and Isothermal Models of Atrazine by Zeolite Prepared from Egyptian Kaolin. *Solid State Sci.* **2011**, *13*, 198–203. [CrossRef]
174. Jodeyri, M.; Haghighi, M.; Shabani, M. Plasmon-assisted demolition of antibiotic using sono-photoreduction decoration of Ag on 2D C3N4 nanophotocatalyst enhanced with acid-treated clinoptilolite. *Ultrason. Sonochem.* **2019**, *54*, 220–232. [CrossRef] [PubMed]

175. Heidari, S.; Haghighi, M.; Shabani, M. Ultrasound assisted dispersion of Bi2Sn2O7-C3N4 nanophotocatalyst over various amount of zeolite Y for enhanced solar-light photocatalytic degradation of tetracycline in aqueous solution. *Ultrason. Sonochem.* **2018**, *43*, 61–72. [CrossRef]
176. Li, H.J.; Wang, W.J.; Chen, L.J.; Wang, L.Q.; Wang, Z.R.; Gao, Z. Photocatalytic Hierarchical Zirconium-Silica Zeolite: One-Pot Synthesis and Its Effective Degradation Mechanism for Tetracycline. *J. Alloys Compd.* **2024**, *1002*, 175279. [CrossRef]
177. Santos, W.D.; Teixeira, M.M.; Campos, I.R.; de Lima, R.B.; Mantilla, A.; Osajima, J.A.; Alcântara, A. Photocatalytic degradation of ciprofloxacin using semiconductor derived from heterostructured ZIF-8-based materials. *Microporous Mesoporous Mater.* **2023**, *359*, 112657. [CrossRef]
178. Barquist, K.; Larsen, S. Chromate adsorption on amine-functionalized nanocrystalline silicalite-1. *Microporous Mesoporous Mater.* **2008**, *116*, 365–369. [CrossRef]
179. Liaquat, I.; Munir, R.; Abbasi, N.A.; Sadia, B.; Muneer, A.; Younas, F.; Noreen, S. Exploring zeolite-based composites in adsorption and photocatalysis for toxic wastewater treatment: Preparation, mechanisms, and future perspectives. *Environ. Pollut.* **2024**, *349*, 123922. [CrossRef] [PubMed]
180. Lee, Y.R.; Soe, J.T.; Zhang, S.; Ahn, J.W.; Park, M.B.; Ahn, W. Synthesis of nanoporous materials via recycling coal fly ash and other solid wastes: A mini review. *Chem. Eng. J.* **2017**, *317*, 821–843. [CrossRef]
181. El-Nahas, S.; Osman, A.I.; Arafat, A.S.; Ala'a, H.; Salman, H. Facile and affordable synthetic route of nano powder zeolite and its application in fast softening of water hardness. *J. Water Process Eng.* **2020**, *33*, 101104. [CrossRef]
182. de Jesús Ruíz-Baltazar, Á. Advancements in nanoparticle-modified zeolites for sustainable water treatment: An interdisciplinary review. *Sci. Total Environ.* **2024**, *946*, 174373. [CrossRef]
183. Irma, M.; Foo, K.Y.; Susilawati, S.; Yusof, E.N.; Nishiyama, N.; Sabar, S. Advancements in zeolite and zeolite-based sorbents: Modification strategies in mitigating nitrogen-containing pollutants from water and wastewater. *Inorg. Chem. Commun.* **2024**, *172*, 113715. [CrossRef]
184. Wibowo, Y.G.; Rianjanu, A.; Taher, T.; Susanti, N.; Karo, P.K.; Supriyatna, Y.I.; Prasetyo, E.; Bahfie, F.; Amin, M.; Syarifuddin, H.; et al. Synthesis, Characterization, and performance test of modified Zeolite-Si/Al nanocomposite from pumice and waste cans using slow pyrolysis process for removal pollutant parameters from hard water. *Environ. Nanotechnol. Monit. Manag.* **2023**, *20*, 100853.

catalysts

MDPI

Article

One-Pot In Situ Synthesis of Porous Vanadium-Doped g-C_3N_4 with Improved Photocatalytic Removal of Pharmaceutical Pollutants

Yafeng Huang †, Rui Pang †, Shanshan Sun, Xiufang Chen *, Fengtao Chen and Wangyang Lu

National & Local Joint Engineering Research Center for Textile Fiber Materials and Processing Technology, College of Material Science and Engineering, Zhejiang Sci-Tech University, Hangzhou 310018, China; huang1786337170@163.com (Y.H.); 19857115160@163.com (R.P.); sunshanshan333biu@163.com (S.S.); cft0923@163.com (F.C.); luwy@zstu.edu.cn (W.L.)

* Correspondence: chenxf@zstu.edu.cn

† These authors contributed equally to this work.

Abstract: The peroxymonosulfate (PMS)-assisted photocatalytic process has shown considerable potential for the treatment of wastewater. g-C_3N_4-based catalysts are widely applied to eliminate organic pollutants in wastewater. However, the bulk catalyst prepared by dicyandiamide has the drawback of a low surface area (10 m^2/g), while the porous catalyst prepared by urea suffers from a low catalyst yield based on urea (3.5%). To address these challenges, a porous V-doped carbon nitride (V/CN) was designed through one-step thermal polymerization using urea and dicyandiamide as the carbon nitride precursor and NH_4VO_3 as the V precursor. When the ratio of urea to dicyandiamide was 10:1, the yield of V/CN was improved, while it maintained a rich porous structure with a specific surface area (64.6 m^2/g). The synergetic effect of V doping and nanosheet and hollow tubular structures facilitated the separation of photogenerated carriers, leading to boosting the photocatalytic activity of g-C_3N_4 in the PMS system. V/CN(10:1) could completely degrade CBZ within 20 min, exhibiting an equivalent catalytic efficiency comparable to that of V/CN prepared by urea (V/UCN), while markedly surpassing both V/DCN and CN prepared by urea alone (UCN) in performance. This study presents an economical and effective approach for the photocatalytic degradation of pharmaceutical pollutants in aquatic environments.

Keywords: V-doped; graphitic carbon nitride; PMS activation; photocatalysis

1. Introduction

With the rapid acceleration of industrialization and urbanization, the emission of harmful and toxic organic pollutants has been steadily increasing, thereby posing threats to water quality and human health [1,2]. Persistent organic pollutants (POPs) constitute a category of organic compounds distinguished by their prolonged persistence in the environment, semi-volatility, and high toxicity [3–6]. These pollutants can be transported over long distances through various environmental media, such as the atmosphere, aquatic systems, and living organisms, thus presenting substantial risks to both human health and ecosystems. Carbamazepine (CBZ), a pharmaceutical compound widely used for the treatment of epilepsy and pain relief, is increasingly released into aquatic environments in large quantities [7–9]. The low biodegradability coupled with the high stability of CBZ in water poses a critical challenge in addressing water pollution. Recently, peroxymonosulfate (PMS)-assisted advanced oxidation processes (AOPs) have garnered much attention within the field of water pollution remediation [10–15]. PMS can be activated by catalysts to generate numerous free radicals with strong redox capabilities, which can degrade various types of refractory organic pollutants. Coupling photocatalytic oxidation technology with PMS oxidation has demonstrated considerable efficacy as a treatment method for pharmaceutical pollutants in water [16,17].

Graphitic carbon nitride (g-C_3N_4), as an inorganic semiconductor material, has garnered significant attention in the field of photocatalytic degradation of organic pollutants due to its non-toxic nature, environmental compatibility, and visible-light responsiveness [18–21]. The bulk g-C_3N_4 is typically synthesized from conventional precursors (e.g., cyanamide, dicyandiamide, and melamine) resulting in a high catalyst yield (40–60% based on the precursors). However, the bulk material exhibits a low photocatalytic efficiency, owing to its low specific surface area, weak absorption of visible light, and rapid recombination of photoexcitons [22–26]. Bulk g-C_3N_4 was found to activate PMS for the CBZ degradation under light irradiation; however, only a low degradation rate of 48% was achieved within 30 min with bulk g-C_3N_4 using PMS. To improve its photocatalytic performance, g-C_3N_4 nanosheets with porous structures were synthesized through using urea as a precursor. This porous catalyst possesses a large surface area (66.3 m^2/g); however, it suffers from a low yield of carbon nitride based on urea (~3.5%), which consequently increases synthesis costs. These problems may hinter their practical applications [27,28]. Therefore, it is essential to develop porous g-C_3N_4-based nanocatalysts with improved yields by optimizing the precursors.

The application of porous g-C_3N_4 in the treatment of pharmaceutical pollutants is also limited by its low efficiency in solar energy utilization and the rapid recombination of photoexcitons [29,30]. Metal doping has emerged as an effective strategy to enhance photocatalytic performance in wastewater treatment, particularly when combined with PMS oxidation. Various metals, including Mn, Fe, and Cu, have been employed to dope into carbon nitride; however, the removal efficiency of these photocatalysts remains insufficient for practical applications [31–36]. In our previous work, vanadium oxide (VO_x) was loaded first onto porous g-C_3N_4 to construct a highly effective photocatalyst for the removal of refractory pharmaceutical pollutants in wastewater using PMS as an oxidant [25]. The superior activity can likely be ascribed to the diverse oxidation states and coordination polyhedral structures exhibited by VO_x species. Nevertheless, a relatively high metal content (8%VO_x) was required to achieve optimized photocatalytic activity. Therefore, it is meaningful to explore the design of V-modified carbon nitride catalysts with lower V content. To date, there have been few investigations into the photocatalytic performance of V-doped materials for the removal of refractory pharmaceutical pollutants in water utilizing PMS as an oxidant [37,38].

In this study, we developed a porous vanadium-doped g-C_3N_4 catalyst (referred to as V/CN) to improve the PMS activation. The catalysts were synthesized through an economical and straightforward thermal polymerization method using urea and dicyandiamide as precursors for carbon nitride, alongside NH_4VO_3 serving as the vanadium source. The morphology and yield of the V/CN catalyst were tuned by varying the ratio of urea to dicyandiamide. We conducted a comprehensive characterization of the morphology and chemical structure of vanadium-doped catalysts using a variety of analytical instruments. Additionally, we investigated in detail how both morphology and V modification influenced the photocatalytic degradation of CBZ in wastewater with PMS. For the first time, V-doped CN catalysts were applied to eliminate organic pharmaceuticals from wastewater via a combination of photocatalysis and PMS oxidation. Furthermore, we proposed a mechanism for CBZ degradation with V/CN as the catalyst. Our findings provide novel insights into designing effective photocatalysts for removing pharmaceutical-contaminated water.

2. Results and Discussion

2.1. Fabrication Process of V/CN Catalysts

The V-doped g-C_3N_4 was fabricated through a one-step thermal polymerization method, using urea and dicyandiamide as precursors for carbon nitride, while ammonium metavanadate (NH_4VO_3) served as the vanadium precursor. The preparation route is illustrated in Scheme 1. In the initial stage, urea, dicyandiamide, and NH_4VO_3 were thoroughly mixed in an aqueous solution. This mixture was stirred at 80 °C to facilitate water removal, and a V-containing complex was obtained, which was subsequently pyrolyzed at 550 °C in air. During this process, the V-containing complex underwent polymerization to form graphitic carbon nitride, with vanadium species being immobilized within the g-C_3N_4 matrix by forming a V–N_x structure simultaneously. Concurrently, abundant gases (e.g., NH_3, NO, and CO) were generated, resulting in significant volume expansion. As a result of these processes, fluffy V-doped carbon nitride nanosheets were produced. The obtained carbon nitride and V-doped samples were denoted as CN(x:y) and V/CN(x:y), where x:y represents the mass ratio of urea to dicyandiamide. The carbon nitride and V-doped samples synthesized from urea alone are referred to as UCN and V/UCN, while those derived solely from dicyandiamide are designated as DCN and V/DCN, respectively. The mass fraction of the initial mass of V to the total mass of V-doped carbon nitride catalyst was about 3%.

Scheme 1. A synthetic route diagram of the V/CN catalyst.

2.2. Characterizations of V/CN Catalysts

The influence of vanadium doping and different precursors for carbon nitride on the catalyst yield based on precursors is presented in Table S1. The yields for UCN, CN(10:1), and DCN were 3.5%, 7.4%, and 62.6%, respectively, while the yields for V/UCN, V/CN(10:1), and V/DCN were 3.7%, 7.9%, and 45%. These results indicated that the introduction of dicyandiamide into urea could significantly enhance the yield of carbon nitride. Furthermore, it appeared that vanadium doping did not have a notable impact on the yield. The impact of vanadium doping and different precursors for carbon nitride on the porous structure of g-C_3N_4-based catalysts was investigated through N_2 sorption measurement. As illustrated in Figure 1a, all samples exhibited typical IV adsorption isotherms accompanied by H3-type hysteresis loops, which implied the presence of mesoporous structures [26,39]. The BJH pore size analysis presented in Figure 1b revealed that pore size distributions for V/DCN, V/UCN, V/CN(10:1), and CN(10:1) were similar; moreover, these samples displayed a wide range of pore sizes, predominantly concentrated between 0 and 60 nm. Specifically, the average pore sizes for CN(10:1) and V/CN(10:1) were found to be 13.5 nm and 12.9 nm, respectively. A summary of the specific surface areas and pore volumes for these catalysts is provided in Table S1. The specific surface areas measured for DCN, UCN, and CN(10:1) were found to be 6.7 m^2/g, 66.3 m^2/g, and 44.7 m^2/g, respectively; notably, the specific surface area of UCN was nearly ten times greater than that of DCN, while that of CN(10:1) increased up to approximately 6.7 times compared with DCN. This indicated that using urea as a precursor facilitated the formation of a nanosheet with porous structure, which enhanced the number of active sites [40]. After vanadium modification, the specific surface areas of V/DCN, V/UCN, and V/CN(10:1) increased to 20.3 m^2/g, 100.6 m^2/g, and 64.6 m^2/g, respectively. This indicated that V doping may have partially interfered with the thermal polymerization process of urea and dicyandiamide, thereby modifying the charge distribution on the carbon nitride plane. This alteration may have reduced the electrostatic repulsion on the surface of carbon nitride and decreased its lateral dimensions [41,42]. Consequently, it could facilitate the formation of smaller nanosheets and more porous structures, which in turn exposed a greater number of active sites. This enhancement would promote substrate diffusion from the solvent to both the inner and outer surfaces of the catalyst, thereby effectively participating in photocatalytic reactions. Notably, when the mass ratio of urea to dicyandiamide was 10:1, the yield of the V-doped sample (7.9%) could increase by one fold, compared with V/UCN (3.7%), but it still maintained a rich porous structure with a high surface area (64.6 m^2/g).

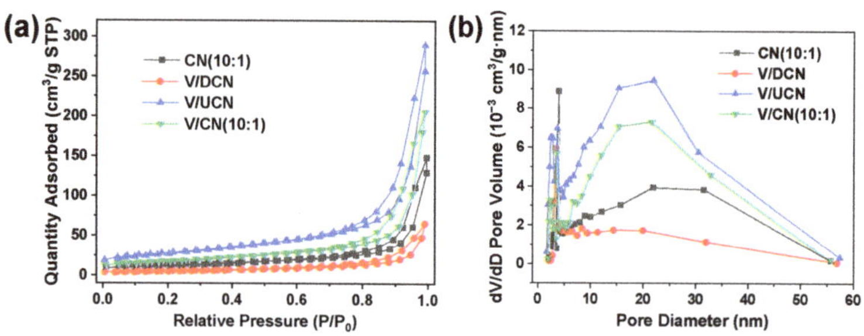

Figure 1. (**a**) N_2 adsorption/desorption curves and (**b**) BJH pore size distribution of CN(10:1), V/DCN, V/UCN, and V/CN(10:1).

The influences of different precursors for carbon nitride on the morphologies of V-doped carbon nitride were characterized using SEM. As illustrated in Figure 2a,b, the V/DCN exhibited a dense structure, composed of numerous particles ranging from tens to hundreds of micrometers. In contrast, Figure 2c,d reveal that V/UCN displayed a porous folded nanosheet stacking morphology characterized by high porosity, resembling UCN. Furthermore, as shown in Figure 2e,f, both lamellar and hollow tubular structures were observed in the V/CN(10:1) sample. The nanosheets with porous structures formed through thermal polymerization of urea were retained; however, part of the lamellae curled to create a more ordered hollow tube structure by the introduction of dicyandiamide [43]. SEM mapping for V/CN(10:1), presented in Figure 2g, confirmed that the elements C, N, O, and V were uniformly distributed throughout the carbon nitride structure. The V content in V/CN(10:1), determined by EDS and inductively coupled plasma optical emission spectroscopy (ICP-OES), was found to be 1.96% (Figure S1) and 2.35%, respectively. Figure 3 presents the TEM images of CN(10:1) and V/CN(10:1). As illustrated in Figure 3a–c, the CN(10:1) sample supported the stacked nanosheets characterized by a porous structure. In contrast, Figure 3d,e reveal that V/CN(10:1) primarily consisted of smaller nanosheet stacks. The HR-TEM image in Figure 3f reflects the presence of the (002) plane of carbon nitride with lattice spacings of about 0.329 nm. Notably, no discernible metal or metal oxide nanoparticles were detected in the HRTEM image, suggesting the absence of metallic V or vanadium oxides within V/CN(10:1). Consequently, combined with SEM mapping results, it can be hypothesized that vanadium may have been incorporated into the carbon nitride framework.

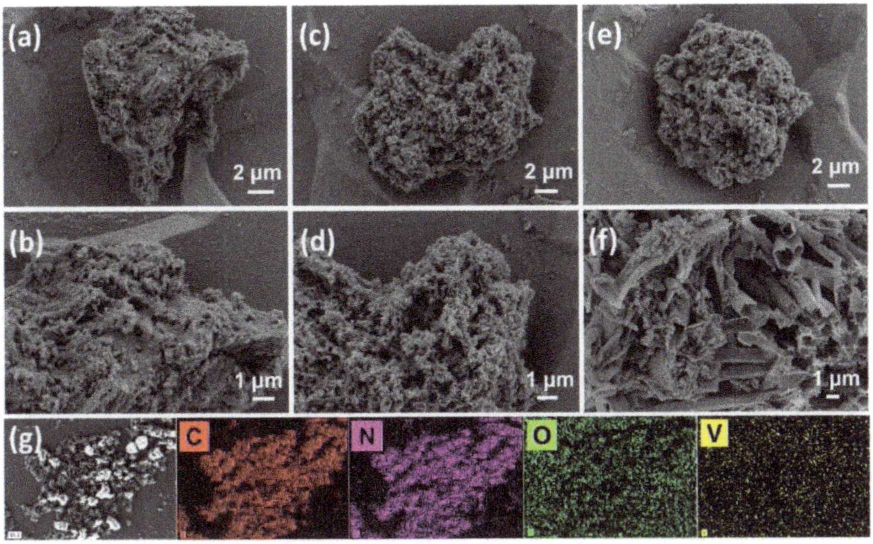

Figure 2. SEM images of (**a**,**b**) V/DCN, (**c**,**d**) V/UCN, and (**e**,**f**) V/CN(10:1) at different magnifications and (**g**) STEM mapping images of V/CN(10:1) (C, N, O, V elements).

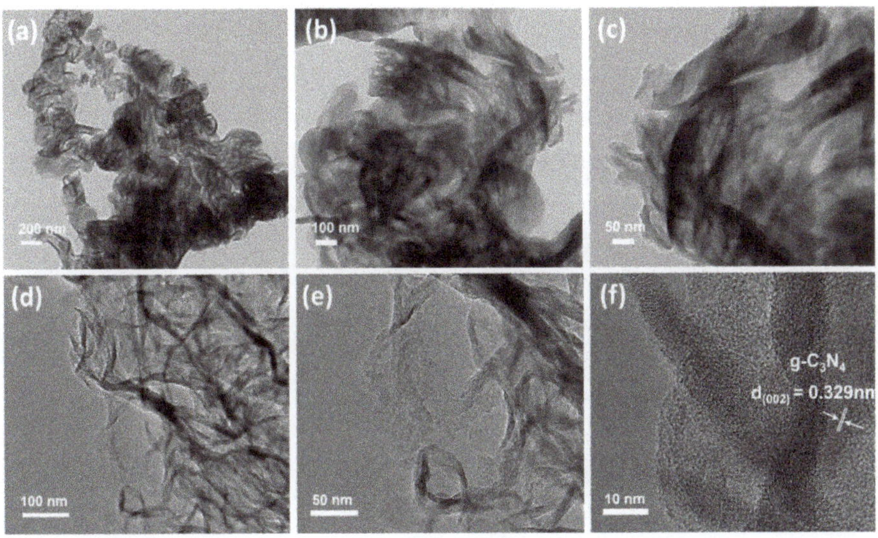

Figure 3. TEM images of (**a**–**c**) CN(10:1) and (**d**,**e**) V/CN(10:1) and (**f**) HRTEM images of V/CN(10:1).

The chemical structure of the V/CN materials was characterized by XRD and FTIR. The XRD patterns for various g-C_3N_4 samples, including V-doped g-C_3N_4, are presented in Figure 4a. Distinct peaks were observed at 2θ = 13.1° and 27.4° in all samples, belonging to the (100) and (002) planes of carbon nitride, respectively [44]. Notably, the intensity of the (002) peak for CN(10:1) and DCN exhibited a significant enhancement compared with UCN. The possible reason was that the introduction of dicyandiamide helped the improvement of the regular stacking of the aromatic rings, thereby leading to an increased degree of crystallization [45]. A similar trend was also noted in the V-doped samples. Furthermore, the intensity of the (002) peak for V/CN(10:1) was weaker than that for

CN(10:1), which could be ascribed to incomplete polymerization within the graphitic layer, as well as disruptions in stacking caused by vanadium doping [46,47]. Importantly, no discernible peaks corresponding to metallic V or VO_x species were detected; this finding further supported the notion that vanadium atoms may be uniformly embedded into the polymer backbone. Figure 4b shows the FT-IR spectra of UCN, DCN, CN(10:1), and V/CN(10:1). All the samples had several similar adsorption bands at 3100–3500, 1200–1650, and 810 cm^{-1}, attributed to the stretching vibrations of amino groups, C=N/C–N–C groups, and the s-triazine ring, respectively [47]. The intensity of the broad peak in the range of 3000–3500 cm^{-1} exhibited a significant increase after V doping, indicating that the amount of amino groups had risen as a result of this doping process.

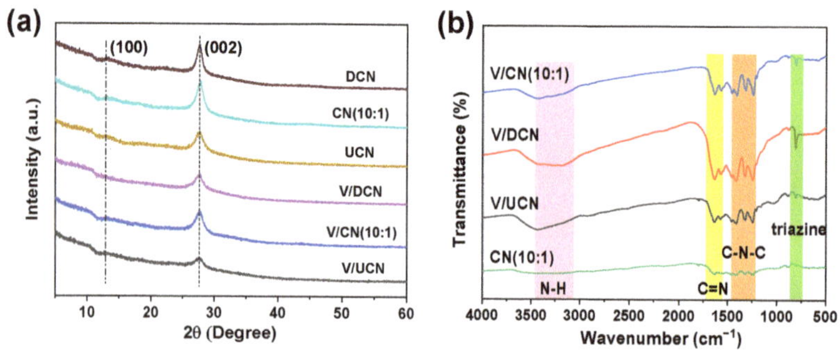

Figure 4. (**a**) XRD patterns and (**b**) FT-IR spectra of different catalysts.

XPS was conducted to elucidate the surface composition and chemical states of the g-C_3N_4-based catalysts. The full spectrum presented in Figure 5a indicates that the g-C_3N_4 samples primarily consisted of C, N, and O elements, whereas V-doped carbon nitride contained C, N, O, and V elements. The C 1s spectra shown in Figure 5b revealed three peaks at 284.6 eV, 286.4 eV, and 288.2 eV for all samples; these corresponded to the graphite C–C bond, the C–O/C–N bond, and the C–N=C bond, respectively [48]. Notably, the atomic ratio of C to N increased from 1.12 for CN(10:1) to 1.14 for V/CN(10:1), suggesting that V doping resulted in a partial loss of nitrogen atoms within the carbon nitride framework, thereby introducing a small amount of nitrogen vacancies. The N 1s spectra depicted in Figure 5c could be deconvoluted into four peaks at 398.7 eV, 400.1 eV, 401 eV, and 404.4 eV; these were attributed to sp^2 N (C–N=C) and sp^3 N–$(C)_3$ groups, amino functional groups (C–N–H), and π-excitation bonds, respectively [49]. In comparison with CN(10:1), both the C–N=C peak and the C–N–H peak in V/CN(10:1) exhibited shifts of 0.1–0.2 V toward higher binding energies. This phenomenon may be ascribed to a decrease in electron density around N atoms due to V–N bond formation [50,51]. Compared with the undoped sample, the V/CN(10:1) sample exhibited a slight increase in the peak intensity of C–N–H groups. This observation may be associated with the formation of smaller nanosheets by V doping. Furthermore, analysis of the V 2p spectra illustrated in Figure 5d revealed four peaks; specifically, two peaks located at 524.4 eV and 516.8 eV were associated with vanadium in the V(V) state, while two other peaks at 523.3 eV and 516.0 eV corresponded to vanadium in the V(IV) state [25]. The ratios of V(V)/V(IV) observed for V/DCN, V/CN(10:1), and V/UCN were approximately 5.1, 2.0, and 3.5, respectively. The aforementioned analyses confirmed that V species existed in the form of V (V) and V (IV). The V atoms were doped within the carbon nitride structure through the V–N bonds.

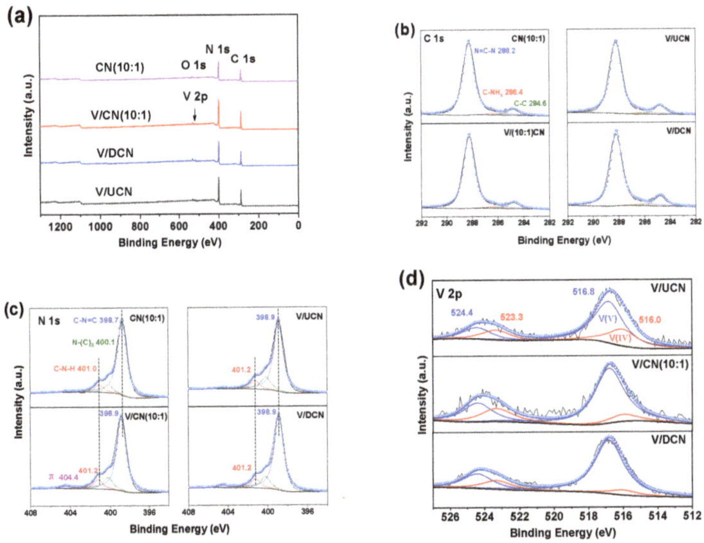

Figure 5. (**a**) XPS full spectrum, (**b**) C 1s spectra, (**c**) N 1s spectra, and (**d**) V 2p spectra of different carbon nitride and V-doped samples.

UV–vis diffuse reflection spectra were utilized to investigate the optical absorption properties of the samples. As shown in Figure 6, all samples displayed significant absorption in the visible light region, with the absorption edge predominantly located around 460 nm. Notably, compared with UCN and CN(10:1), the absorption edge of DCN displayed a red shift. This phenomenon may be ascribed to the quantum size effect associated with nanosheets in UCN and CN(10:1). Additionally, a color change from light yellow to yellow was observed for UCN, CN(10:1), and DCN. Vanadium doping boosted obviously the light absorption intensity of V/DCN within the wavelength range of 400–750 nm, thereby improving its light harvesting efficiency. Additionally, vanadium modification led to a slight increase in light absorption intensity within the range of 438–600 nm for V/CN(10:1), which may be attributed to the formation of V–N_x bonds. Furthermore, compared with CN(10:1), a slight blue shift in the absorption edge was observed in V/CN(10:1), indicating that the nanosheets in the vanadium-doped sample were smaller [51].

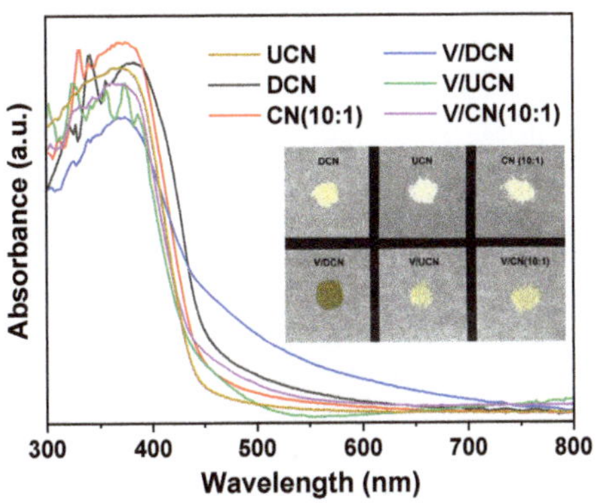

Figure 6. Ultraviolet-visible diffuse reflection spectra of different carbon nitrides and V-doped carbon nitrides.

Transient photocurrent (TPC) analyses were conducted on g-C_3N_4-based catalysts to investigate the behaviors of photogenerated electron–hole separation and transfer. Figure 7a illustrates the photocurrent responses of various carbon nitride and V-doped samples. The current density of V/CN(10:1) was found to be 3.4, 1.4, and 2.1 times higher than those of CN(10:1), V/UCN, and V/DCN, respectively. This enhancement in photocurrent density confirmed that V doping and nanosheet and hollow tubular structures significantly promoted the rapid separation of charge carriers. It was likely that the interaction between nitrogen and vanadium atoms through V–N_x bonds enhanced both the e^-/h^+ separation efficiency and charge transfer rate. Furthermore, the larger specific surface area of V/CN(10:1) along with its nanosheet and hollow tube structure provided abundant active sites for interaction with reactants while facilitating their transport through interconnected channels within the catalyst. Figure 7b presents the electrochemical impedance spectroscopy (EIS) Nyquist plots of the samples. Generally, the radius of the Nyquist plot was indicative of the charge transfer resistance; a larger arc radius corresponded to higher charge transfer resistance within the material [52]. Notably, there was no significant difference in the radii among DCN, UCN, and CN(10:1), suggesting that their charge transfer rates were comparable. In contrast, V/CN(10:1) displayed an arc with the smallest radius, signifying more efficient electron–hole separation and expedited charge transfer. The result was in good agreement with TPC analysis. These results implied the excellent catalytic performance of the V/CN(10:1) catalyst.

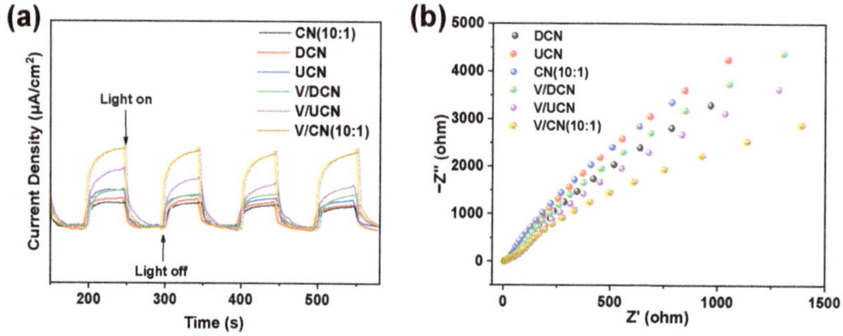

Figure 7. (**a**) TPC response. (**b**) EIS Nyquist plots of different carbon nitride and V-doped samples.

2.3. *Photocatalytic Performance Evaluation*

The photocatalytic degradation of CBZ was employed as a reaction model to investigate the catalytic properties of g-C_3N_4-based materials. Initially, the catalytic performance of V/CN(10:1) catalysts was assessed in different reaction systems. As illustrated in Figure 8a, only 1.3% of CBZ was removed within 30 min using the V/CN(10:1) catalyst in the dark, indicating its low adsorption capacity for CBZ. However, under simulated sunlight conditions, the degradation rate of CBZ increased to 31.7% with the same catalyst over a period of 30 min, demonstrating that V/CN(10:1) exhibited moderate photocatalytic performance. In contrast, when PMS was utilized without any catalyst, the degradation rates of CBZ were recorded at 18.8% and 20.4% after 30 min under dark and light conditions, respectively. These results indicated that only a small amount of CBZ underwent degradation, due to PMS alone, even when coupled with light irradiation. When employing V/CN(10:1) as a catalyst alongside PMS as an oxidant, almost complete degradation of CBZ occurred within 20 min under simulated sunlight irradiation. This finding confirmed that V/CN(10:1) was effective in activating PMS for the photodegradation of CBZ under light irradiation.

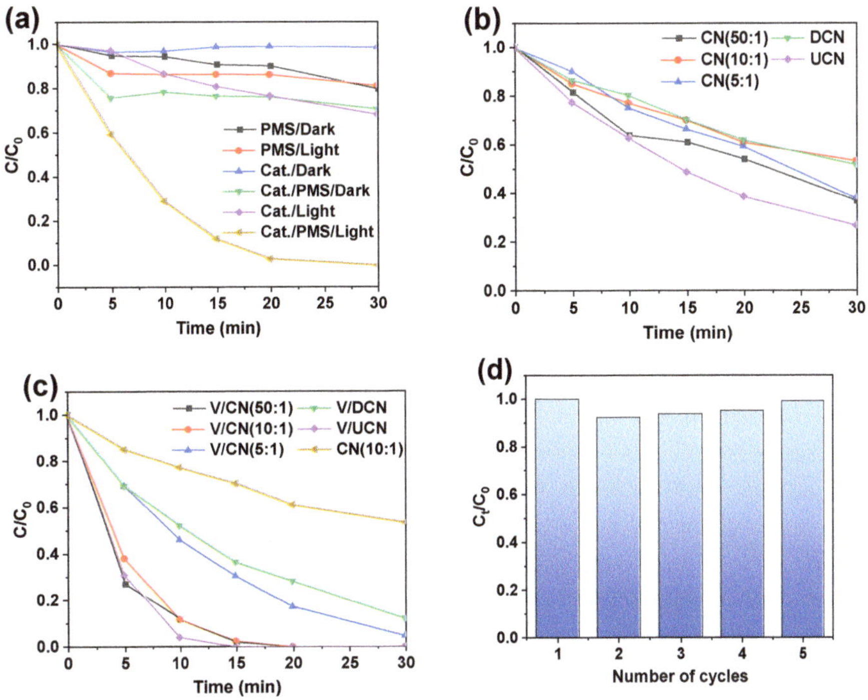

Figure 8. (**a**) Degradation performance of CBZ by V/CN(10:1) under different reaction conditions. Effects of different proportions of urea and dicyandiamide on (**b**) the CN/PMS/light system and (**c**) the V/CN/PMS/light system. (**d**) Recycling test of V/CN(10:1) for CBZ degradation. Conditions: [CBZ] = 25 μM; [PMS] = 0.2 mM; [catalyst] = 0.1 g/L; initial [pH] = 7.0; [T] = 25 °C.

In order to determine the optimal ratio of urea to dicyandiamide, we investigated the impact of this ratio on the photocatalytic performance of g-C_3N_4 and V-doped samples. As illustrated in Figure 8b, when DCN was employed as a catalyst, approximately 30% of CBZ was converted within 30 min under 0.2 mM PMS and light irradiation. Increasing the mass ratio of urea to dicyandiamide generally enhanced the photocatalytic activity of carbon nitride, with UCN exhibiting the highest level of activity. The modification of vanadium was found to significantly improve the photocatalytic performance of carbon nitride. As shown in Figure 8c, under similar reaction conditions, V/DCN achieved a CBZ degradation rate of 71.6% within 20 min. When the ratio of urea to dicyandiamide increased from 5:1 to 10:1, the degradation rates for V-doped carbon nitride rose dramatically to 82.6%, and 100%, respectively. Notably, further increasing this ratio to either 50:1 or pure urea (1:0) did not yield significant enhancements in photocatalytic efficiency; both samples were able to completely degrade CBZ within 20 min. Although V/UCN demonstrated outstanding photocatalytic activity, its low yield posed challenges for industrial applications. Considering both catalyst yield and photocatalytic performance comprehensively, it was concluded that V/CN(10:1) represented the optimal catalyst for treating CBZ in aqueous environments. Furthermore, we observed that V/CN(10:1) exhibited good catalytic recyclability, as depicted in Figure 8d; there was no noticeable decline in photocatalytic activity throughout five cycles during our experiments.

Figure S2 investigates the effects of various reaction parameters, including PMS dosage, catalyst dosage, initial pH, and anionic species, on the catalytic performance of V/CN(10:1). As illustrated in Figure S2a, when the catalyst dosage was increased from 0.1 g/L to 0.3 g/L, a slight enhancement in degradation efficiency was observed; however, CBZ could still be degraded within 20 min with the low catalyst dosage of 0.1 g/L. Considering cost-effectiveness, a catalyst dosage of 0.1 g/L was selected for this study. From Figure S2b, it is evident that increasing the PMS dosage from 0.05 mM to 0.2 mM significantly improved the catalytic activity from 89.8% to nearly 100% within 20 min. This suggested that elevating the PMS concentration enhanced the catalyst activity at a low level of PMS concentration. However, further increases in PMS dosage to 0.3 mM and 0.4 mM did not yield any additional improvement in degradation rate; thus, a PMS concentration of 0.2 mM was chosen for subsequent experiments. Figure S2c demonstrates that the V/CN(10:1) catalyst exhibited excellent catalytic performance across a pH range of 3–7. The degradation efficiency decreased under alkaline conditions, which could be attributed to the following factors: (1) When the pH of the solution exceeded 9.4, PMS dissociated into sulfate radical anions ($SO_5{}^{2-}$), which exhibited lower oxidation potential compared with the peroxymonosulfate ion ($HSO_5{}^{-}$). (2) Under alkaline conditions, PMS decomposed into sulfate ions ($SO_4{}^{2-}$) according to Equation (1). (3) Excessive OH^- in the system led to a negatively charged catalyst surface, thereby hindering effective interactions between the catalyst and both PMS and CBZ [53,54]. The influence of various anions, such as Cl^-, $SO_4{}^{2-}$, and $H_2PO_4{}^{-}$, on photocatalytic activity was also examined. The results are presented in Figure S2d. Notably, no significant decrease in catalytic efficiency was detected within the V/CN(10:1)/PMS/light catalytic system in the presence of these anionic species (Cl^-, $SO_4{}^{2-}$ and $H_2PO_4{}^{-}$). These findings indicated that the V/CN(10:1) possessed good adaptability under acidic and weakly alkaline conditions, as well as in solutions containing different inorganic salts.

$$2HSO_5^- + 2OH^- \rightarrow 2H_2O + 2SO_4^{2-} + O_2 \quad (1)$$

Total organic carbon (TOC) analysis was used as an effective method to assess deep oxidation activity of V-doped catalyst. As illustrated in Figure 9, after 60 min of reaction under the set conditions, the TOC removal rate of CN(10:1) was 27.1%, while the TOC removal rates of V/DCN, V/UCN, and V/CN(10:1) were 45.1%, 59.45%, and 52.9%, respectively. The TOC removal of V/CN(10:1) was increased by 95% compared with pure g-C_3N_4. The results demonstrated that V doping and porous structure significantly enhanced the mineralization efficiency of carbon nitride, facilitating the conversion of macromolecular organic compounds into CO_2.

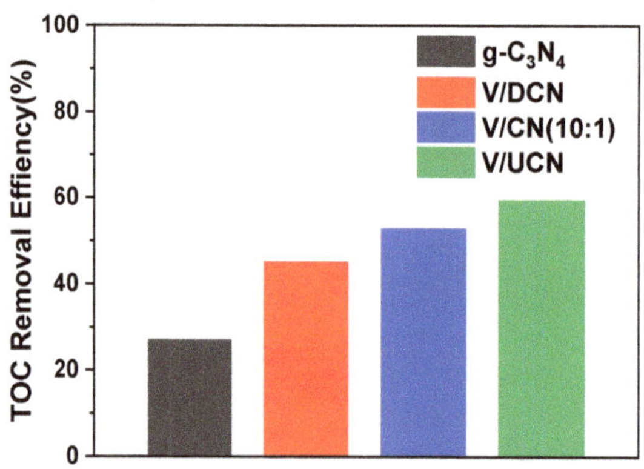

Figure 9. TOC removal rate of the catalyst/PMS/light system after 60 min. Conditions: [CBZ] = 25 μM, [PMS] = 0.2 mM, [catalyst] = 0.1 g/L, initial [pH] = 7.0, T= 40 °C.

2.4. Reaction Mechanism

Various trapping agents were introduced into the system to investigate the impact of reactive oxygen species (ROS) on the degradation process of CBZ. Different trapping agents, including potassium iodide (KI), tert-butanol (TBA), p-benzoquinone (p-BQ), ethanol (EtOH), and histidine (L-His), were added into the V/CN(10:1)/PMS/light system [26,55], with the experimental results illustrated in Figure 10a. It is known that TBA serves as the primary trapping agent for •OH, while EtOH functions as the main trapping agent for both $SO_4•^-$ and •OH radicals. The degradation rate of CBZ was partially inhibited by TBA, indicating that •OH was one of the principal active species involved in the photocatalytic process. This inhibitory effect was further intensified with the addition of an equivalent concentration of EtOH to the reaction system, suggesting that $SO_4•^-$ also played a role in the photocatalytic reaction. It is important to note that neither TBA nor EtOH completely suppressed CBZ degradation; thus, other active species were likely generated within the photocatalytic system. The degradation rate of CBZ also decreased significantly upon the addition of KI, indicating that h^+ was one of the primary active species involved in the photocatalytic process. Similarly, adding L-His and p-BQ led to notable reductions in degradation rates, suggesting that singlet oxygen ($^1O_2$) and superoxide radicals ($O_2•^-$) also played critical roles in this photocatalytic mechanism. Therefore, within the V/CN(10:1)/PMS/light system, $SO_4•^-$, •OH, h^+, $^1O_2$, and $O_2•^-$ emerged as major contributors to CBZ degradation processes.

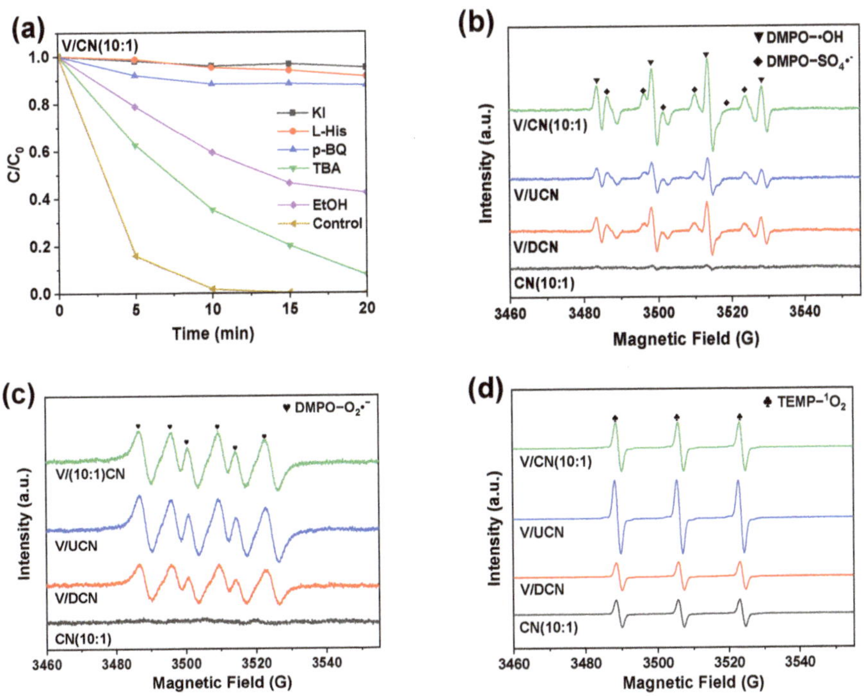

Figure 10. (**a**) Degradation performance of V/CN(10:1) with different capture agents (KI, TBA, p-BQ, EtOH, and L-His) under simulated sunlight irradiation; (**b**) DMPO spin-trapping ESR spectra in aqueous solution; (**c**) DMPO spin-trapping ESR spectra in methanol solution; and (**d**) TEMP spin-trapping ESR spectra in aqueous solution in the presence of V/CN(10:1), V/UCN, V/DCN, and CN(10:1) under simulated sunlight irradiation. Conditions: [CBZ] = 25 μM, [catalyst] = 0.1 g/L, [PMS] = 0.2 mM, [KI] = 0.1 mM, [EtOH] = [TBA] = 10 mM, [p-BQ] = [L-His] = 1 mM.

The generation of reactive radicals during the catalytic process in the V/CN/PMS/light system was further investigated using electron spin resonance (ESR) techniques. 5,5-dimethyl-1-pyrroline N-oxide (DMPO) was employed to detect •OH and SO_4•$^-$ radicals in aqueous solutions, as well as O_2•$^-$ radicals in methanol solutions. Additionally, $^1O_2$ was detected using 2,2,6,6-tetramethyl-4-piperidinol (TEMP) in aqueous solutions [56]. As illustrated in Figure 10b, the V-doped samples exhibited quadruple state peaks for DMPO-•OH with an intensity ratio of 1:2:2:1 and six characteristic peaks for DMPO-SO_4•$^-$ with a signal intensity ratio of 1:1:1:1:1:1. The signal intensity of V/CN(10:1) was significantly higher than that observed for V/UCN and V/DCN. In contrast, CN(10:1) displayed weak signals for both •OH and SO_4•$^-$. These results indicated that V/CN(10:1) effectively promoted the activation of PMS to generate SO_4•$^-$ and •OH radicals during photocatalysis. The characteristic signal peaks of DMPO-O_2•$^-$ can be seen in the g-C_3N_4-based samples shown in Figure 10c. Notably, the detectable signal intensity for CN(10:1) was very weak; however, the O_2•$^-$ radical signals generated by V/CN(10:1) and V/UCN were considerably stronger than those from CN(10:1). As depicted in Figure 10d, all samples exhibited characteristic triple signal peaks for TEMP-$^1O_2$ with a signal intensity ratio of 1:1:1. The intensity of the $^1O_2$ peaks for V/CN(10:1) was slightly weaker than that for V/UCN but was 1.78 times as much as that for CN(10:1). Combining the findings from the ESR techniques and the trapping agent experiments revealed that SO_4•$^-$, •OH, $^1O_2$, O_2•$^-$, and h^+ collectively contributed to the remarkable catalytic activity of V/CN(10:1) for CBZ degradation. Moreover, compared with CN(10:1) and V/DCN, V/CN(10:1) demonstrated enhanced activation capability toward PMS, resulting in increased production of SO_4•$^-$, •OH, $^1O_2$, and O_2•$^-$ radicals, thereby leading to its superior photocatalytic performance.

Based on the results of the above analysis, a possible mechanism for the degradation of CBZ in the V/CN(10:1)/PMS/light system was proposed. The schematic is shown in Scheme 2. When V/CN(10:1) was irradiated by light, the electrons were excited to conduction band, and a pair of charge carriers were generated on the surface of the catalyst (Equation (2)). The photogenerated e^- reacted with dissolved oxygen in water to produce O_2•$^-$ (Equation (3)). The hole h^+ oxidized H_2O to form •OH, which directly participated in the degradation of CBZ (Equation (4)), wherein HSO_5^- could be oxidized by V(V) to form SO_5•$^-$, which was reduced to V(IV) (Equation (5)). In addition, V(IV) further reduced HSO_5^- to form SO_4•$^-$, which was oxidized to V(V) (Equation (6)). The cyclic redox reaction of V(V) and V(IV) within the V/CN(10:1) ensured the generation of massive SO_5•$^-$ and SO_4•$^-$ during the photocatalytic reaction. Furthermore, the SO_5•$^-$ reacted with water to produce $^1O_2$ (Equation (8)). The •OH were also generated through the reaction of SO_4•$^-$ with water (Equation (7)). The reactive radicals generated by the V/CN(10:1) ultimately oxidized CBZ to CO_2 and H_2O (Equation (9)) [25].

$$V/CN(10:1) + h\nu \rightarrow h^+ + e^- \quad (2)$$

$$O_2 + e^- \rightarrow O_2^- \quad (3)$$

$$h^+ + H_2O \rightarrow \cdot OH + H^+ \quad (4)$$

$$V^{5+} + HSO_5^- \rightarrow V^{4+} + SO_5^- + H^+ \quad (5)$$

$$V^{4+} + HSO_5^- \rightarrow V^{5+} + SO_4^- + OH^- \quad (6)$$

$$SO_4^- + H_2O \rightarrow HSO_4^- + \cdot OH \quad (7)$$

$$SO_5^- + H_2O \rightarrow HSO_4^- + \quad (8)$$

$$SO_4^- / O_2^- / {}^1O_2 / h^+ / \cdot OH \rightarrow Intermediate + CO_2 + H_2O \quad (9)$$

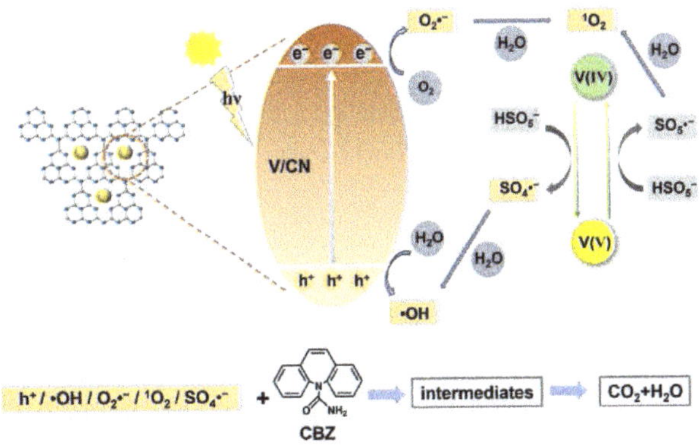

Scheme 2. Proposed mechanism of the V/CN(10:1)/PMS/light system on CBZ degradation.

2.5. CBZ Degradation Pathway

To elucidate the oxidation pathway of CBZ, ultra-performance liquid chromatography coupled with high-definition mass spectrometry (UPLC/HDMS) in positive ion mode was employed to detect the intermediates generated during the CBZ degradation by V/CN(10:1) in the presence of PMS. Table S2 summarizes the molecular formulas, retention times, and relative molecular masses of potential intermediates. The plausible degradation pathways are illustrated in Figure 11. Initially, PMS was activated by V/CN(10:1) under light conditions, leading to the generation of abundant ROS. The olefinic double bonds of CBZ underwent oxidation by ROS, resulting in the formation of P1 (m/z = 252.1). Subsequently, P1 could be hydrolyzed to yield both P2 (m/z = 270.1) and P3 (m/z = 270.1). Additionally, P1 may also transform into P4 (m/z = 252.1) through ring opening and further oxidation by ROS. P2 experienced a series of transformations including oxidation, rearrangement, and deamidation of its amide group (-CONH$_2$), ultimately producing P5 (m/z = 180.1). Ultimately, compounds P3, P4, and P5 were further degraded into smaller organic molecules that were subsequently mineralized into carbon dioxide and water.

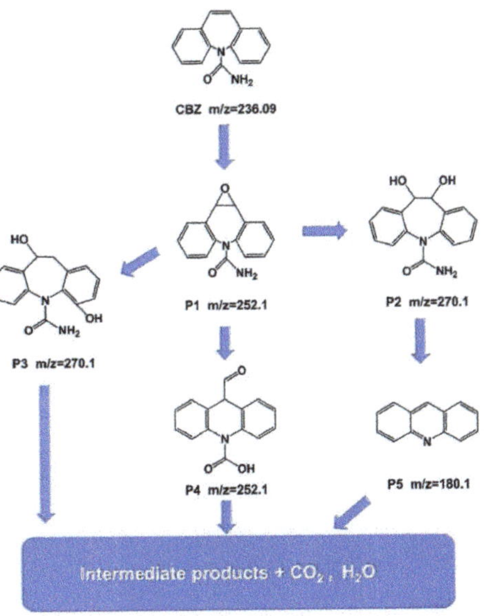

Figure 11. A proposed degradation pathway of CBZ in the V/CN(10:1)/PMS/light system.

3. Experimental Section

3.1. Materials

Dicyandiamide ($C_2H_4N_4$) and 5,5-dimethyl-1-pyrroline N-oxide (DMPO, $C_6H_{11}NO$) was purchased from Beijing J&K Technology Co., Ltd., Beijing, China. Carbamazepine (CBZ, $C_{15}H_{12}N_2O$), ammonium metavanadate (NH_4VO_3), sodium chloride (NaCl), sodium sulphate (Na_2SO_4), sodium carbonate (Na_2CO_3), sodium dihydrogen phosphate (NaH_2PO_4), sodium hydroxide (NaOH), sodium chloroacetate ($ClCH_2COONa$), p-Benzoquinone (p-BQ, $C_6H_4O_2$), L-Histidine (L-His, $C_6H_9N_3O_2$), potassium iodide (KI), and 2,2,6,6-Tetramethylpiperidine (TEMP, $C_9H_{20}N_2$) were purchased from Aladdin Biochemical Technology Co., Ltd., Shanghai, China. Urea ($CO(NH_2)_2$), tert-butanol (TBA, $C_4H_{10}O$), and ethanol (EtOH, C_2H_6O) were purchased from Maclean's Biochemical Technology Co., Ltd., Shanghai, China.

3.2. Synthesis of V-Doped g-C_3N_4 Catalyst

Briefly, 10 g of urea, a specified amount of dicyandiamide, and NH_4VO_3 were dissolved in 15 mL of deionized water. The solution was then stirred at 80 °C for 12 h to ensure complete removal of water. Subsequently, the mixture was transferred to a covered ceramic crucible and heated in a muffle furnace at 550 °C for 2 h. The resulting product was designated as V/CN(x:y), where x:y denotes the mass ratio of urea to dicyandiamide. The products synthesized solely from urea and dicyandiamide were referred to as UCN and DCN, respectively. Following doping with V atoms, the products were named V/UCN and V/DCN accordingly. The mass fraction of the initial mass of V relative to the total mass of the V-doped carbon nitride catalyst is approximately 3%.

3.3. Characterization

The chemical structures of the synthesized g-C_3N_4-based catalysts were characterized using Fourier transform infrared spectroscopy (FT-IR; Nicolet is50, Thermo Fisher Scientific, Waltham, MA, USA), X-ray diffraction (XRD; D8 Discover, Bruker, Karlsruhe, Germany), and X-ray photoelectron spectroscopy (XPS; Thermo Scientific K-Alpha, Thermo Fisher Scientific, Waltham, MA, USA). The morphology of the catalysts was analyzed via field emission scanning electron microscopy coupled with energy-dispersive X-ray spectroscopy (SEM-EDS; Gemini-500, Carl Zeiss AG, Oberkochen, Germany) and transmission electron microscopy (TEM; JEM-2100, JEOL, Tokyo, Japan). The Brunauer–Emmett–Teller (BET) surface area and pore volume of the prepared catalysts were determined from nitrogen adsorption isotherms at 77 K using a surface area and porosity analyzer (BET; APSP 2460, Micromeritics, Norcross, GA, USA). The optical absorption spectra of the as-prepared samples were obtained through UV-Vis diffuse reflectance spectroscopy (DRS; Lambda 950, PerkinElmer, Waltham, MA, USA). Ultra-performance liquid chromatography/high-definition mass spectrometry (UPLC/HDMS) in positive ion mode was employed to detect intermediates generated within the photocatalytic system. Additionally, electron paramagnetic resonance spectroscopy (EPR, A300, Bruker, Karlsruhe, Germany) was utilized to identify various reactive radical species such as $SO_4\bullet^-$, $\bullet OH$, $O_2\bullet^-$, and $^1O_2$ captured by DMPO or TEMP in the photocatalytic system.

3.4. Photocatalytic Tests

The catalytic activity of the catalyst was investigated through the degradation of carbamazepine (CBZ). All reactions were conducted in 40 mL glass vials. Following the addition of 2 mg of catalyst powder to 20 mL of a CBZ solution (25 μM), the mixture was sonicated for 10 min to achieve adsorption equilibrium. Subsequently, PMS was introduced, and the vials were placed on a shaker within a solar simulation device (Q-SUN Xe-1, Q-Lab,

Cleveland, OH, USA) for photocatalytic reaction. The concentration of CBZ at various reaction times was analyzed using ultra-high-performance liquid chromatography (HPLC, Waters, Milford, MA, USA). An acetonitrile-water eluent with a volume ratio of 40:60 and a flow rate of 0.30 mL min^{-1} served as the mobile phase. The photo-degradation efficiency was evaluated by calculating C/C_0, where C_0 represents the initial concentration of CBZ and C denotes the remaining concentration at specific time intervals. In both acidic and alkaline systems, pH adjustments were made prior to irradiation by adding either 0.1 M H_2SO_4 or NaOH. The mineralization rate was determined using a total organic carbon analyzer (TOC-LCPH; Shimadzu, Kyoto, Japan).

4. Conclusions

In this study, vanadium-doped carbon nitride was synthesized through a one-step thermal polymerization process, utilizing urea and dicyandiamide as precursors for carbon nitride and NH_4VO_3 as the vanadium source. The effects of the urea-to-dicyandiamide ratio and vanadium doping on the catalyst yield, structure, and photocatalytic performance were systematically investigated. The findings indicated that when the urea-to-dicyandiamide ratio was set at 10:1, the yield of the V-doped sample was doubled compared with that of V/UCN. This sample maintained a nanosheet and hollow tubular structures with a high surface area. It was likely that vanadium was incorporated into the carbon nitride framework in the form of $V–N_x$ bonds within V/CN(10:1). Consequently, V/CN(10:1) exhibited superior photocatalytic activity for CBZ degradation, compared with both CN(10:1) and V/DCN; its performance was comparable to that of V/UCN. Furthermore, this catalyst had good catalytic reusability, in which the degradation efficiency remained above 95% after five reaction cycles. During the photocatalytic process, key active species such as $SO_4\bullet^-$, $\bullet OH$, $^1O_2$, $O_2\bullet^-$, and h^+ played pivotal roles in CBZ degradation. Compared with CN(10:1) and V/DCN, more $SO_4\bullet^-$, $\bullet OH$, $^1O_2$, and $O_2\bullet^-$ radicals could be produced by the activation of PMS by V/CN(10:1), thereby contributing to its excellent photocatalytic performance. The degradation pathway within the V/CN/PMS/light system was elucidated through the detection of intermediates produced during catalytic reactions using UPLC-MS. This study established a foundation for the development of efficient photocatalysts aimed at treating pharmaceutical pollutants in water.

Supplementary Materials: The following supporting information can be downloaded at https://www.mdpi.com/article/10.3390/catal15030206/s1, Figure S1: EDS spectrum of V/CN(10:1). Figure S2: The photocatalytic activity of V/CN(10:1) under (a) different catalyst concentrations, (b) different concentrations of PMS, (c) different pH conditions, and (d) different inorganic salt ions. Table S1: Catalyst yield based on precursors and BET surface area and total pore volume of various samples. Table S2: Main information on intermediates in degradation systems.

Author Contributions: Original draft preparation, investigation, and validation. Y.H., R.P. and S.S.; Review, editing, and funding acquisition, X.C.; Data curation and validation, F.C. and W.L. All authors have read and agreed to the published version of the manuscript.

Funding: This research was funded by the Baima Lake Laboratory Joint Funds of the Zhejiang Provincial Natural Science Foundation of China (LBMHY24E060003), the Scientific Research Foundation of Zhejiang Sci-Tech University (19212450-Y), and the Fundamental Research Funds of Zhejiang Sci-Tech University (23212112-Y).

Data Availability Statement: The authors can confirm that all relevant data are included in the article.

Conflicts of Interest: The authors do not have any financial or non-financial interests that are directly or indirectly related to the work submitted for publication.

References

1. Dong, K.; Shen, C.; Yan, R.; Liu, Y.; Zhuang, C.; Li, S. Integration of Plasmonic Effect and S-Scheme Heterojunction into $Ag/Ag_3PO_4/C_3N_5$ Photocatalyst for Boosted Photocatalytic Levofloxacin Degradation. *Acta Phys.-Chim. Sin.* **2024**, *40*, 2310013. [CrossRef]
2. Xu, P.; Zhang, R.; Gong, J.; Luo, Y.; Zhuang, Y.; Zhang, P. S-Scheme $WO_3/SnIn_4S_8$ Heterojunction for Water Purification: Enhanced Photocatalytic Performance and Mechanism. *Catalysts* **2023**, *13*, 1450. [CrossRef]
3. Ahmad, S.; Liu, L.; Zhang, S.; Tang, J. Nitrogen-Doped Biochar (N-Doped BC) and Iron/Nitrogen Co-Doped Biochar (Fe/N Co-Doped BC) for Removal of Refractory Organic Pollutants. *J. Hazard. Mater.* **2023**, *446*, 130727. [CrossRef] [PubMed]
4. Shen, T.; Tian, K.; Cao, M.; Li, L.; Shi, F.; Qu, J.; Zheng, Q.; Zhang, G. Application of Nickel Foam Supported Cu–MnO_2 in Microwave Enhanced Fenton-like Process for p-Nitrophenol Removal: Degradation, Synergy and Mechanism Insight. *J. Clean. Prod.* **2023**, *397*, 136442. [CrossRef]
5. Stella, R.; Sreevani, I.; Gurugubelli, T.; Ravikumar, R.; Koutavarapu, R. Enhanced Solar Light-Driven Photocatalytic Degradation of Tetracycline Using Fe^{3+}-Doped CdO/ZnS Nanocomposite: Mechanistic Insights and Performance Evaluation. *Catalysts* **2023**, *13*, 1312. [CrossRef]
6. Zhang, X.; Pan, Y.; Wang, Y.; Wu, T.; Shao, B.; He, Q.; Zhou, L.; Li, T.; Liu, S.; Huang, X.; et al. MOF Derived $MnFeO_x$ Supported on Carbon Cloth as Electrochemical Anode for Peroxymonosulfate Electro-Activation and Persistent Organic Pollutants Degradation. *Chem. Eng. J.* **2024**, *481*, 148646. [CrossRef]
7. Wang, Y.; Huang, Y.; Chen, Y.; Dou, L.; Ren, Y.; Li, N.; Lai, B.; Tan, B. Cobalt Doped g-C_3N_4 Activated Peroxymonosulfate for Organic Pollutant Degradation: Alterations in Cobalt Species and Reactive Oxygen Species. *Chemosphere* **2024**, *369*, 143763. [CrossRef]
8. Lai, B.; Wu, P.; Li, B.; Liu, L.; Wang, T.; Zhu, N.; Dang, Z. Activation of Peroxomonosulfate by Manganese-Containing Non-Homogeneous Catalysts Prepared via a Soft Template Calcination Strategy for Efficient Degradation of Carbamazepine. *Chem. Eng. J.* **2024**, *500*, 157199. [CrossRef]
9. Ying, Y.; Liang, S.; Zhang, F.; Xu, X.; Qian, C.; Jiang, L.; Zhou, J.; Wan, Y.; Wang, L.; Yao, Y. Accelerated Fe^{3+}/Fe^{2+} Cycle in Mo_2C-Based Fe Catalyst to Promote Peroxymonosulfate Activation. *Chemosphere* **2024**, *367*, 143380. [CrossRef]
10. Shen, Z.; Fan, L.; Yang, S.; Yao, Y.; Chen, H.; Wang, W. Fe-Based Carbonitride as Fenton-like Catalyst for the Elimination of Organic Contaminants. *Environ. Res.* **2021**, *198*, 110486. [CrossRef]
11. Zhen, J.; Sun, J.; Xu, X.; Wu, Z.; Song, W.; Ying, Y.; Liang, S.; Miao, L.; Cao, J.; Lv, W.; et al. M−N_3 Configuration on Boron Nitride Boosts Singlet Oxygen Generation via Peroxymonosulfate Activation for Selective Oxidation. *Angew. Chem. Int. Ed.* **2024**, *63*, e202402669. [CrossRef]
12. Yan, H.; Zhen, J.; Yao, Y. Fe^{3+} and H_2O_2 Assisted Dopamine Rapid Polymerization on Melamine Foam to Activate PMS for Organic Pollutant Degradation. *Environ. Sci.-Water Res. Technol.* **2024**, *10*, 2698–2708. [CrossRef]
13. Wang, M.; Wang, Y.; Sun, J.; Zhen, J.; Lv, W. Layered Double Hydroxide/Carbonitride Heterostructure with Potent Combination for Highly Efficient Peroxymonosulfate Activation. *Chemosphere* **2023**, *313*, 137394. [CrossRef]
14. Honarmandrad, Z.; Sun, X.; Wang, Z.; Naushad, M.; Boczkaj, G. Activated Persulfate and Peroxymonosulfate Based Advanced Oxidation Processes (AOPs) for Antibiotics Degradation—A Review. *Water Resour. Ind.* **2023**, *29*, 100194. [CrossRef]
15. Hassani, A.; Scaria, J.; Ghanbari, F.; Nidheesh, P.V. Sulfate Radicals-Based Advanced Oxidation Processes for the Degradation of Pharmaceuticals and Personal Care Products: A Review on Relevant Activation Mechanisms, Performance, and Perspectives. *Environ. Res.* **2023**, *217*, 114789. [CrossRef]
16. Deng, Y.; Ye, Y.-X.; He, Y.; Xu, J.; Ke, Z.; Zhang, X.; Ouyang, G.; Yang, X. Highly Effective Activation of Peroxymonosulfate via Oxygen-Coordinated Single-Atom Iron for Water Decontamination. *Chem. Eng. J.* **2024**, *485*, 149782. [CrossRef]
17. Chen, Z.; Meng, G.; Han, Z.; Li, H.; Chi, S.; Hu, G.; Zhao, X. Interfacial Anchoring Cobalt Species Mediated Advanced Oxidation: Degradation Performance and Mechanism of Organic Pollutants. *J. Colloid Interface Sci.* **2025**, *679*, 67–78. [CrossRef]
18. Wang, C.; Xie, Q.; Shen, W.; Chen, Z.; Hong, X.; Fu, Y.; Si, Y. Synthesis of Highly Crystallized g-C_3N_4 by Regulating Staged Gaseous Intermediates for Hydrogen Production. *J. Mater. Res.* **2023**, *38*, 3214–3226. [CrossRef]
19. Xie, Q.; Wang, C.; Shen, W.; Fu, Y.; Si, Y. Regulating Intermediate Concentration to Synthesize Highly Crystalline g-C_3N_4 under Spontaneous Ultrahigh Pressure. *ChemNanoMat* **2023**, *9*, e202300102. [CrossRef]
20. Zhuang, C.; Chang, Y.; Li, W.; Li, S.; Xu, P.; Zhang, H.; Zhang, Y.; Zhang, C.; Gao, J.; Chen, G.; et al. Light-Induced Variation of Lithium Coordination Environment in g-C_3N_4 Nanosheet for Highly Efficient Oxygen Reduction Reactions. *ACS Nano* **2024**, *18*, 5206–5217. [CrossRef]
21. Chen, L.; Maigbay, M.A.; Li, M.; Qiu, X. Synthesis and Modification Strategies of g-C_3N_4 Nanosheets for Photocatalytic Applications. *Adv. Powder Mater.* **2024**, *3*, 100150. [CrossRef]
22. Song, X.-L.; Chen, L.; Gao, L.-J.; Ren, J.-T.; Yuan, Z.-Y. Engineering g-C_3N_4 Based Materials for Advanced Photocatalysis: Recent Advances. *Green Energy Environ.* **2024**, *9*, 166–197. [CrossRef]

23. Li, Z.; Lv, H.; Tong, K.; He, Y.; Zhai, C.; Yun, Y.; Zhu, M. Modulating the Precursors of Carbon Nitride to Boost Local Electron Delocalization for H_2O_2 Photosynthesis to Remove Oxytetracycline and Its Antibiotic Resistant Genes. *Appl. Catal. B Environ.* **2024**, *345*, 123690. [CrossRef]
24. Wang, D.; Zhang, X.; Zhang, H.; Song, P.; Yang, P. Effect of Precursors on Cu Particle Distribution in g-C_3N_4 Nanosheets towards Efficient Photocatalytic Degradation and H_2 Generation. *Int. J. Hydrogen Energy* **2024**, *68*, 463–471. [CrossRef]
25. Cheng, L.; Sun, S.; Chen, X.; Chen, F.; Chen, X.; Lu, W. Convenient Fabrication of Ultrafine VO_x Decorated on Porous g-C_3N_4 for Boosting Photocatalytic Degradation of Pharmaceuticals with Peroxymonosulfate. *Surf. Interfaces* **2023**, *42*, 103300. [CrossRef]
26. Peng, X.; Chen, X.; Pang, R.; Cheng, L.; Chen, F.; Lu, W. The Impact of Polymerization Atmosphere on the Microstructure and Photocatalytic Properties of Fe-Doped g-C_3N_4 Nanosheets. *Catalysts* **2024**, *14*, 520. [CrossRef]
27. Wu, X.; Tan, L.; Chen, G.; Kang, J.; Wang, G. g-C_3N_4-Based S-Scheme Heterojunction Photocatalysts. *Sci. China-Mater.* **2024**, *67*, 444–472. [CrossRef]
28. Hou, S.; Gao, X.; Lv, X.; Zhao, Y.; Yin, X.; Liu, Y.; Fang, J.; Yu, X.; Ma, X.; Ma, T.; et al. Decade Milestone Advancement of Defect-Engineered g-C_3N_4 for Solar Catalytic Applications. *Nano-Micro Lett.* **2024**, *16*, 70. [CrossRef]
29. Cong, X.; Li, A.; Guo, F.; Qin, H.; Zhang, X.; Wang, W.; Xu, W. Construction of CdS@g-C_3N_4 Heterojunction Photocatalyst for Highly Efficient Degradation of Gaseous Toluene. *Sci. Total Environ.* **2024**, *913*, 169777. [CrossRef]
30. Deng, J.; Zeng, Y.; Almatrafi, E.; Liang, Y.; Wang, Z.; Wang, Z.; Song, B.; Shang, Y.; Wang, W.; Zhou, C.; et al. Advances of Carbon Nitride Based Atomically Dispersed Catalysts from Single-Atom to Dual-Atom in Advanced Oxidation Process Applications. *Coord. Chem. Rev.* **2024**, *505*, 215693. [CrossRef]
31. Wei, F. The Future of Carbon Catalysis, Energy, Material and Engineering. *Carbon Future* **2024**, *1*, 9200006. [CrossRef]
32. Fan, Y.; Lin, W. Rational Design of Multifunctional Framework Materials for Sustainable Photocatalysis. *Carbon Future* **2024**, *1*, 9200018. [CrossRef]
33. Chen, X.; Cheng, L.; Yang, Y.; Chen, X.; Chen, F.; Lu, W. Construction of High-Density Fe Clusters Embedded in a Porous Carbon Nitride Catalyst with Effectively Selective Transformation of Benzene. *ACS Sustain. Chem. Eng.* **2023**, *11*, 1518–1526. [CrossRef]
34. Kalidasan, K.; Mallapur, S.; Munirathnam, K.; Nagarajaiah, H.; Reddy, M.B.M.; Kakarla, R.R.; Raghu, A.V. Transition Metals-Doped g-C_3N_4 Nanostructures as Advanced Photocatalysts for Energy and Environmental Applications. *Chemosphere* **2024**, *352*, 141354. [CrossRef]
35. Zhang, C.; Pan, D.; Zhang, Y.; Lin, L.; Wang, Y.; Zhou, M.; Li, Z.; Xu, S. Reasonably Design of Hollow Spherical g-C_3N_4/$Mn_{0.25}Cd_{0.75}S$ Heterojunction for Efficient Photocatalytic Hydrogen Production and Tetracycline Degradation. *J. Environ. Chem. Eng.* **2024**, *12*, 111956. [CrossRef]
36. Cao, G.; Shen, Z.; Cui, J.; Yu, M.; Li, W. Bifunctional Activation of Peroxymonosulfate over CuS/g-C_3N_4 Composite for Efficient Degradation of Tetracycline Antibiotics. *Chem. Eng. J.* **2024**, *483*, 149082. [CrossRef]
37. Shentu, Q.; Wu, Z.; Song, W.; Pan, S.; Zhou, Z.; Lv, W.; Song, C.; Yao, Y. Carbon Doped Boron Nitride Nanosheet as Efficient Metal-Free Catalyst for Peroxymonosulfate Activation: Important Role of B-N-C Moieties. *Chem. Eng. J.* **2022**, *446*, 137274. [CrossRef]
38. Wang, L.; Rao, L.; Ran, M.; Shentu, Q.; Wu, Z.; Song, W.; Zhang, Z.; Li, H.; Yao, Y.; Lv, W.; et al. A Polymer Tethering Strategy to Achieve High Metal Loading on Catalysts for Fenton Reactions. *Nat. Commun.* **2023**, *14*, 7841. [CrossRef]
39. Sun, S.; Peng, X.; Guo, X.; Chen, X.; Liu, D. Boosting Solvent-Free Aerobic Oxidation of Benzylic Compounds into Ketones over Au-Pd Nanoparticles Supported by Porous Carbon. *Catalysts* **2024**, *14*, 158. [CrossRef]
40. Chen, X.; Yang, Y.; Xiao, Q.; Guo, X.; Chen, F.; Liu, X.; Lu, W. Controlling Transformation of Sorbitol into Glycols over Ru-WOx Modifed Biomass-derived N-doped Carbon. *J. Mater. Sci.* **2024**, *59*, 8186. [CrossRef]
41. Sun, L.; Yang, M.; Huang, J.; Yu, D.; Hong, W.; Chen, X. Freestanding Graphitic Carbon Nitride Photonic Crystals for Enhanced Photocatalysis. *Adv. Funct. Mater.* **2016**, *26*, 4943–4950. [CrossRef]
42. Wang, K.; Jiang, L.; Xin, T.; Li, Y.; Wu, X.; Zhang, G. Single-Atom V-N Charge-Transfer Bridge on Ultrathin Carbon Nitride for Efficient Photocatalytic H2 Production and Formaldehyde Oxidation under Visible Light. *Chem. Eng. J.* **2022**, *429*, 132229. [CrossRef]
43. Guo, F.; Chen, Z.; Shi, Y.; Cao, L.; Cheng, X.; Shi, W.; Chen, L.; Lin, X. A Ragged Porous Hollow Tubular Carbon Nitride towards Boosting Visible-Light Photocatalytic Hydrogen Production in Water and Seawater. *Renew. Energy* **2022**, *188*, 1–10. [CrossRef]
44. Chen, X.; Zhang, J.; Fu, X.; Antonietti, M.; Wang, X. Fe-g-C_3N_4-Catalyzed Oxidation of Benzene to Phenol Using Hydrogen Peroxide and Visible Light. *J. Am. Chem. Soc* **2009**, *131*, 11658–11659. [CrossRef]
45. Wan, S.; Xu, J.; Cao, S.; Yu, J. Promoting Intramolecular Charge Transfer of Graphitic Carbon Nitride by Donor–Acceptor Modulation for Visible-light Photocatalytic H_2 Evolution. *Interdiscip. Mater.* **2022**, *1*, 294–308. [CrossRef]
46. Lee, Y.-J.; Jeong, Y.J.; Cho, I.S.; Park, S.-J.; Lee, C.-G.; Alvarez, P.J.J. Facile Synthesis of N Vacancy g-C_3N_4 Using Mg-Induced Defect on the Amine Groups for Enhanced Photocatalytic •OH Generation. *J. Hazard. Mater.* **2023**, *449*, 131046. [CrossRef]
47. Zhang, L.; Liu, D.; Guan, J.; Chen, X.; Guo, X.; Zhao, F.; Hou, T.; Mu, X. Metal-free g-C_3N_4 Photocatalyst by Sulfuric Acid Activation for Selective Aerobic Oxidation of Benzyl Alcohol under Visible Light. *Mater. Res. Bull.* **2014**, *59*, 84–92. [CrossRef]

48. Sun, H.; Guo, F.; Pan, J.; Huang, W.; Wang, K.; Shi, W. One-Pot Thermal Polymerization Route to Prepare N-Deficient Modified g-C_3N_4 for the Degradation of Tetracycline by the Synergistic Effect of Photocatalysis and Persulfate-Based Advanced Oxidation Process. *Chem. Eng. J.* **2021**, *406*, 126844. [CrossRef]
49. Liang, Z.; Xue, Y.; Wang, X.; Zhou, Y.; Zhang, X.; Cui, H.; Cheng, G.; Tian, J. Co Doped MoS_2 as Cocatalyst Considerably Improved Photocatalytic Hydrogen Evolution of g-C_3N_4 in an Alkalescent Environment. *Chem. Eng. J.* **2021**, *421*, 130016. [CrossRef]
50. Chen, X.; Xiao, Q.; Yang, Y.; Dong, B.; Zhao, Z. Constructing Polyphosphazene Microsphere-Supported Pd Nanocatalysts for Efffcient Hydrogenation of Quinolines under Mild Conditions. *Catalysts* **2024**, *14*, 345. [CrossRef]
51. Zhang, L.; Cheng, L.; Hu, Y.; Xiao, Q.; Chen, X.; Lu, W. Robust Co_3O_4 Nanocatalysts Supported on Biomass-derived Porous N-doped Carbon Toward Low-Pressure Hydrogenation of Furfural. *Front. Mater. Sci.* **2023**, *17*, 230645. [CrossRef]
52. Li, M.; Feng, C.; Shi, X.; Chen, X.; Yang, Y. Breaking the activity-selectivity trade-off in photocatalytic semihydrogenation of alkynes over palladium nanoparticles with phosphate modiffcation. *Appl. Catal. B Environ. Energy* **2025**, *365*, 124948. [CrossRef]
53. Yan, Y.; Zhang, H.; Wang, W.; Li, W.; Ren, Y.; Li, X. Synthesis of Fe^0/Fe_3O_4@porous Carbon through a Facile Heat Treatment of Iron-Containing Candle Soots for Peroxymonosulfate Activation and Efficient Degradation of Sulfamethoxazole. *J. Hazard. Mater.* **2021**, *411*, 124952. [CrossRef] [PubMed]
54. Luo, J.; Bo, S.; Qin, Y.; An, Q.; Xiao, Z.; Zhai, S. Transforming Goat Manure into Surface-Loaded Cobalt/Biochar as PMS Activator for Highly Efficient Ciprofloxacin Degradation. *Chem. Eng. J.* **2020**, *395*, 125063. [CrossRef]
55. Yang, T.; Fan, S.; Li, Y.; Zhou, Q. Fe-N/C Single-Atom Catalysts with High Density of Fe-N_x Sites toward Peroxymonosulfate Activation for High-Efficient Oxidation of Bisphenol A: Electron-Transfer Mechanism. *Chem. Eng. J.* **2021**, *419*, 129590. [CrossRef]
56. Gan, P.; Zhang, Z.; Hu, Y.; Li, Y.; Ye, J.; Tong, M.; Liang, J. Insight into the Role of Fe in the Synergetic Effect of Persulfate/Sulfite and Fe_2O_3@g-C_3N_4 for Carbamazepine Degradation. *Sci. Total Environ.* **2022**, *819*, 152787. [CrossRef]

catalysts

MDPI

Review

Red- and Black-Phosphorus-Based Materials for Photocatalytic Water Disinfection

Dongyang He, Jingyuan Dong, Ya-Nan Zhang *, Sui Zhang, Yi-Nan Zhang and Jiao Qu *

School of Environment, Northeast Normal University, Changchun 130117, China; hedy443@nenu.edu.cn (D.H.); dongjy725@nenu.edu.cn (J.D.); zhangsui888@nenu.edu.cn (S.Z.); zhangyinan@nenu.edu.cn (Y.-N.Z.)
* Correspondence: zhangyn912@nenu.edu.cn (Y.-N.Z.); quj100@nenu.edu.cn (J.Q.)

Abstract: Given the growing threat of pathogens in drinking water, exploring novel semiconductor photocatalysts for effective water disinfection has become a critical area of research. Recently, red and black phosphorus have attracted significant attention due to their unique physicochemical properties, making them well suited for photocatalytic water disinfection. Particularly, recent studies have demonstrated that red and black phosphorus can be combined with other materials to improve the photocatalytic disinfection performance of the resulting composites. Herein, we present the first review of red- and black-phosphorus-based materials for photocatalytic water disinfection. This article begins by presenting the fundamental principle of photocatalytic water disinfection before reviewing the latest developments in the application of red- and black-phosphorus-based photocatalysts for this purpose. In conclusion, it provides a summary and proposes potential avenues for future research in this field.

Keywords: photocatalytic water disinfection; red phosphorus; black phosphorus; inactivation mechanism

1. Introduction

Pathogens in drinking water present a serious threat to public health, responsible for a range of waterborne diseases, including diarrhea, gastroenteritis, hepatitis, meningitis, typhoid, and encephalitis [1]. Globally, these diseases claim over 2 million lives annually, with one child under the age of 5 succumbing to preventable diarrheal cases every 2 min [2]. As a result, there is a pressing need to develop effective water disinfection methods for pathogen-contaminated water. While traditional techniques such as chlorination, ozonation, and UV irradiation have been widely implemented in water treatment for decades, they are often accompanied by secondary pollution and substantial operational costs. These limitations underscore the necessity for innovative approaches that align with the principle of safe and sustainable development.

From both economic and environmental perspectives, photocatalytic water disinfection employing semiconductor-based photocatalysts is widely regarded as a "green" and sustainable alternative to conventional disinfection methods. Upon absorbing photons with energy equal to or exceeding its bandgap energy (E_g), a semiconductor photocatalyst generates electron-hole pairs [3]. When the conduction-band and valence-band potentials of the semiconductor are suitably positioned, the photo-excited electrons and holes can independently react with water and dissolved O_2, yielding reactive oxygen species (ROS) such as superoxide radical ($\bullet O_2^-$), hydroxyl radical ($\bullet OH$), hydrogen peroxide (H_2O_2), and singlet oxygen ($^1O_2$) [4]. These ROS serve as highly reactive oxidants, capable of inactivating pathogens by inflicting irreparable damage on their essential biomolecules [5].

https://doi.org/10.3390/catal15030218

The development of low-cost and high-performance semiconductor photocatalysts that pose no secondary pollution risks is pivotal for the practical implementation of photocatalytic water disinfection. Over the past few decades, substantial progress has been achieved in developing various photocatalysts for the photocatalytic inactivation of pathogens. TiO_2 is the most studied photocatalytic material; its bandgap, however, is generally too large to harvest most of the spectrum of visible light [6]. Reducing the bandgap of TiO_2 to enable its visible-light activation typically requires intricate synthesis strategies, such as non-metal doping, metal deposition, heterostructure coupling, and advanced structural engineering [7]. Nevertheless, the complex nature of these methods introduces substantial practical challenges, which significantly impede the widespread application of TiO_2. Thus, numerous efforts have been devoted to developing novel visible-light-driven photocatalysts for water disinfection [8].

Among these, two P allotropes, red phosphorus and black phosphorus, have recently attracted growing attention. In addition to their inherent ability to harness visible light for water disinfection, these two P allotropes can also be integrated with other semiconductor materials to form composites with markedly improved photocatalytic disinfection performances. Furthermore, red and black phosphorus exhibit excellent biocompatibility, which minimizes environmental and health risks while effectively addressing the secondary pollution issues often associated with metal-based photocatalysts [9–11]. However, a comprehensive review focusing on the application of red- and black-phosphorus-based materials for photocatalytic water disinfection is yet to be available. Given the growing interest in these two P allotropes and their promising potential in photocatalytic microbial inactivation, it is both timely and necessary to summarize the current progress in this field. This article first outlines the fundamental principle of photocatalytic water disinfection, followed by a summary of recent advancements in red- and black-phosphorus-based photocatalysts for this application. Finally, it provides a summary and future research directions in this field.

2. Basic Principle of Photocatalytic Water Disinfection

Photocatalytic water disinfection refers to the use of ROS generated through photocatalytic reactions to attack and damage the essential biomolecules of pathogenic microorganisms, thereby achieving their inactivation [12]. Specifically, when a semiconductor is exposed to light with energy equal to or greater than its E_g, valence-band electrons are excited, transitioning to the semiconductor's conduction band, where they form photo-generated electrons. Simultaneously, corresponding photo-generated holes are created in the semiconductor's valence band [13]. Collectively, these electrons and holes are referred to as photo-generated charge carriers. These charge carriers are inherently unstable and tend to recombine, releasing energy as heat. However, unrecombined photo-generated charge carriers can migrate to the semiconductor's surface, where they participate in various photocatalytic reactions.

Photo-generated electrons exhibit electron-donating characteristics, thereby displaying reductive behavior, while photo-generated holes possess electron-accepting tendencies, manifesting an oxidative property. The redox capabilities of these charge carriers are governed by the semiconductor's band structure [14]. It is widely accepted that a more negative conduction-band potential corresponds to improved reductive capabilities for photo-generated electrons, while a more positive valence-band potential correlates with enhanced oxidative abilities for photo-generated holes. Multiple ROS can be generated during photocatalytic reactions (Figure 1a). The generation of $\bullet O_2^-$ typically occurs via the reduction of O_2 by photo-generated electrons. The necessary condition for this process is that the semiconductor's conduction-band potential is more negative than the redox potential for $O_2/\bullet O_2^-$ (−0.33 V versus NHE) [15]. Furthermore, $\bullet O_2^-$ can serve as a precursor for the generation of additional ROS. For instance, photo-generated holes can oxidize $\bullet O_2^-$ to yield $^1O_2$, while photo-generated electrons can reduce $\bullet O_2^-$ to form H_2O_2 [16,17]. •OH is typically produced by the oxidation of H_2O by photo-generated holes, under the condition that the semiconductor's valence-band potential exceeds the redox potential for $H_2O/\bullet OH$ (+2.72 V versus NHE) [18]. As for H_2O_2, three distinct pathways can lead to its formation, including the two-step single-electron reduction of O_2, the one-step two-electron reduction of O_2, and the two-hole oxidation of water [19–21].

Microbial cellular components, such as proteins, nucleic acids, and lipids, are composed of organic substances, and the aforementioned ROS can clearly target and disrupt them. Taking bacteria as an example, proteins are essential components for bacterial cell walls and membranes, and the reaction rate constant for protein decomposition by $^1O_2$ is exceptionally high. Therefore, $^1O_2$ can effectively disrupt bacterial outer structures [22]. H_2O_2 has the ability to oxidize and damage bacterial cell walls and membranes. And, due to its relatively long lifespan, it can also diffuse into the cytoplasm of bacterial cells, where it oxidizes intracellular components [4]. •OH has a reaction energy of 402.8 mJ/mol, which is higher than the bond energies of various chemical bonds in organic matter. As a result, •OH can rapidly decompose the organic substances that make up bacteria. It should be noted that $\bullet O_2^-$, $^1O_2$, and •OH have short lifespans and typically require direct contact between the semiconductor material and microorganisms to exert their disinfecting effects, often referred to as direct ROS. In contrast, H_2O_2 has a much longer lifespan than other ROS, allowing it to exert its disinfecting effect without the need of direct contact between the semiconductor material and microorganisms, known as indirect ROS [23]. Through the synergistic action of these ROS, multiple critical biological functions of bacteria are comprehensively disrupted (Figure 1b). This includes cellular structure damage, energy metabolism inhibition, antioxidant enzyme system dysregulation, and genetic material degradation, ultimately leading to irreversible bacterial inactivation [24].

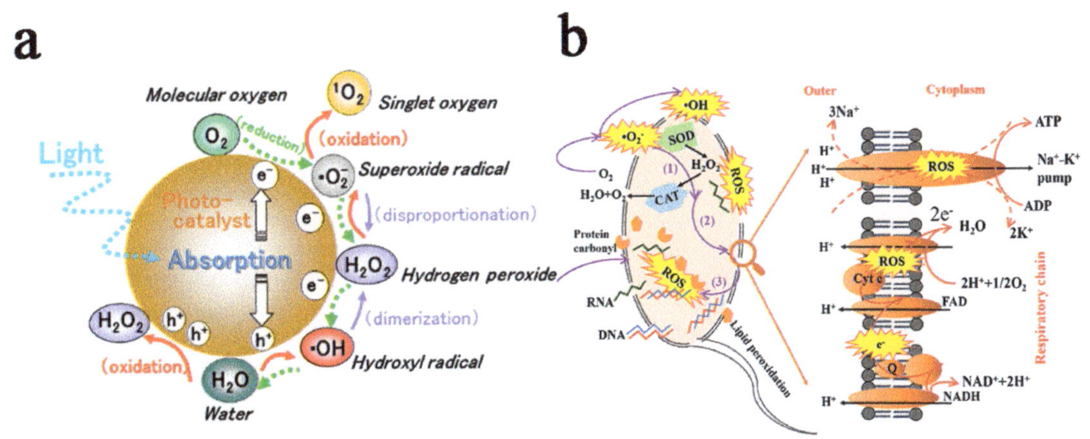

Figure 1. (**a**) Schematic illustration of ROS generation by a semiconductor through photocatalytic reactions. Reprinted with permission from ref. [21]. Copyright 2017 American Chemical Society. (**b**) ROS attacks on multiple critical biological functions of bacteria. Reprinted with permission from ref. [24]. Copyright 2019 American Chemical Society.

3. Red- and Black-Phosphorus-Based Photocatalysts for Water Disinfection

3.1. Red-Phosphorus-Based Antimicrobial Photocatalysts

P constitutes approximately 0.1% of the Earth's crust, making it one of the most abundant elements on the Earth [25]. Two allotropes of P, including red phosphorus and black phosphorus, have been applied in photocatalytic water disinfection. Red phosphorus, owing to its well-established preparation methods and low production costs, has been extensively studied in the field of photocatalytic water disinfection. Xia et al. are the first to investigate the feasibility of using red phosphorus for photocatalytic water disinfection [26]. Starting with commercial red phosphorus, they employ a hydrothermal method to remove its oxide layers, thereby obtaining a red-phosphorus-based photocatalyst for their study. The resulting product is amorphous red phosphorus, which appears as microparticles ranging from 2 to 10 μm in size. Its E_g is determined to be 1.42 eV, indicating its effective utilization of visible light. The valence- and conduction-band potentials of the synthesized red phosphorus are found to be 0.8 and −0.62 V versus NHE, respectively, suggesting that the ROS in its photocatalytic system are predominantly generated via reduction pathways. The authors observe a complete 7-log inactivation of *E. coli* after 90 min of visible-light irradiation, while, when excited by sunlight, the time required for complete bacterial inactivation is reduced to 60 min (Figure 2a). They subsequently conduct a scavenger study along with a steady-state ROS concentration test, revealing that $\bullet O_2^-$ is the predominant ROS responsible for *E. coli* inactivation (Figure 2b). Finally, the authors discover that the attack of ROS disrupts the energy metabolism of *E. coli* cells, compromising both the integrity and permeability of the cell membranes. This leads to the leakage of intracellular components, the impairment of the antioxidant enzyme system, and the degradation of genetic materials, ultimately culminating in the death of *E. coli* (Figure 2c). This work provides valuable insights for the application of red phosphorus in photocatalytic water disinfection. In another work, Athira et al. conduct a combined study of DFT calculations and experiments to investigate the band structure of commercial red phosphorus [27]. Their findings indicate that the E_g of commercial red phosphorus is approximately 2 eV, enabling it to absorb a significant portion of visible light. Furthermore, the conduction-band potential of commercial red phosphorus is relatively negative, which is sufficient for photo-generated electrons to reduce O_2, thereby generating $\bullet O_2^-$ and H_2O_2. However, its valence-band potential is insufficient to directly oxidize H_2O to produce •OH, and, thus, •OH is likely to be generated through a reduction pathway. Subsequently, they explore the photocatalytic bactericidal efficacy of commercial red phosphorus under sunlight irradiation. The results reveal that a 50 min photocatalytic treatment completely inactivates *E. coli*. Commercially available red phosphorus is typically classified as amorphous. When subjected to high-temperature annealing, it undergoes a structural transition, giving rise to fibrous red phosphorus. Roshith et al. synthesize a fibrous red phosphorus photocatalyst with an urchin-like morphology via a solid-state method [28]. This material exhibits an E_g value of 1.9 eV, with valence- and conduction-band potentials well suited for $\bullet O_2^-$ and H_2O_2 generation, underscoring its strong capability for visible-light-driven ROS production. Under sunlight irradiation, the synthesized fibrous red phosphorus achieves a complete 8-log inactivation of *E. coli* within 30 min of photocatalytic treatment. Notably, this performance ranks as one of the best among various photocatalysts reported to date, highlighting its exceptional disinfection potential.

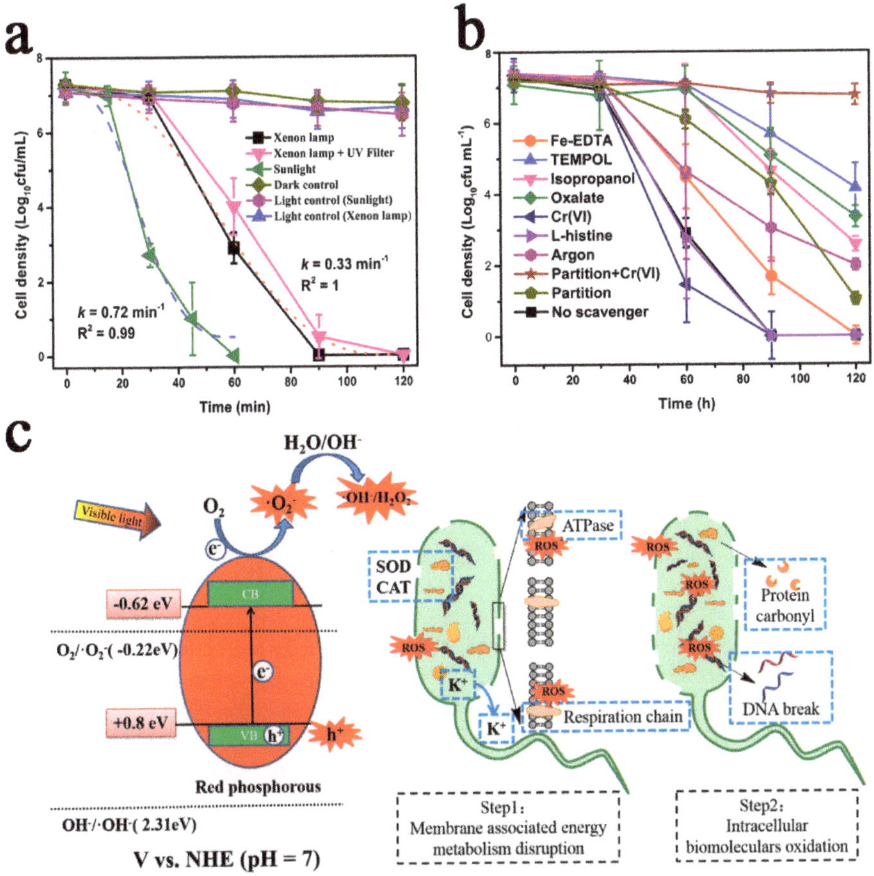

Figure 2. (**a**) Visible-light and sunlight photocatalytic disinfection activities of red phosphorus. (**b**) Photocatalytic disinfection efficiency of red phosphorus in the presence of various scavengers. (**c**) Illustration depicting the proposed bactericidal mechanism of red phosphorus under visible-light irradiation. Reprinted with permission from ref. [26]. Copyright 2015 American Chemical Society.

The three studies mentioned above utilize a photocatalytic system in which light is directed into a slurry containing red phosphorus particles. However, such an approach can only demonstrate the feasibility of using red phosphorus for photocatalytic water disinfection at the laboratory scale. Roshith et al., for the first time, immobilize red phosphorus on the inner walls of a quartz millicapillary using a solid-state method to construct an optofluidic reactor operated in a continuous-flow manner [29]. By adjusting the flow rate of the bacterial solution through the reactor, its residence time can be controlled, enabling the effective regulation of disinfection efficiency. Specifically, under sunlight illumination, a 6.7-log inactivation of *E. coli*, corresponding to a disinfection efficiency of >99.9999%, can be achieved with a residence time of 14 min. When the residence time reaches 28 min, no viable *E. coli* can be detected. Finally, the authors conduct a multicycle test (six consecutive runs), observing that, even with a residence time of 2 min, each cycle can achieve a disinfection efficiency of >95% for *E. coli*, demonstrating the significant potential of the developed reactor for practical applications.

Beyond its intrinsic photocatalytic disinfection capability, red phosphorus can be integrated with other materials, thereby augmenting the photocatalytic disinfection performance of the resulting composites. Wang et al. pioneer this line of research, employing a one-step "green" sonochemical method at room temperature to simultaneously treat graphitic carbon nitride and red phosphorus [30]. During this process, bulk graphitic carbon nitride is exfoliated into few-layer nanosheets, and red phosphorus is downsized and anchored onto the graphitic carbon nitride nanosheets, forming a close-knit metal-free heterojunction photocatalyst. On one hand, red phosphorus exhibits responsiveness to long-wavelength photons up to 700 nm, significantly extending the light absorption range of the heterojunction (Figure 3a). On the other hand, the results of the band structure analysis reveal the establishment of a type-I band alignment at the heterojunction interface (Figure 3b). In this configuration, although both photo-induced charge carriers migrate from graphitic carbon nitride to red phosphorus, the differential migration rates of electrons and holes across the interface facilitate the efficient separation of these charge carriers. As a result, the photocatalytic disinfection capability of the heterojunction is significantly enhanced compared with that of pure graphitic carbon nitride and red phosphorus. Under visible-light irradiation for 80 min, the photocatalytic system achieves a 7-log reduction in *E. coli* concentration (Figure 3c). This disinfection efficiency represents one of the highest performances achieved by previously developed metal-free photocatalysts. In another study, Chen et al. fabricate a 0D/1D heterojunction photocatalyst by modifying carbon nitride nanotubes with red phosphorus nanodots (Figure 3d) [31]. This heterojunction demonstrates an exceptional photocatalytic disinfection performance, achieving a 7-log inactivation of *S. aureus* within 30 min under white LED irradiation, and completely inactivating *E. coli* within only 25 min (Figure 3e,f). Through a scavenger study, the authors investigate the contributions of different ROS to bacterial inactivation and determine that $\bullet O_2^-$ and H_2O_2 predominantly drive the disinfection process. Interestingly, in this work, red phosphorus nanodots exhibit a negligible photocatalytic disinfection ability, which further underscores the critical role of red phosphorus in enhancing photocatalytic disinfection performance when properly integrated into a composite system.

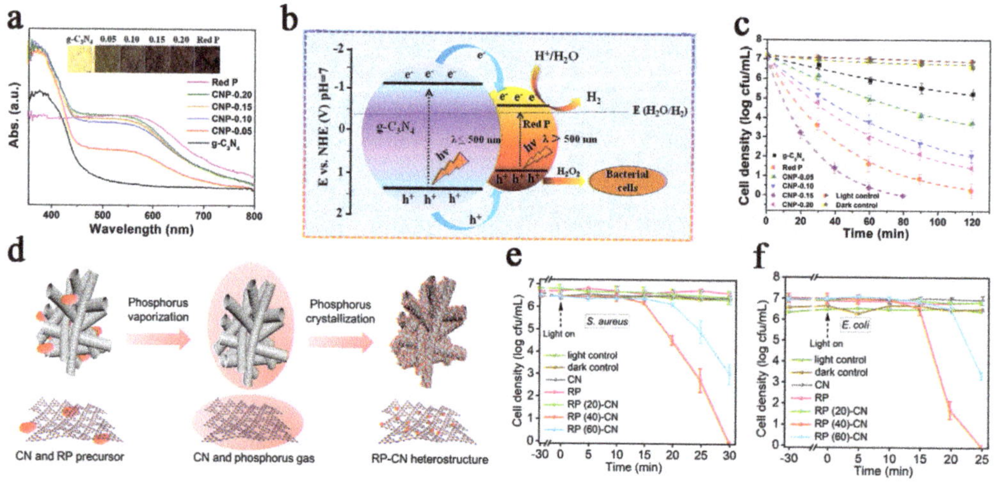

Figure 3. (**a**) UV-vis diffuse reflectance spectra of red phosphorus, graphitic carbon nitride, and heterojunction materials. (**b**) Electronic band structures of red phosphorus and graphitic carbon nitride. (**c**) Visible-light photocatalytic disinfection activities of the samples. Reprinted with permission from ref. [30]. Copyright 2018 Elsevier. (**d**) Schematic diagram illustrating the synthetic process for preparing the heterojunction photocatalyst. Photocatalytic disinfection activities of the prepared samples against (**e**) *S. aureus* and (**f**) *E. coli* under white LED irradiation. Reprinted with permission from ref. [31]. Copyright 2022 American Chemical Society.

Within the spectrum of widely researched photocatalytic antimicrobial materials, wide-bandgap semiconductors such as TiO_2 and ZnO have garnered the most extensive research attention. However, these materials are fundamentally constrained by their limited ability to effectively harness solar light. For instance, anatase TiO_2 exemplifies this limitation with an E_g value of 3.2 eV, restricting its absorption to UV light with wavelengths shorter than 387.5 nm [32]. Since UV light accounts for only about 4% of the total solar energy, visible light, which represents 43% of the solar spectrum, remains substantially underutilized by such material [33,34]. To enhance the solar energy utilization of these wide-bandgap semiconductor materials, they can be integrated with narrow-bandgap semiconductors to broaden the spectral response range of the resulting composites. Liu et al. deposit a red phosphorus nanolayer onto 1D TiO_2 nanofibers using a vacuum ampoule strategy, resulting in the formation of a TiO_2/red phosphorus (TiO_2@RP) core–shell heterostructure (Figure 4a) [35]. Red phosphorus, with an absorption edge extending up to 700 nm, effectively broadens the spectral utilization range of the heterojunction. Moreover, the synergistic interaction between the red phosphorus nanolayer and TiO_2 core significantly enhances the charge carrier transport and separation efficiency. Collectively, the synthesized TiO_2@RP core–shell heterojunction demonstrates a photocatalytic disinfection activity far superior to that of red phosphorus or TiO_2 alone. Under visible-light irradiation, a 7-log reduction in *E. coli* and *S. aureus* can be achieved within 25 and 30 min, respectively (Figure 4b,c). In another study, to overcome the limitations of ZnO in visible-light utilization and its high rate of photo-generated charge carrier recombination, Li et al. first deposit red phosphorus onto a Ti plate via chemical vapor deposition, followed by the deposition of a ZnO nanofilm on the red phosphorus layer using atomic layer deposition, thereby constructing a heterostructured red phosphorus/ZnO thin film [36]. The most stable geometric configuration of the heterojunction, formed by the interface between the red phosphorus (001) plane and the ZnO (002) plane, exhibits pronounced charge redistribution at the interfacial region. This effective interfacial charge transfer, coupled with the enhanced separation efficiency of photo-generated electron-hole pairs, significantly promotes ROS generation. Consequently, this heterojunction exhibits an excellent photocatalytic antibacterial performance. As shown in Figure 4d, under simulated solar light, $99.63 \pm 0.24\%$ of *S. aureus* can be inactivated within 20 min, while $99.72 \pm 0.12\%$ of *E. coli* can be inactivated within 10 min. When using real sunlight as the light source, a $99.92 \pm 0.02\%$ inactivation of *S. aureus* is achieved at 15 min, and a $99.86 \pm 0.10\%$ inactivation of *E. coli* is achieved at 7 min. Based on the results of control experiments, the authors eliminate the effects of thermal influence and direct UV disinfection, demonstrating that the inactivation of bacteria is primarily attributed to visible-light-driven photocatalysis. This finding further indicates that the integration of ZnO with red phosphorus significantly broadens the light absorption range of the resulting composite. Additionally, to assess the indoor-light-activated disinfection performance of the heterojunction, antibacterial experiments are conducted under LED light. At 20 min, the disinfection rate for *S. aureus* is $93.95 \pm 1.21\%$, while at 10 min, it reaches $96.44 \pm 1.24\%$ for *E. coli*. Interestingly, using the bacterial imprinting method, the authors visually demonstrate that even the faint LED light emitted from a mobile phone can activate the prepared heterojunction photocatalyst to exert a photocatalytic disinfection effect (Figure 4e). This work provides a new perspective for enhancing the visible-light photocatalytic water disinfection activity of wide-bandgap semiconductors through the use of red phosphorus.

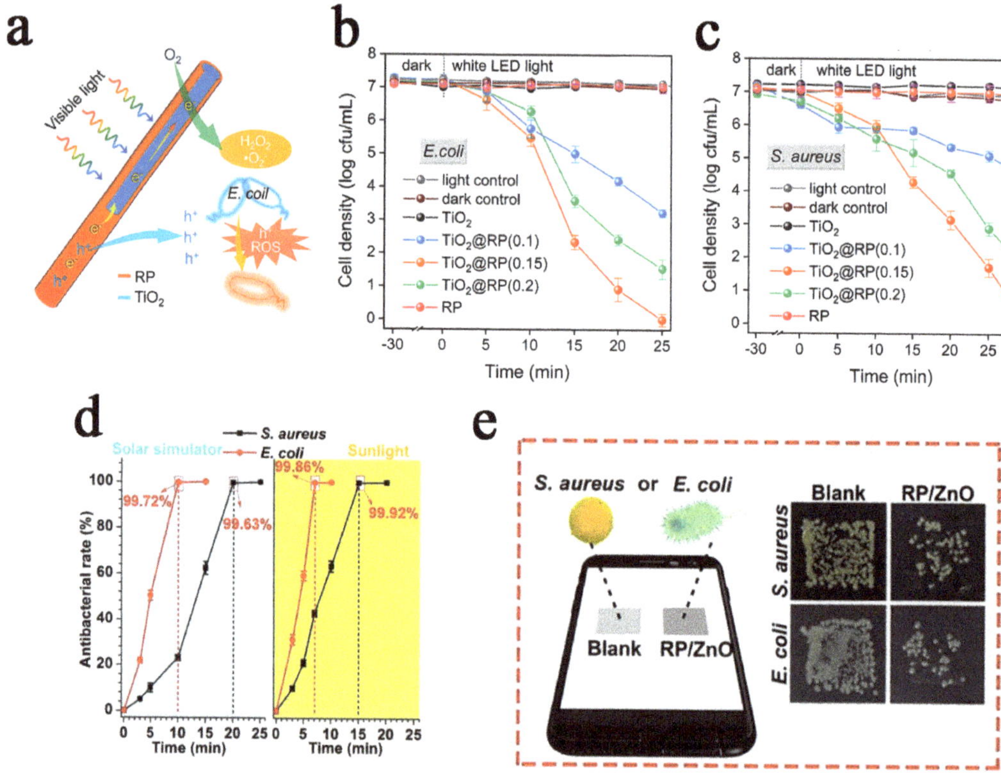

Figure 4. (**a**) Proposed photocatalytic reaction process of the prepared TiO_2@RP core–shell heterostructure. Photocatalytic disinfection activities of the prepared samples against (**b**) *E. coli* and (**c**) *S. aureus* under visible-light irradiation. Reprinted with permission from ref. [35]. Copyright 2021 Elsevier. (**d**) Photocatalytic disinfection efficiency of red phosphorus/ZnO thin film against *S. aureus* and *E. coli* under simulated solar light and real sunlight. (**e**) Photocatalytic disinfection on a small piece of screen protector film coated with red phosphorus/ZnO under LED light from a phone. Reprinted with permission from ref. [36]. Copyright 2019 Wiley-VCH.

Carbonaceous nanomaterials have been demonstrated to be highly effective co-catalysts for coupling with semiconductor materials, as their exceptional electronic properties can significantly suppress the recombination of photo-generated electron-hole pairs. Zhang et al. first fabricate a crystalline red phosphorus film on a Ti plate via chemical vapor deposition, followed by the deposition of a graphene oxide layer to enhance its photoelectrochemical properties (Figure 5a,b) [37]. Through a range of characterization techniques, including an open-circuit potential test, transient photocurrent response analysis, electrochemical impedance spectroscopy, and photoluminescence spectroscopy, it is confirmed that the incorporation of a graphene oxide layer effectively enhances the generation of photo-induced charge carriers while facilitating their efficient transfer and separation. As a result, the photocatalytic antibacterial performance of the synthesized red phosphorus/graphene oxide film is significantly enhanced compared with that of red phosphorus film. Under simulated sunlight, red phosphorus film achieves an antibacterial efficiency of 78.68% against *S. aureus* after 20 min of irradiation, which is markedly enhanced to 98.28% upon graphene oxide deposition. Moreover, the synthesized red phosphorus/graphene oxide film exhibits a broad-spectrum antibacterial activity, demonstrating an excellent disinfection performance against other bacteria, such as *E. coli*. After 15 min of irradiation, an antibacterial efficacy of 99.91% can be achieved. This work highlights the critical role of carbonaceous nanomaterials in enhancing the photocatalytic antibacterial activity of red phosphorus. Recently, our group has developed a ternary heterojunction comprising magnetic carbon nanotubes, red phosphorus, and graphitic carbon nitride [38]. Given that red phosphorus has a less-positive valence-band potential and a less-negative conduction-band potential compared to graphitic carbon nitride, a type-I heterojunction is formed at the red phosphorus/graphitic carbon nitride interface, effectively facilitating charge carrier separation. Furthermore, the incorporated magnetic carbon nanotubes not only serve as an efficient electron shuttle to further suppress charge carrier recombination, but also endow the composite with a magnetic property, enabling convenient recovery and reuse (Figure 5c,d). Our findings reveal that the ternary heterojunction exhibits markedly elevated generation abilities for H_2O_2, $•O_2^-$, and •OH compared to graphitic carbon nitride and red phosphorus/graphitic carbon nitride photocatalysts. This substantial increase in ROS generation undoubtedly plays a pivotal role in its superior photocatalytic antibacterial performance. For the ternary heterojunction, a complete 7-log reduction in *S. aureus* cell density is achieved within 45 min of visible-light illumination, as verified by both plate counting and live/dead fluorescence assays (Figure 5e). This gives a first-order disinfection rate constant of 0.14 min^{-1}, approximately 9.3 times higher than that of graphitic carbon nitride. Notably, this heterojunction exhibits a superior performance compared to the widely used P25 TiO_2 under real-sunlight irradiation, achieving a complete 7-log inactivation of *S. aureus* within 60 min (Figure 5f). The corresponding first-order disinfection rate constant is determined to be 0.067 min^{-1}, approximately 13.9 times higher than that of P25 TiO_2. Subsequently, we systematically investigate the inactivation process of *S. aureus* and reveal that the photocatalytic treatment of the heterojunction induces bacterial inactivation through multiple pathways. These include the membrane damage at the structural and metabolic levels, the leakage of intracellular components, the penetration of extracellular ROS into the cytoplasm, and the decomposition of genomic DNA, ultimately leading to the inactivation of bacteria (Figure 5g). Notably, the synthesized heterojunction catalyst exhibits exceptional antimicrobial activities against a diverse array of pathogenic microorganisms, including *E. coli*, *S. aureus*, *E. faecalis*, *P. aeruginosa*, tetracycline-resistant *E. coli*, ampicillin-resistant *E. coli*, and *C. parvum*, underscoring its broad-spectrum disinfection capability. More importantly, the disinfection efficiency of the prepared heterojunction catalyst is compared with that of several conventional water disinfection techniques at a scale of 5 L. While UV irradiation and boiling exhibit higher disinfection rates for *S. aureus*, they are energy-intensive, with UV irradiation requiring 68.57 J/log and boiling demanding 237,000 J/log. Furthermore, bacterial reactivation is observed after UV irradiation and chlorination treatment, whereas the photocatalytic treatment based on the synthesized heterojunction inactivates *S. aureus* more thoroughly. This work offers valuable insights into the design and fabrication of magnetic red-phosphorus-based antimicrobial photocatalysts.

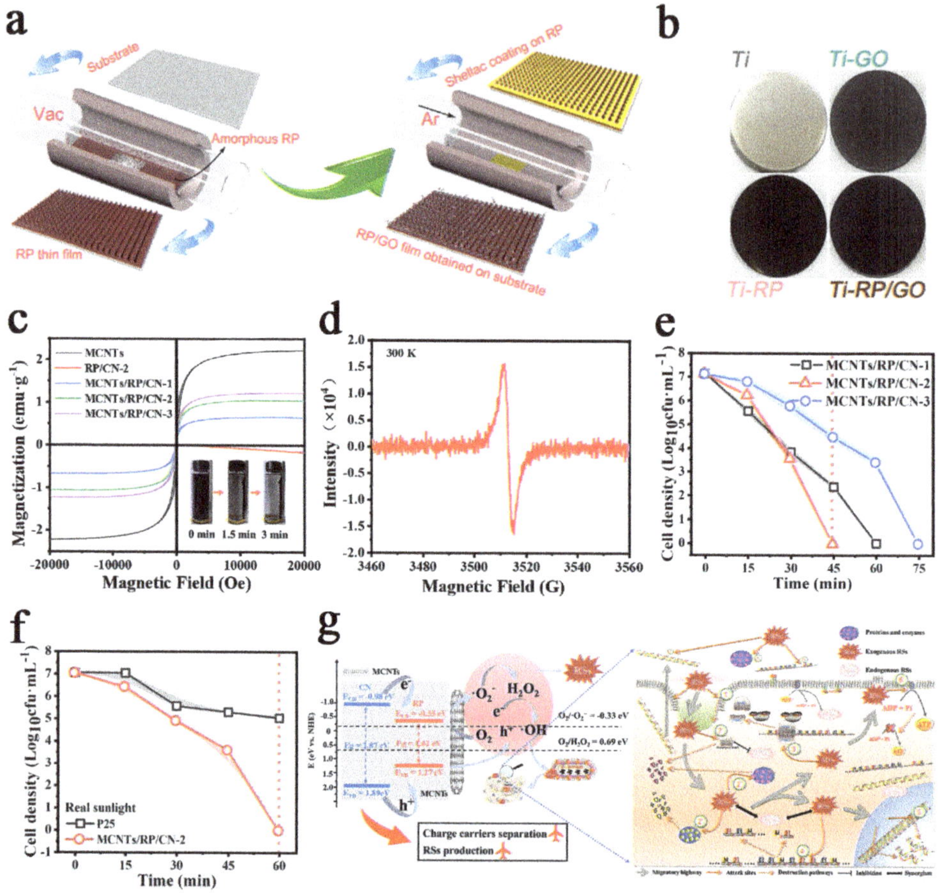

Figure 5. (**a**) Schematic diagram of the synthesis process of the red phosphorus/graphene oxide film. (**b**) Photographic images of the samples. Reprinted with permission from ref. [37]. Copyright 2020 Elsevier. (**c**) Hysteresis loops of the samples used to determine their saturation magnetization values. The inset illustrates the magnetic property of the ternary heterojunction for its convenient separation from water. (**d**) Local self-spin test using electron spin resonance spectroscopy for magnetic carbon nanotubes. (**e**) Photocatalytic disinfection activity of the ternary heterojunction against *S. aureus* under visible-light irradiation. (**f**) Photocatalytic disinfection activities of the samples against *S. aureus* under real sunlight. (**g**) Illustration summarizing the proposed disinfection mechanism of the ternary heterojunction under visible-light irradiation. Reprinted with permission from ref. [38]. Copyright 2023 Elsevier.

3.2. Black-Phosphorus-Based Antimicrobial Photocatalysts

The discovery of black phosphorus was made over a century [25]. It has gained attention for its superior thermodynamic stability compared to other P allotropes in recent years. Owing to its exceptional physicochemical properties, black phosphorus has now been recognized as a promising candidate for photocatalytic applications. These properties include (1) a tunable layer-dependent bandgap ranging from 0.3 to 2.0 eV; (2) a broad light absorption range extending from UV to the near-infrared region; (3) a large specific surface area in its few-layer form, offering abundant active sites for photocatalytic reactions; and (4) an exceptionally high hole mobility of up to 1000 cm^2 V^{-1} S^{-1} [16,39]. Our group pioneer the utilization of black phosphorus as a metal-free cocatalyst, establishing it as a promising alternative to noble metals for photocatalytic bacterial inactivation [40]. We synthesize a heterojunction photocatalyst by integrating black phosphorus with graphitic carbon nitride and, for the first time, apply it to photocatalytic bacterial inactivation. A series of characterizations confirm the successful synthesis of the composite and its enhanced visible-light absorption without disrupting the intrinsic structure of graphitic carbon nitride. Furthermore, photoluminescence and transient photocurrent response measurements reveal that this integration can effectively suppress charge carrier recombination, thereby extending the lifetime of photo-generated electrons and holes. This arises from the formation of a type-I heterojunction between black phosphorus and graphitic carbon nitride, where photo-generated electron-hole pairs in graphitic carbon nitride migrate to black phosphorus at different rates, effectively suppressing electron-hole recombination and extending their lifetimes to facilitate enhanced ROS generation. Thus, the overall ROS generation ability of the heterojunction is significantly greater compared with that of graphitic carbon nitride. As a result, this heterojunction exhibits a remarkable antibacterial performance, achieving a complete 7-log inactivation of *E. coli* within 60 min under visible-light irradiation. This disinfection efficiency is seven times higher than that of graphitic carbon nitride. Bacterial scanning electron microscope observation and K^+ leakage experiment confirm that the inactivation mechanism for *E. coli* is driven by ROS-induced membrane damage and intracellular content leakage. Additionally, the heterojunction demonstrates its stability, maintaining a 99.97% *E. coli* inactivation rate after three consecutive cycles. Furthermore, we conduct an acute toxicity assessment using *Vibrio fischeri*, demonstrating that the bioluminescence intensity before and after photocatalysis remains virtually unchanged, confirming the non-toxic and environmentally friendly nature of the developed heterojunction. Our study offers insights into the design of purely metal-free heterostructure photocatalysts based on black phosphorus for efficient photocatalytic water disinfection.

Although the type-I charge transfer demonstrated above can enhance the separation efficiency of electron-hole pairs thermodynamically, this improvement comes at the expense of reduced redox abilities for charge carriers, which is detrimental to the following photocatalytic reactions. Building on this, we recently fabricated an S-scheme heterojunction photocatalyst by integrating black phosphorus with BiOBr [41]. In this band configuration, BiOBr, serving as an oxidation-type component, possesses a more positive valence-band potential, while black phosphorus, acting as a reduction-type component, features a more negative conduction-band potential. Due to the Fermi level difference between these two materials, when black phosphorus and BiOBr are in close contact, electrons in black phosphorus spontaneously transfer to BiOBr, establishing an internal electric field oriented from black phosphorus to BiOBr. Under visible-light irradiation, this internal electric field facilitates the transfer of photo-generated electrons in the conduction band of BiOBr to recombine with the photo-induced holes in the valence band of black phosphorus. Thus, the remaining electrons and holes possess stronger redox abilities and extended lifetimes, thereby enhancing ROS generation. Consequently, the synthesized S-scheme heterojunction exhibits a significantly enhanced photocatalytic disinfection performance, achieving a complete 7-log inactivation of *E. coli* after 7.5 min of visible-light irradiation. This study offers new insights into the application of black-phosphorus-based S-scheme heterojunction photocatalysts for efficient water disinfection.

4. Summary and Future Research Directions

In recent years, systematic research has been conducted on the design, synthesis, and photocatalytic disinfection performance of red- and black-phosphorus-based materials. On one hand, red phosphorus, including its amorphous and fibrous forms, has demonstrated its intrinsic photocatalytic water disinfection ability. Moreover, photocatalytic reactors based on red phosphorus have been developed, further highlighting its potential for practical water disinfection applications. Notably, red phosphorus can be integrated with other materials to enhance the photocatalytic disinfection performance of the resulting composites. This enhancement primarily arises from two aspects: (i) the formation of heterojunctions by integrating semiconductors with matching band structures, which facilitates charge carrier separation, and (ii) the intrinsic strong light absorption of red phosphorus, which broadens the spectral response of the obtained composites. Additionally, red phosphorus is well suited for coupling with carbonaceous nanomaterials, which can further improve charge carrier separation. Certain carbonaceous materials also exhibit a magnetic property, endowing the resulting composites with magnetic responsiveness, thereby enabling convenient recovery and reuse. On the other hand, black phosphorus has also attracted significant attention in photocatalytic water disinfection due to its unique physicochemical properties. Research efforts have primarily focused on the construction of black-phosphorus-based heterojunction photocatalysts, including type-I and S-scheme configurations, to enhance charge carrier separation, thereby increasing the generation of ROS within the photocatalytic systems.

Although significant progress has been made, several challenges persist, warranting further investigation in future studies. First, one of the key advantages of red and black phosphorus lies in their metal-free nature, which offers benefits such as environmental friendliness, non-toxicity, and a composition based on the abundantly available P element. However, many existing red- and black-phosphorus-based materials incorporate metal components, thereby diminishing the unique advantages of their metal-free nature. Future research should focus on the design and development of purely metal-free photocatalysts based on red and black phosphorus. This may involve tailoring their intrinsic morphological structures or selecting metal-free semiconductors with complementary band structures to construct heterojunctions, while also exploring the potential of the resulting materials for photocatalytic water disinfection. Notably, while metal-free materials are generally considered to pose a lower risk of secondary pollution compared to metal-based catalysts, it remains essential to establish conclusive evidence. Future studies should, therefore, investigate whether the synthesized materials release substances with potential ecological risks.

Moreover, most existing studies primarily focus on evaluating the photocatalytic disinfection performance of red- and black-phosphorus-based materials in pure water or saline solution. However, the composition of authentic water matrices is far more complex. Natural organic matter, carbonate species, and other background constituents in authentic water matrices can significantly reduce the effectiveness of photocatalytic disinfection by scavenging ROS, which is one of the key limitations for the large-scale application of these red- and black-phosphorus-based materials. Additionally, certain halide ions in these matrices may react with ROS, generating halogen radicals whose impact on the efficiency of photocatalytic water disinfection remains unclear [42]. Therefore, future research should focus on assessing the disinfection effectiveness of red- and black-phosphorus-based materials in authentic water matrices, with a detailed evaluation of how common water constituents affect the process of photocatalytic disinfection. Additionally, it is essential to evaluate the production cost and long-term stability when considering the application of these materials in real-world water disinfection.

Author Contributions: Writing—original draft preparation, D.H.; investigation, J.D., S.Z. and Y.-N.Z. (Ya-Nan Zhang); writing—review and editing, Y.-N.Z. (Yi-Nan Zhang) and J.Q. All authors have read and agreed to the published version of the manuscript.

Funding: This work is financially supported by National Natural Science Foundation of China (42130705, 42377048, 42407082, 42422702, and 22176030), China Postdoctoral Science Foundation (2022M720700), and Fundamental Research Funds for the Central Universities (2412023QD027).

Conflicts of Interest: The authors declare no competing financial interests.

References

1. Chu, C.; Ryberg, E.C.; Loeb, S.K.; Suh, M.J.; Kim, J.H. Water Disinfection in Rural Areas Demands Unconventional Solar Technologies. *Acc. Chem. Res.* **2019**, *52*, 1187–1195. [CrossRef] [PubMed]
2. Prüss-Ustün, A.; Bartram, J.; Clasen, T.; Colford, J.M.; Cumming, O.; Curtis, V.; Bonjour, S.; Dangour, A.D.; De France, J.; Fewtrell, L.; et al. Burden of disease from inadequate water, sanitation and hygiene in low- and middle-income settings: A retrospective analysis of data from 145 countries. *Trop. Med. Int. Health* **2014**, *19*, 894–905. [CrossRef] [PubMed]
3. He, D.; Liu, C.; Zhang, Y.-N.; Yang, H.; Qu, J. Efficient water disinfection accelerated by polymerization-degree-controlled graphitic carbon nitride under visible light. *J. Environ. Chem. Eng.* **2024**, *12*, 112247. [CrossRef]
4. Liu, C.; He, D.; Yang, H.; Zhang, K.; Zhou, X.; Zhang, T.; Qu, J. Selective generation of H_2O_2 by Cu-modified graphitic carbon nitride for rapid water disinfection via intracellular Fenton reaction. *Chem. Eng. J.* **2023**, *477*, 146835. [CrossRef]
5. Yang, H.; He, D.; Fan, L.; Cheng, F.; Zhou, X.; Zhou, T.; Liu, C.; Wang, C.; Zhang, Y.n.; Qu, J. Visible-Light-Induced Rapid Elimination of Antibiotic Resistance Contaminations Using Graphitic Carbon Nitride Tailored with Carrier Confinement Domains. *Adv. Mater.* **2024**, *37*, 2410221. [CrossRef] [PubMed]
6. Wu, T.; Liu, B.; Liu, C.; Wan, J.; Yang, A.; Liu, K.; Shi, F.; Zhao, J.; Lu, Z.; Chen, G.; et al. Solar-driven efficient heterogeneous subminute water disinfection nanosystem assembled with fingerprint MoS_2. *Nat. Water* **2023**, *1*, 462–470. [CrossRef]
7. Liu, C.; Kong, D.; Hsu, P.C.; Yuan, H.; Lee, H.W.; Liu, Y.; Wang, H.; Wang, S.; Yan, K.; Lin, D.; et al. Rapid water disinfection using vertically aligned MoS_2 nanofilms and visible light. *Nat. Nanotechnol.* **2016**, *11*, 1098–1104. [CrossRef]
8. You, J.; Guo, Y.; Guo, R.; Liu, X. A review of visible light-active photocatalysts for water disinfection: Features and prospects. *Chem. Eng. J.* **2019**, *373*, 624–641. [CrossRef]
9. Zhang, L.-S.; Wong, K.-H.; Yip, H.-Y.; Hu, C.; Yu, J.C.; Chan, C.-Y.; Wong, P.-K. Effective Photocatalytic Disinfection of *E. coli* K-12 Using AgBr-Ag-Bi_2WO_6 Nanojunction System Irradiated by Visible Light: The Role of Diffusing Hydroxyl Radicals. *Environ. Sci. Technol.* **2010**, *44*, 1392–1398. [CrossRef] [PubMed]
10. Hu, X.; Hu, C.; Peng, T.; Zhou, X.; Qu, J. Plasmon-Induced Inactivation of Enteric Pathogenic Microorganisms with Ag-AgI/Al_2O_3 under Visible-Light Irradiation. *Environ. Sci. Technol.* **2010**, *44*, 7058–7062. [CrossRef]
11. Xia, D.; Ng, T.W.; An, T.; Li, G.; Li, Y.; Yip, H.Y.; Zhao, H.; Lu, A.; Wong, P.-K. A Recyclable Mineral Catalyst for Visible-Light-Driven Photocatalytic Inactivation of Bacteria: Natural Magnetic Sphalerite. *Environ. Sci. Technol.* **2013**, *47*, 11166–11173. [CrossRef] [PubMed]
12. Xia, P.; Cao, S.; Zhu, B.; Liu, M.; Shi, M.; Yu, J.; Zhang, Y. Designing a 0D/2D S-Scheme Heterojunction over Polymeric Carbon Nitride for Visible-Light Photocatalytic Inactivation of Bacteria. *Angew. Chem. Int. Ed. Engl.* **2020**, *59*, 5218–5225. [CrossRef]
13. Sánchez-Albores, R.; Cano, F.J.; Sebastian, P.J.; Reyes-Vallejo, O. Microwave-assisted biosynthesis of ZnO-GO particles using orange peel extract for photocatalytic degradation of methylene blue. *J. Environ. Chem. Eng.* **2022**, *10*, 108924. [CrossRef]
14. Xu, Q.; Zhang, L.; Cheng, B.; Fan, J.; Yu, J. S-Scheme Heterojunction Photocatalyst. *Chem* **2020**, *6*, 1543–1559. [CrossRef]
15. Wu, Z.; Shen, J.; Li, W.; Li, J.; Xia, D.; Xu, D.; Zhang, S.; Zhu, Y. Electron self-sufficient core-shell BiOCl@Fe-BiOCl nanosheets boosting Fe(III)/Fe(II) recycling and synergetic photocatalysis-Fenton for enhanced degradation of phenol. *Appl. Catal. B Environ.* **2023**, *330*, 122642. [CrossRef]
16. He, D.; Jin, D.; Cheng, F.; Zhang, T.; Qu, J.; Zhou, Y.; Yuan, X.; Zhang, Y.N.; Peijnenburg, W. Development of a metal-free black phosphorus/graphitic carbon nitride heterostructure for visible-light-driven degradation of indomethacin. *Sci. Total Environ.* **2022**, *804*, 150062. [CrossRef] [PubMed]
17. Wang, T.; Jiang, Z.; An, T.; Li, G.; Zhao, H.; Wong, P.K. Enhanced Visible-Light-Driven Photocatalytic Bacterial Inactivation by Ultrathin Carbon-Coated Magnetic Cobalt Ferrite Nanoparticles. *Environ. Sci. Technol.* **2018**, *52*, 4774–4784. [CrossRef] [PubMed]
18. He, D.; Yang, H.; Jin, D.; Qu, J.; Yuan, X.; Zhang, Y.-N.; Huo, M.; Peijnenburg, W.J.G.M. Rapid water purification using modified graphitic carbon nitride and visible light. *Appl. Catal. B Environ.* **2021**, *285*, 119864. [CrossRef]
19. Zhang, X.; Su, H.; Cui, P.; Cao, Y.; Teng, Z.; Zhang, Q.; Wang, Y.; Feng, Y.; Feng, R.; Hou, J.; et al. Developing Ni single-atom sites in carbon nitride for efficient photocatalytic H_2O_2 production. *Nat. Commun.* **2023**, *14*, 7115. [CrossRef] [PubMed]
20. Teng, Z.; Zhang, Q.; Yang, H.; Kato, K.; Yang, W.; Lu, Y.-R.; Liu, S.; Wang, C.; Yamakata, A.; Su, C.; et al. Atomically dispersed antimony on carbon nitride for the artificial photosynthesis of hydrogen peroxide. *Nat. Catal.* **2021**, *4*, 374–384. [CrossRef]

21. Nosaka, Y.; Nosaka, A.Y. Generation and Detection of Reactive Oxygen Species in Photocatalysis. *Chem. Rev.* **2017**, *117*, 11302–11336. [CrossRef] [PubMed]
22. Cho, M.; Lee, J.; Mackeyev, Y.; Wilson, L.J.; Alvarez, P.J.J.; Hughes, J.B.; Kim, J.-H. Visible Light Sensitized Inactivation of MS-2 Bacteriophage by a Cationic Amine-Functionalized C_{60} Derivative. *Environ. Sci. Technol.* **2010**, *44*, 6685–6691. [CrossRef]
23. Teng, Z.; Yang, N.; Lv, H.; Wang, S.; Hu, M.; Wang, C.; Wang, D.; Wang, G. Edge-Functionalized g-C_3N_4 Nanosheets as a Highly Efficient Metal-free Photocatalyst for Safe Drinking Water. *Chem* **2019**, *5*, 664–680. [CrossRef]
24. Xiao, K.; Wang, T.; Sun, M.; Hanif, A.; Gu, Q.; Tian, B.; Jiang, Z.; Wang, B.; Sun, H.; Shang, J.; et al. Photocatalytic Bacterial Inactivation by a Rape Pollen-MoS_2 Biohybrid Catalyst: Synergetic Effects and Inactivation Mechanisms. *Environ. Sci. Technol.* **2019**, *54*, 537–549. [CrossRef]
25. Liu, H.; Du, Y.; Deng, Y.; Ye, P.D. Semiconducting black phosphorus: Synthesis, transport properties and electronic applications. *Chem. Soc. Rev.* **2015**, *44*, 2732–2743. [CrossRef]
26. Xia, D.; Shen, Z.; Huang, G.; Wang, W.; Yu, J.C.; Wong, P.K. Red Phosphorus: An Earth-Abundant Elemental Photocatalyst for "Green" Bacterial Inactivation under Visible Light. *Environ. Sci. Technol.* **2015**, *49*, 6264–6273. [CrossRef] [PubMed]
27. Athira, T.K.; Roshith, M.; Kadrekar, R.; Arya, A.; Kumar, M.S.; Anantharaj, G.; Gurrala, L.; Saranyan, V.; Babu, T.G.S.; Darbha, V.R.K. Evaluation of photocatalytic activity of commercial red phosphorus towards the disinfection of *E. coli* and reduction of Cr (VI) under direct sunlight. *Mater. Res. Express* **2020**, *7*, 104002. [CrossRef]
28. Roshith, M.; Kumar, M.S.; Nanda Kumar, A.K.; Ramasubramanian, S.; Stanley, J.; Satheesh Babu, T.G.; Ravi Kumar, D.V. Urchin-like fibrous red phosphorus as an efficient photocatalyst for solar-light-driven disinfection of *E. coli*. *J. Photochem. Photobiol. A Chem.* **2019**, *384*, 112034. [CrossRef]
29. Roshith, M.; Pathak, A.; Nanda Kumar, A.K.; Anantharaj, G.; Saranyan, V.; Ramasubramanian, S.; Satheesh Babu, T.G.; Ravi Kumar, D.V. Continuous flow solar photocatalytic disinfection of *E. coli* using red phosphorus immobilized capillaries as optofluidic reactors. *Appl. Surf. Sci.* **2021**, *540*, 148398. [CrossRef]
30. Wang, W.; Li, G.; An, T.; Chan, D.K.L.; Yu, J.C.; Wong, P.K. Photocatalytic hydrogen evolution and bacterial inactivation utilizing sonochemical-synthesized g-C_3N_4/red phosphorus hybrid nanosheets as a wide-spectral-responsive photocatalyst: The role of type I band alignment. *Appl. Catal. B Environ.* **2018**, *238*, 126–135. [CrossRef]
31. Chen, J.; Zhu, Y.; Yang, X.; Ye, W.; Liu, J.; Butenko, D.S.; Lu, P.; Meng, P.; Xu, Y.; Yang, D.; et al. Red Phosphorus Nanodot-Decorated Polymeric Carbon Nitride Nanotubes for Visible-Light-Driven Photocatalytic Bacterial Inactivation. *ACS Appl. Nano Mater.* **2022**, *5*, 862–870. [CrossRef]
32. Pelaez, M.; Nolan, N.T.; Pillai, S.C.; Seery, M.K.; Falaras, P.; Kontos, A.G.; Dunlop, P.S.M.; Hamilton, J.W.J.; Byrne, J.A.; O'Shea, K.; et al. A review on the visible light active titanium dioxide photocatalysts for environmental applications. *Appl. Catal. B Environ.* **2012**, *125*, 331–349. [CrossRef]
33. Moniz, S.J.A.; Shevlin, S.A.; Martin, D.J.; Guo, Z.-X.; Tang, J. Visible-light driven heterojunction photocatalysts for water splitting-a critical review. *Energy Environ. Sci.* **2015**, *8*, 731–759. [CrossRef]
34. Zhang, S.; Duan, S.; Chen, G.; Meng, S.; Zheng, X.; Fan, Y.; Fu, X.; Chen, S. $MoS_2/Zn_3In_2S_6$ composite photocatalysts for enhancement of visible light-driven hydrogen production from formic acid. *Chin. J. Catal.* **2021**, *42*, 193–204. [CrossRef]
35. Liu, J.; Zhu, Y.; Chen, J.; Butenko, D.S.; Ren, J.; Yang, X.; Lu, P.; Meng, P.; Xu, Y.; Yang, D.; et al. Visible-light driven rapid bacterial inactivation on red phosphorus/titanium oxide nanofiber heterostructures. *J. Hazard. Mater.* **2021**, *413*, 125462. [CrossRef] [PubMed]
36. Li, J.; Liu, X.; Tan, L.; Liang, Y.; Cui, Z.; Yang, X.; Zhu, S.; Li, Z.; Zheng, Y.; Yeung, K.W.K.; et al. Light-Activated Rapid Disinfection by Accelerated Charge Transfer in Red Phosphorus/ZnO Heterointerface. *Small Methods* **2019**, *3*, 1900048. [CrossRef]
37. Zhang, Q.; Liu, X.; Tan, L.; Cui, Z.; Li, Z.; Liang, Y.; Zhu, S.; Yeung, K.W.K.; Zheng, Y.; Wu, S. An UV to NIR-driven platform based on red phosphorus/graphene oxide film for rapid microbial inactivation. *Chem. Eng. J.* **2020**, *383*, 123088. [CrossRef]
38. Yang, H.; He, D.; Zhang, T.; Liu, C.; Cheng, F.; Zhou, Y.; Zhang, Y.-n.; Qu, J. Magnetic carbon nanotubes/red phosphorus/graphitic carbon nitride heterojunction for highly efficient visible-light photocatalytic water disinfection. *Chem. Eng. J.* **2023**, *466*, 143309. [CrossRef]
39. Ran, J.; Guo, W.; Wang, H.; Zhu, B.; Yu, J.; Qiao, S.Z. Metal-Free 2D/2D Phosphorene/g-C_3N_4 Van der Waals Heterojunction for Highly Enhanced Visible-Light Photocatalytic H_2 Production. *Adv. Mater.* **2018**, *30*, 1800128. [CrossRef] [PubMed]
40. He, D.; Zhang, Z.; Xing, Y.; Zhou, Y.; Yang, H.; Liu, H.; Qu, J.; Yuan, X.; Guan, J.; Zhang, Y.-n. Black phosphorus/graphitic carbon nitride: A metal-free photocatalyst for "green" photocatalytic bacterial inactivation under visible light. *Chem. Eng. J.* **2020**, *384*, 123258. [CrossRef]

41. Zhang, K.; Zhang, Y.; Zhang, D.; Liu, C.; Zhou, X.; Yang, H.; Qu, J.; He, D. Efficient photocatalytic water disinfection by a novel BP/BiOBr S-scheme heterojunction photocatalyst. *Chem. Eng. J.* **2023**, *468*, 143581. [CrossRef]
42. Yang, H.; He, D.; Fan, L.; Cheng, F.; Zhou, Y.; Lei, Y.; Zhang, Y.-N.; Yang, X.; Qu, J. Evaluating the Impact of $Cl_2^{\bullet-}$ Generation on Antibiotic-Resistance Contamination Removal via UV/Peroxydisulfate. *Environ. Sci. Technol.* **2024**, *58*, 5578–5588. [CrossRef] [PubMed]

MDPI AG
Grosspeteranlage 5
4052 Basel
Switzerland
Tel.: +41 61 683 77 34

Catalysts Editorial Office
E-mail: catalysts@mdpi.com
www.mdpi.com/journal/catalysts

www.ingramcontent.com/pod-product-compliance
Lightning Source LLC
Chambersburg PA
CBHW061512210926
51850CB00008B/1895
9783725861811